Next Generation Photovoltaics
High efficiency through full spectrum utilization

T0132559

Series in Optics and Optoelectronics

Series Editors:　**R G W Brown**, University of Nottingham, UK
　　　　　　　　E R Pike, Kings College, London, UK

Other titles in the series

Applications of Silicon–Germanium Heterostructure Devices
C K Maiti and G A Armstrong

Optical Fibre Devices
J-P Goure and I Verrier

Optical Applications of Liquid Crystals
L Vicari (ed)

Laser-Induced Damage of Optical Materials
R M Wood

Forthcoming titles in the series

High Speed Photonic Devices
N Dagli (ed)

Diode Lasers
D Sands

High Aperture Focussing of Electromagnetic Waves and Applications in Optical
Microscopy
C J R Sheppard and P Torok

Other titles of interest

Thin-Film Optical Filters (Third Edition)
H Angus Macleod

Series in Optics and Optoelectronics

Next Generation Photovoltaics
High efficiency through full spectrum utilization

Edited by

Antonio Martí and Antonio Luque

Istituto de Energia Solar—ETSIT,
Universidad Politécnica de Madrid, Spain

CRC Press
Taylor & Francis Group
Boca Raton London New York

CRC Press is an imprint of the
Taylor & Francis Group, an **informa** business

First published 2004 by IOP Publishing Ltd

Published 2020 by CRC Press
Taylor & Francis Group
6000 Broken Sound Parkway NW, Suite 300
Boca Raton, FL 33487-2742

First issued in paperback 2020

© 2004 by Taylor & Francis Group, LLC
CRC Press is an imprint of Taylor & Francis Group, an Informa business

No claim to original U.S. Government works

ISBN 13: 978-0-367-57847-3 (pbk)
ISBN 13: 978-0-7503-0905-9 (hbk)

**Visit the Taylor & Francis Web site at
http://www.taylorandfrancis.com**

**and the CRC Press Web site at
http://www.crcpress.com**

British Library Cataloguing-in-Publication Data

A catalogue record for this book is available from the British Library.

Library of Congress Cataloging-in-Publication Data are available

Cover Design: Victoria Le Billon

Typeset in LᵀEX 2$_\varepsilon$ by Text 2 Text Limited, Torquay, Devon

Contents

Preface   xi

1 Non-conventional photovoltaic technology: a need to reach goals
Antonio Luque and Antonio Martí   1
  1.1   Introduction   1
  1.2   On the motivation for solar energy   2
  1.3   Penetration goals for PV electricity   7
  1.4   Will PV electricity reach costs sufficiently low to permit a wide
       penetration?   9
  1.5   The need for a technological breakthrough   14
  1.6   Conclusions   17
       References   18

2 Trends in the development of solar photovoltaics
Zh I Alferov and V D Rumyantsev   19
  2.1   Introduction   19
  2.2   Starting period   20
  2.3   Simple structures and simple technologies   21
  2.4   Nanostructures and 'high technologies'   23
  2.5   Multi-junction solar cells   28
  2.6   From the 'sky' to the Earth   34
  2.7   Concentration of solar radiation   35
  2.8   Concentrators in space   43
  2.9   'Non-solar' photovoltaics   44
  2.10 Conclusions   47
       References   48

3 Thermodynamics of solar energy converters
Peter Würfel   50
  3.1   Introduction   50
  3.2   Equilibria   50
       3.2.1   Temperature equilibrium   51
       3.2.2   Thermochemical equilibrium   52

3.3   Converting chemical energy into electrical energy:
      the basic requirements for a solar cell   57
3.4   Concepts for solar cells with ultra high efficiencies   59
      3.4.1   Thermophotovoltaic conversion   60
      3.4.2   Hot carrier cell   60
      3.4.3   Tandem cells   60
      3.4.4   Intermediate level cells   61
      3.4.5   Photon up- and down-conversion   61
3.5   Conclusions   62
      References   63

4   Tandem cells for very high concentration
A W Bett   **64**
4.1   Introduction   64
4.2   Tandem solar cells   66
      4.2.1   Mechanically stacked tandem cells   67
      4.2.2   Monolithic tandem cells   72
      4.2.3   Combined approach: mechanical stacking of monolithic
              cells   77
4.3   Testing and application of monolithic dual-junction concentrator
      cells   77
      4.3.1   Characterization of monolithic concentrator solar cells 77
      4.3.2   Fabrication and characterization of a test module   80
      4.3.3   FLATCON module   82
      4.3.4   Concentrator system development   83
4.4   Summary and perspective   85
      Acknowledgments   87
      References   88

5   Quantum wells in photovoltaic cells
C Rohr, P Abbott, I M Ballard, D B Bushnell, J P Connolly,
N J Ekins-Daukes and K W J Barnham   **91**
5.1   Introduction   91
5.2   Quantum well cells   91
5.3   Strain compensation   94
5.4   QWs in tandem cells   96
5.5   QWCs with light trapping   97
5.6   QWCs for thermophotovoltaics   99
5.7   Conclusions   102
      References   103

**6   The importance of the very high concentration in third-generation
      solar cells**
Carlos Algora   **108**
6.1   Introduction   108

	6.2	Theory	109
		6.2.1 How concentration works on solar cell performance	109
		6.2.2 Series resistance	112
		6.2.3 The effect of illuminating the cell with a wide-angle cone of light	115
		6.2.4 Pending issues: modelling under real operation conditions	118
	6.3	Present and future of concentrator third-generation solar cells	120
	6.4	Economics	122
		6.4.1 How concentration affects solar cell cost	122
		6.4.2 Required concentration level	124
		6.4.3 Cost analysis	126
	6.5	Summary and conclusions	134
		Note added in press	136
		References	136

7 Intermediate-band solar cells
A Martí, L Cuadra and A Luque **140**

	7.1	Introduction	140
	7.2	Preliminary concepts and definitions	142
	7.3	Intermediate-band solar cell: model	148
	7.4	The quantum-dot intermediate-band solar cell	150
	7.5	Considerations for the practical implementation of the QD-IBSC	155
	7.6	Summary	160
		Acknowledgments	162
		References	162

8 Multi-interface novel devices: model with a continuous substructure
Z T Kuznicki **165**

	8.1	Introduction	165
	8.2	Novelties in Si optoelectronics and photovoltaics	167
		8.2.1 Enhanced absorbance	168
		8.2.2 Enhanced conversion	168
	8.3	Active substructure and active interfaces	169
	8.4	Active substructure by ion implantation	170
		8.4.1 Hetero-interface energy band offset	173
		8.4.2 Built-in electric field	174
		8.4.3 Built-in strain field	176
		8.4.4 Defects	178
	8.5	Model of multi-interface solar cells	178
		8.5.1 Collection efficiency and internal quantum efficiency	181
		8.5.2 Generation rate	181
		8.5.3 Carrier collection limit	181
		8.5.4 Surface reservoir	182
		8.5.5 Collection zones	183
		8.5.6 Impurity band doping profile	184

8.5.7 Uni- and bipolar electronic transport in a multi-interface emitter 184
8.5.8 Absorbance in presence of a dead zone 186
8.5.9 Self-consistent calculation 187
8.6 An experimental test device 189
8.6.1 Enhanced internal quantum efficiency 190
8.6.2 Sample without any carrier collection limit (CCL) 191
8.7 Concluding remarks and perspectives 192
Acknowledgments 193
References 194

9 Quantum dot solar cells
A J Nozik **196**
9.1 Introduction 196
9.2 Relaxation dynamics of hot electrons 199
9.2.1 Quantum wells and superlattices 201
9.2.2 Relaxation dynamics of hot electrons in quantum dots 206
9.3 Quantum dot solar cell configuration 214
9.3.1 Photoelectrodes composed of quantum dot arrays 216
9.3.2 Quantum dot-sensitized nanocrystalline TiO$_2$ solar cells 216
9.3.3 Quantum dots dispersed in organic semiconductor polymer matrices 217
9.4 Conclusion 218
Acknowledgments 218
References 218

10 Progress in thermophotovoltaic converters
Bernd Bitnar, Wilhelm Durisch, Fritz von Roth, Günther Palfinger, Hans Sigg, Detlev Grützmacher, Jens Gobrecht, Eva-Maria Meyer, Ulrich Vogt, Andreas Meyer and Adolf Heeb **223**
10.1 Introduction 223
10.2 TPV based on III/V low-bandgap photocells 224
10.3 TPV in residential heating systems 225
10.4 Progress in TPV with silicon photocells 227
10.4.1 Design of the system and a description of the components 227
10.4.2 Small prototype and demonstration TPV system 228
10.4.3 Prototype heating furnace 230
10.4.4 Foam ceramic emitters 231
10.5 Design of a novel thin-film TPV system 235
10.5.1 TPV with nanostructured SiGe photocells 240
10.6 Conclusion 243
Acknowledgments 243
References 243

11 Solar cells for TPV converters
V M Andreev   **246**
11.1 Introduction   246
11.2 Predicted efficiency of TPV cells   247
11.3 Germanium-based TPV cells   251
11.4 Silicon-based solar PV cells for TPV applications   254
11.5 GaSb TPV cells   256
11.6 TPV cells based on InAs- and GaSb-related materials   260
    11.6.1 InGaAsSb/GaSb TPV cells   261
    11.6.2 Sub-bandgap photon reflection in InGaAsSb/GaSb TPV
        cells   263
    11.6.3 Tandem GaSb/InGaAsSb TPV cells   263
    11.6.4 TPV cells based on low-bandgap InAsSbP/InAs   264
11.7 TPV cells based on InGaAs/InP heterostructures   266
11.8 Summary   268
    Acknowledgments   269
    References   269

12 Wafer-bonding and film transfer for advanced PV cells
C Jaussaud, E Jalaguier and D Mencaraglia   **274**
12.1 Introduction   274
12.2 Wafer-bonding and transfer application to SOI structures   274
12.3 Other transfer processes   277
12.4 Application of film transfer to III–V structures and PV cells   279
    12.4.1 HEMT InAlAs/InGaAs transistors on films transferred
        onto Si   280
    12.4.2 Multi-junction photovoltaic cells with wafer bonding
        using metals   281
    12.4.3 Germanium layer transfer for photovoltaic applications   281
12.5 Conclusion   283
    References   283

13 Concentrator optics for the next-generation photovoltaics
P Benítez and J C Miñano   **285**
13.1 Introduction   285
    13.1.1 Desired characteristics of PV concentrators   286
    13.1.2 Concentration and acceptance angle   287
    13.1.3 Definitions of geometrical concentration and optical
        efficiency   288
    13.1.4 The effective acceptance angle   290
    13.1.5 Non-uniform irradiance on the solar cell:
        How critical is it?   296
    13.1.6 The PV design challenge   305
    13.1.7 Non-imaging optics: the best framework for concentrator
        design   309

13.2 Concentrator optics overview   312
       13.2.1 Classical concentrators   312
       13.2.2 The SMS PV concentrators   314
13.3 Advanced research in non-imaging optics   319
13.4 Summary   320
       Acknowledgments   321
       Appendix: Uniform distribution as the optimum illumination   321
       References   322

**Appendix: Conclusions of the Third-generation PV workshop
for high efficiency through full spectrum utilization**   **326**

Index   **328**

11 Solar cells for TPV converters
   V M Andreev   **246**
   11.1 Introduction   246
   11.2 Predicted efficiency of TPV cells   247
   11.3 Germanium-based TPV cells   251
   11.4 Silicon-based solar PV cells for TPV applications   254
   11.5 GaSb TPV cells   256
   11.6 TPV cells based on InAs- and GaSb-related materials   260
        11.6.1 InGaAsSb/GaSb TPV cells   261
        11.6.2 Sub-bandgap photon reflection in InGaAsSb/GaSb TPV
                cells   263
        11.6.3 Tandem GaSb/InGaAsSb TPV cells   263
        11.6.4 TPV cells based on low-bandgap InAsSbP/InAs   264
   11.7 TPV cells based on InGaAs/InP heterostructures   266
   11.8 Summary   268
        Acknowledgments   269
        References   269

12 Wafer-bonding and film transfer for advanced PV cells
   C Jaussaud, E Jalaguier and D Mencaraglia   **274**
   12.1 Introduction   274
   12.2 Wafer-bonding and transfer application to SOI structures   274
   12.3 Other transfer processes   277
   12.4 Application of film transfer to III–V structures and PV cells   279
        12.4.1 HEMT InAlAs/InGaAs transistors on films transferred
                onto Si   280
        12.4.2 Multi-junction photovoltaic cells with wafer bonding
                using metals   281
        12.4.3 Germanium layer transfer for photovoltaic applications   281
   12.5 Conclusion   283
        References   283

13 Concentrator optics for the next-generation photovoltaics
   P Benítez and J C Miñano   **285**
   13.1 Introduction   285
        13.1.1 Desired characteristics of PV concentrators   286
        13.1.2 Concentration and acceptance angle   287
        13.1.3 Definitions of geometrical concentration and optical
                efficiency   288
        13.1.4 The effective acceptance angle   290
        13.1.5 Non-uniform irradiance on the solar cell:
                How critical is it?   296
        13.1.6 The PV design challenge   305
        13.1.7 Non-imaging optics: the best framework for concentrator
                design   309

Chapter 1

Non-conventional photovoltaic technology: a need to reach goals

Antonio Luque and Antonio Martí
Istituto de Energía Solar, Universidad Politécnica de Madrid
ETSI Telecomunicación, Ciudad Universitaria s/n, 28040,
Madrid, Spain

1.1 Introduction

This book is the result of a workshop celebrated in the splendid mountain residence of the Polytechnic University of Madrid next to the village of Cercedilla, near Madrid. There, a group of specialists gathered under the initiative of the Energy R&D programme of the European Commission, to discuss the feasibility of new forms for effectively converting solar energy into electricity. This book collects together the contributions of most of the speakers.

Among the participants we were proud to count the Nobel Laureate Zhores Alferov who, in the early 1980s, invented the modern III–V heterojunction solar cells, Hans Queisser who, in the early 1960s, together with the late Nobel Laureate William Shockley established the physical limits of photovoltaic (PV) conversion and Martin Green, the celebrated scientist, who, after having established records of efficiency for the now common silicon cell, hoisted the banner for the need for a 'third generation of solar cells' able to overcome the limitations of the present technological effort in PV. Together they closed the workshop.

This chapter will present the opening lecture that presented the motivation for the gathering. The thesis of this document is that present technology, despite the current impressive growth in PV, will be unlikely to reach the low cost level that is necessary for it to replace a large proportion of fuel-based electricity production. As a consequence, new forms of solar energy conversion must be developed to fulfil society's expectations for it.

1

We want immediately to state that our thesis is not to be considered to be in conflict with the PV industry as a whole nor with the mainstream of PV development. On the contrary, we think that the support of present PV technology and the expansion of the industry based on it is a must for any further step forward in the development of solar energy conversion. Furthermore, we cannot totally discard the notion that it might reach the necessary prices and goals. When talking about the future, we can only talk about likely scenarios and about recommended actions to ensure that we help towards building a sustainable future.

Accordingly, this chapter will present the stresses that advise us of the necessity for the development of renewable energies (and, among them, solar electricity), the volume of installation that will be necessary to mitigate such stresses and the forecast exercises that allow us to support our thesis (i.e. that incumbent forms of PV are probably unable to reach the necessary costs for achieving the goals of penetration defined as relevant). Then, the ways to change this situation will be briefly sketched and forecasted. For additional information other authors in this book will explain the different options in more detail. However, the collective conclusions reached at the end of the Workshop are presented in an appendix.

1.2   On the motivation for solar energy

The most obvious reason for supporting the development of a new form of energy is the exhaustion of existing ones. Will this situation occur, at least within the next half-century? Let us look at the answer given by the Royal Dutch/Shell Group [1], the big oil corporation:

> Coal will not become scarce within this timescale, though resources are concentrated in a few countries and will become increasingly complex and distant from markets. Costs of exploiting and using them will eventually affect coal's competitiveness.

> Oil production has long been expected to peak. Some think this is now imminent. But a scarcity of oil supplies—including unconventional sources and natural gas liquids—is very unlikely before 2025. This could be extended to 2040 by adopting known measures to increase vehicle efficiency and focusing oil demand on this sector. Technology improvements are likely to outpace rising depletion costs for at least the next decade, keeping new supplies below $20 per barrel. The costs of bio-fuels and gas to liquids should both fall well below $20 per barrel of oil equivalent over the next two decades, constraining oil prices.

> Gas resource uncertainty is significant. Scarcity could occur as early as 2025, or well after 2050. Gas is considered by many to be more scarce than oil, constraining expansion. But the key issue is whether there

can be timely development of the infrastructure to transport remote gas economically.

Nuclear energy expansion has stalled in OECD countries, not only because of safety concerns but because new nuclear power is uncompetitive. Even with emission constraints, the liberalisation of gas and power markets means this is unlikely to change over the next two decades. Further ahead, technology advances could make a new generation of nuclear supplies competitive.

Renewable energy resources are adequate to meet all potential energy needs, despite competing with food and leisure for land use. But widespread use of solar and wind will require new forms of energy storage. Renewable energy has made few inroads into primary energy supply. Although the costs of wind and photovoltaic sources have fallen dramatically over the past two decades, this is also true for conventional energy (direct quotations to Shell report reproduced here with permission from Shell International Limited, 2001).

Thus, in summary, no global energy shortage is expected to appear in the next 50 years but for Shell:

Demographics, urbanisation, incomes, market liberalisation and energy demand are all important factors in shaping the energy system but are not likely to be central to its evolution. By contrast, the availability of energy resources and, in particular, potential oil scarcity in the second quarter of the century, followed by gas some time later, will transform the system. What will take the place of oil—an orderly transition to bio-fuels in advanced internal combustion engines or a step-change to new technologies and new fuels?

Therefore, we can expect an important transformation in the energy system caused, to a large extent, by oil scarcity. It is true that this scarcity will affect the energy used for transport, which is not at the moment electric, more directly while here we are dealing with electric energy. However, this may well not be the situation when fuel cells have been developed and penetrate, to an important extent, the transportation system. In any case, the transformation of the energy system will certainly affect electricity production, in the sense of extending its proportion in the final use of energy.

However, the second reason, perhaps publicly perceived as the most important today, for public support of the development of new forms of energy is sustainability. According to the Intergovernmental Panel of Climatic Change (IPCC) [2] in its Third Assessment Report (TAR), based on models corresponding to six scenarios (plus an additional one corresponding to the preceding Second Assessment Report (SAR)), they present a number of statements that are considered robust findings:

Most of observed warming over last 50 years (is) likely due to increases in greenhouse gas concentrations due to human activities. (See figure 1.1.)

CO_2 concentrations increasing over the 21st century (are) virtually certain to be mainly due to fossil-fuel emission. (See figure 1.2.)

Global average surface temperature during 21st century (is) rising at rates very likely without precedent during last 10000 years. (See figure 1.1.)

An additional feature of climatic change is associated with the inertia of the climatic system. Even if we immediately stop the emission of greenhouse gases, the quantity of CO_2 will continue to rise as well as the temperature. However, the social system also has inertia and reductions in greenhouse gas emission cannot occur immediately. A semi-qualitative diagram is presented in figure 1.3. For instance, attempts to stabilize the concentration of CO_2 in the atmosphere require actions to reduce the emission of greenhouse gases to well below the present level. The earlier we start to reduce the emission level, the lower the level of stabilization achieved will be but stabilization will still take one to three centuries. Temperature will stabilize even more slowly and the rise in the sea level due to thermal expansion and ice melting will take millennia.

In contrast, the reduction in 'greenhouse gases' other than CO_2 is easier and can be achieved within decades after the emissions are curbed.

An agreed model has been used to determine the conditions which would lead to a fixed final CO_2 concentration in the atmosphere (stabilization). Based on this, the IPCC TAR states that

stabilization of atmospheric CO_2 concentrations at 450, 650, or 1000 ppm would require global anthropogenic CO_2 emissions to drop below year 1990 levels, within a few decades, about a century, or about 2 centuries, respectively, and continue to decrease steadily thereafter to a small fraction of current emissions. Emissions would peak in about 1 to 2 decades (450 ppm) and roughly a century (1000 ppm) from the present.

Reaching these goals requires a form of energy production virtually free from CO_2 release. Only nuclear power and renewable energies have this characteristic. The large extent of this necessary reduction implies that such sources must eventually be fully developed, both in cost and storage capability.

Other characteristics of the coming climate are, according to the IPCC's official opinion in its TAR,

Nearly all land areas very likely to warm more than the global average, with more hot days and heat waves and fewer cold days and cold waves.

Hydrological cycle more intense. Increase in globally averaged precipitation and more intense precipitation events very likely over many areas.

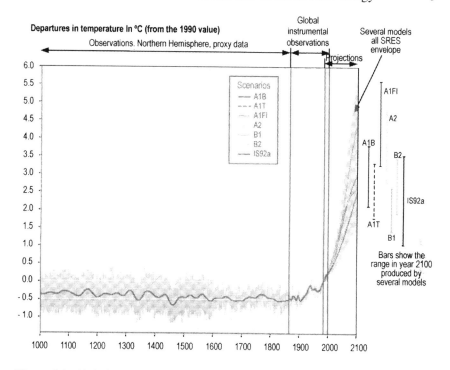

Figure 1.1. Variations in the Earth's surface temperature: years 1000 to 2100. From year 1000 to year 1860 variations in average surface temperature of the Northern Hemisphere are shown (corresponding data from the Southern Hemisphere not available) reconstructed from proxy data (tree rings, corals, ice cores and historical records). Thereafter instrumental data are used. Scenarios A are economically oriented, (A1FI, fossil fuel intensive, AT non-fossil, AB balanced), scenarios B ecologically oriented. Index 1 represents global convergence; index 2, a culture of diversity. The scenario IS92 was used in the SAR. For instance, scenario B1 which is ecologically oriented in a converging world is the most effective to mitigate the temperature increase (© Intergovernmental Panel on Climate Change. Reproduced with permission).

> Increased summer drying and associated risk of drought likely over most mid-latitude continental interior.

But undertaking the ambitious task of stabilizing the CO_2 content is only worthwhile if the consequences of the climatic change are adverse enough. In this respect, the IPCC TAR, while recognizing that the extent of the adverse and favourable effects cannot yet be quantified, advances the following 'robust findings':

> Projected climate change will have beneficial and adverse effects on both environmental and socio-economic systems, but the larger the

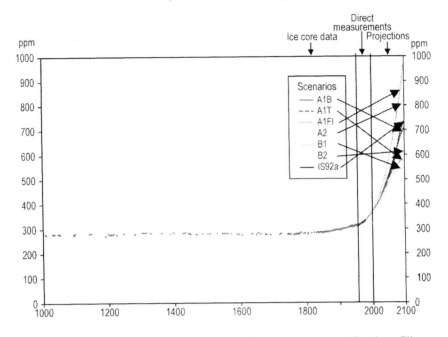

Figure 1.2. Atmospheric CO_2 concentrations (© Intergovernmental Panel on Climate Change. Reproduced with permission).

changes and the rate of change in climate, the more the adverse effects predominate.

The adverse impacts of climate change are expected to fall disproportionately upon developing countries and the poor persons within countries.

Ecosystems and species are vulnerable to climate change and other stresses (as illustrated by observed impacts of recent regional temperature changes) and some will be irreversibly damaged or lost.

In some mid to high latitudes, plant productivity (trees and some agricultural crops) would increase with small increases in temperature. Plant productivity would decrease in most regions of the world for warming beyond a few °C.

Many physical systems are vulnerable to climate change (e.g., the impact of coastal storm surges will be exacerbated by sea-level rise and glaciers and permafrost will continue to retreat).

In summary, a climatic change has already been triggered by human activity. Nature has always possessed a fearsome might. We might rightly say that we are awakening her wrath. By mid-century, the consequences, while certainly

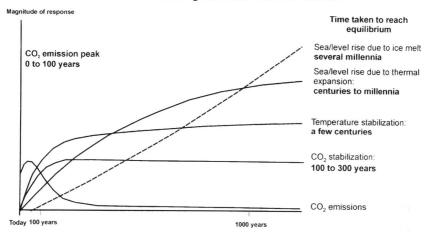

**CO$_2$ concentration, temperature and sea level
continue to rise long after emissions are reduced**

Magnitude of response

Time taken to reach
equilibrium

CO$_2$ emission peak
0 to 100 years

Sea/level rise due to ice melt
several millennia

Sea/level rise due to thermal
expansion:
centuries to millennia

Temperature stabilization:
a few centuries

CO$_2$ stabilization:
100 to 300 years

CO$_2$ emissions

Today 100 years

1000 years

Figure 1.3. Generic illustration of the inertia effects on CO$_2$ concentration, the temperature and sea level rise. Note that stabilization requires a substantial reduction in CO$_2$ emissions, well below its present levels. In the long term, the use of non-polluting energies is a must to reach stabilization (© Intergovernmental Panel on Climate Change. Reproduced with permission).

not pleasant, might perhaps not be sufficiently dramatic globally but they will become so in the centuries to come if we do not immediately initiate a vigorous programme of climatic change mitigation. Intergenerational solidarity requests us to start acting now.

1.3   Penetration goals for PV electricity

In this section we are going to present some results from the Renewable Intensive Global Energy Supply (RIGES) scenario. This scenario was commissioned by the United Nations Solar Energy Group on Environment and Development as part of a book [3] intended to be an input to the 1992 Rio de Janeiro Conference on the Environment and Development. This supply scenario was devised to respond to one of the demand scenarios prepared by the Response Strategies Working Group of the IPCC (who also presented its own supply scenario). The chosen IPCC demand scenario was the one called 'Accelerated Policies'.

In this demand scenario, the growth of Gross Domestic Product (GDP) is assumed to be high in all of the 11 regions into which the scenario is divided. It is, thus, a socially acceptable scenario in which the growth of the poorest is not sacrificed to environmental concerns. Advanced measures in energy efficiency are also assumed.

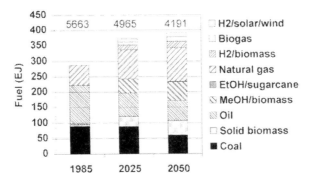

Figure 1.4. Fuel supply in RIGES. The number above the columns gives the carbon emission as CO_2 in Mt of C (elaborated with data from appendix A in [3]).

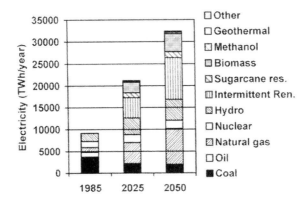

Figure 1.5. Electricity supply in RIGES. The fuels used for electricity generation are included in figure 1.4 (elaborated with data in appendix A from [3]).

In all the IPCC demand scenarios, not only the 'Accelerated Policies' one, much of the final use energy is provided in the form of electricity; therefore, it is electricity that experiences the highest growth while other fuels grow more moderately. All scenarios extend until 2050.

The results of the RIGES scenario separate other fuels from electricity. This avoids any discussion of how to translate the electricity from renewable sources (like hydroelectricity) into an 'equivalent' primary energy that contributes to the final use of the energy without affecting its production. These results are presented in figures 1.4 and 1.5.

The first result to note is that an increase in energy use can be obtained, with an intensive use of renewable energy sources, together with a decrease in CO_2 releases from 5663 Mt of carbon in 1985 to 4191 Mt in 2050. This decrease in CO_2 releases is obtained thanks to a moderate increase of only 36% in primary energy consumption, and an extensive use of renewable energies, that in 2050, will reach 41% of the total fuels used. The large proportion of natural gas, with its large content of hydrogen instead of carbon as the combustible element, also helps this result to be reached.

At the same time, the increments in the final use of the energy are largely satisfied by the 3.5-fold increment in electricity consumption supplied in 2050, mainly by renewable sources (62%) with fossil fuels providing 31%, the remaining 7% being nuclear and geothermal (the latter only 0.6%), that do not release appreciable quantities of CO_2. It is of interest to note that intermittent renewable sources, namely solar and wind power, amount to 30% of the total quantity of electricity generated and constitute the largest contribution to the global electricity supply.

But does this picture constitute a prediction of the energy situation by the middle of the 21st century? Not at all! Scenarios like this represent a set of self-consistent variables that may constitute a picture of the reality but there are other sets of parameters representing alternative and equally possible pictures. However, there are many more pictures with non-self-consistent sets of parameters that cannot occur. The study of scenarios tries to discard such impossible patterns and to focus on the self-consistent ones.

1.4   Will PV electricity reach costs sufficiently low to permit a wide penetration?

Reaching the penetration level assigned in the preceding scenario exercise implies that PV electricity has to reduce its cost to levels that makes it possible for it to compete with other electricity production technologies. Indeed, an energy technology is not adopted on cost considerations alone. Its choice has largely to do with why this technology is more convenient than the competing technologies. Modularity and image (which leads to generous public support for its installation), not price, are the origin of the impressive growth that PV sales have experienced in recent years. But prices must come closer to those of other technologies for any real massive penetration to be viable.

In figure 1.6 we present the evolution of PV module sales. We have witnessed, in the last five years, an explosive growth that almost nobody dared to foresee. The continuous curve represents an annual growth rate of 30%. The broken curve represents the model described later.

We have modelled the growth of the PV module market and the evolution of PV prices [4]. On one side, we have considered the learning curve that states that, for many goods, prices are reduced in a similar proportion every time the cumulated production of the good is doubled (the ratio of prices is the inverse of

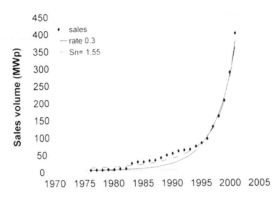

Figure 1.6. Annual sales of photovoltaic modules and model interpolations (from [4]. ©
John Wiley & Sons Ltd. Reproduced with permission).

the ratio of cumulated markets raised to the power n),

$$\frac{p}{p_0} = \left[1 + \frac{\int_0^t m \, dt}{M_0} \right]^{-n} \tag{1.1}$$

p being the price and m the annual market at time t (p_0 and m_0 are the
corresponding values at the initial time of consideration). M_0 is the accumulated
market at the initial time of consideration.

In the case of PV, the price reduction is 17.5% ($n = 0.277$) in constant
dollars every time production doubles. This law allows us to forecast the price of
the modules at any future moment if we know the cumulated sales at this moment
or, alternatively, if we know the annual sales.

In many studies the annual increase in sales is considered to be constant, i.e.
the sales each year are considered to be those of the previous year multiplied by a
constant. This is what has been done to achieve the continuous curve in figure 1.6;
in this case, the annual rate of growth has been taken as 30%. However, we have
preferred to link this growth to an economic variable. This is the demand elasticity
S defined as the opposite of the logarithmic derivative of the annual market with
respect to the price (or the ratio of the relative increment of the annual market for
a very small relative decrement in the price):

$$S = -\frac{p}{m} \frac{dm}{dp}. \tag{1.2}$$

The broken curve in figure 1.6 represents this model when adjusted for best fitting
with the real market data ($Sn = 1.55$). The fit is better than the exponential model.
Combining equations (1.1) and (1.2) leads to

$$\left(\frac{m}{m_0} \right)^{\frac{1}{Sn}} = 1 + \frac{\int_0^t m \, dt}{M_0}. \tag{1.3}$$

For *constant Sn*, the solution is

$$m = m_0 \left[1 - t \frac{m_0}{M_0}(Sn - 1) \right]^{-Sn/(Sn-1)} . \qquad (1.4)$$

This equation shows an asymptote for $t = M_0/[m_0(Sn - 1)]$. This asymptotic behaviour means that the market's rate of growth increases every year and this, in fact, has been observed in recent years. However, this cannot last for long. In fact, with the previously mentioned data, the asymptote is located in 2009 ($t = 0$ is 1998). It is clear that Sn cannot be taken as a constant. In fact, there is no reason for it to be so. While there is much empirical evidence for many products that n is constant as long as there are no drastic changes in technology and this is the case for PV where 90% or more of the market is dominated by flat crystalline silicon modules. However, there is no rule that sets S as a constant. Consequently, S has been considered to be variable according to the following simplified pattern:

$$
\begin{aligned}
&\text{if} \quad (pm < C_s(t) \quad \text{and} \quad p_c < p) \quad \text{then} \quad S = S_i \\
&\text{if} \quad (C_s(t) \le pm \quad \text{and} \quad p_c < p) \quad \text{then} \quad S = S_s \qquad (1.5)\\
&\text{if} \quad (p \le p_c) \quad \text{then} \quad S = S_c
\end{aligned}
$$

where $p(t)$ is the module price. The meaning of this expression is that S takes a high initial value S_i when the total annual expenditure pm in PV modules is below a certain threshold $C_s(t)$, then, when this threshold is reached, S decreases to a stagnation value S_s. Finally, if a certain price of competence p_c is reached, S takes another high value S_c of competence.

The explanation of these conditions is as follows. $S = S_i$ today because people are willing to buy PV modules regardless of their high price as they find one or several convenient characteristics in PV electricity. This has always been so, as is rightly stressed by Shell in its cited report [1]:

> A technology that offers superior or new qualities, even at higher costs, can dramatically change lifestyles and related energy use. Widespread introduction of electricity in the early twentieth century prompted fundamental changes in production processes, business organization and patterns of life. Coal-fired steam engines powered the early stages of industrialisation, replacing wood, water and wind. The internal combustion engine provided vastly superior personal transport, boosting oil consumption.

One such superior quality is certainly a sense of freedom and solidarity and, to no lesser extent, image. PV is a clean technology that gives prestige to its owner (whether an individual or a corporation), more than many other sumptuary expenditures. Furthermore, it is modular. The general expenditure to enjoy this good is not very high. It can be afforded in many homes and you can 'do it yourself' so boosting the sense of freedom from large utility corporations.

Furthermore, the government may satisfy the wishes of the population concerning clean energy with low total cost but high symbolic value. For a stand-alone technology, it is generally reliable and easy to handle, thus reducing maintenance greatly with respect to the alternatives. For developing rural areas, it adds to the preceding advantages the approbation of donor organizations that often support rural development.

However, this generally favourable public acceptance will change when the operating costs really start to affect the economy. Then the opposition to delivering funds for this expensive alternative will increase and any increase in the market will require a real price reduction, i.e. S_s will be lower.

Again, when due to experience, the price has been reduced sufficiently so as to compete with the incumbent electricity generator, the situation will change and S_c will increase because the advantages of PV electricity will no longer be hampered by the price drawback. Yet this model is not intended to study this competition phase, only to detect in its onset—a final vertical asymptotic behaviour—the end of the validity of this study.

An interesting result is that it is virtually independent of the value selected for S_s (as long as $S_s n < 0.45$) and S_c (as long as $S_c n > 1.4$), which are the values of S to be used for the long-term future. For the short-term future, the use of the historic value of S_i seems justified. This leads us to an apparently obvious conclusion: the future markets of PV modules, in monetary terms, will amount, for a long period, to what society is willing to pay for a good that is purchased by its unique characteristics and one which is not competing with any other one equivalent.

To simplify, the level of expenditure that society is willing to pay wordwide for PV modules is assumed to be

$$C_s(t) = C_{s0}(1 + \kappa t) \tag{1.6}$$

which is growing at the rate of the total GDP of the industrialized countries as forecast in RIGES, C_{s0} being parametrized and the parameter κ taking the value $\kappa = 0.056$ year^{-1}. Of course, many other patterns are possible but a proper parametrization will cause them to be within the limits studied.

We present in figure 1.7 the growth of the market for several values of the parameters. The value of $C_{s0} = 5$ billion dollars corresponds to devoting to PV 0.1% of the GDP of the industrialized countries. It is assumed that only one-third of this amount, i.e. five billion dollars, is devoted to the purchase of modules. Additional curves have been drawn with C_{s0} twice and half the preceding value.

The evolution of prices is represented in figure 1.8. Note that the price considered by us [5] to be necessary for competition with conventional electricity, 0.35\$ Wp^{-1}, is not reached until 2050. As for the 1\$ Wp^{-1} barrier, in the most optimistic assumption in our study, it is reached in 2012, for an annual market of 18 GWp and, in the most pessimistic, it is reached in 2027 for an annual market of 7 GWp. This study does not foresee that it can be reached within this decade, as is the goal of some R&D programmes.

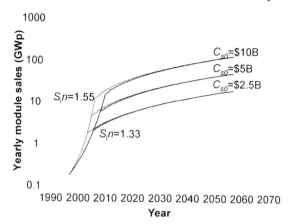

Figure 1.7. Annual module sales, in power units, for several values of the parameters. Note the good predictive behaviour of the model so far. In 1998, when the market was 159 MWp, the model predicted 362 MWp for 2001. The recorded market has been 381 MWp (from [4]. © John Wiley & Sons Ltd. Reproduced with permission).

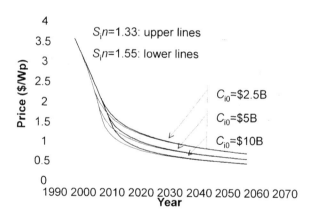

Figure 1.8. Prices predicted by the model for several values of the parameters. The competition price, assumed to be 0.35$ Wp^{-1}, is not reached within the period of study (from [4]. © John Wiley & Sons Ltd. Reproduced with permission).

This relatively disappointing price evolution is due to the low learning curve or rate, which, as we have already said, in PV is only 17.5% in constant dollars. It is higher than that of wind power, 15%, but much smaller than semiconductor memories, some 32%.

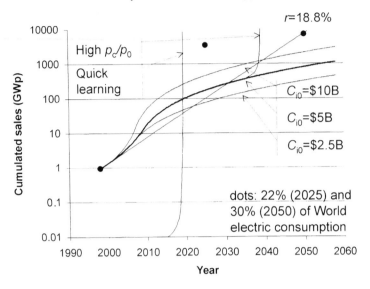

Figure 1.9. Cumulated sales of installed PV capacity based on the model described in the text (from [4]. © John Wiley & Sons Ltd. Reproduced with permission).

1.5   The need for a technological breakthrough

With the help of the indicated model, we draw in figure 1.9 the cumulated market that, given the fast growth of PV and the long expected lifecycle of the modules, is almost the same as that of PV installed power. In this diagram we have also indicated (by dots) the installed PV capacity necessary to provide (in good climates) the annual intermittent electricity programmed in RIGES. We observe that our model leads, depending on the value of C_{s0}, to 4.5–29.1% in the amount programmed for 2050. Furthermore, it is more expensive than incumbent electricity sources.

   Of course, for a sufficiently high value of C_{s0}, the required cumulated sales would be reached. This value is 20 billion dollars which should be compared with the 5 billion of our central case. This would imply devoting up to 0.4% of the GDP of the industrialized countries to the development of PV electricity. In this case, the price of competence $p_c = 0.35\$ \ Wp^{-1}$ is also achieved by about 2050. The question is whether society is willing to support so heavily for so long a cost-ineffective technology that will eventually become cost-effective.

   Even with less support, PV electricity might become cost effective if, for some reason, modules of $0.7\$ \ Wp^{-1}$, instead of the $0.35\$ \ Wp^{-1}$ considered so far, can lead to cost- effective generating plants. This might happen by a misjudgement on our part, if we have fixed too stringent a condition for the cost-competitive module price or, what is the same, by a modification in the costs of commercialization that permit higher cost modules for a given price of the installed PV generator but also by an undesired increase in the global price of electricity in constant dollars. However, the situation is expected to be the opposite: prices of incumbent electricity will decrease, as they have done historically.

If 0.7\$ Wp^{-1} modules become competitive, the cumulated market will follow the pattern of the curve labelled 'high p_c/p_0'. In this case, the price for competition would be reached by 2038 in the central case represented in figure 1.9. Even for the lower case (not drawn), the price of competence would be reached by 2045, before the end of the half of the century.

However, relying on cost reduction by experience is a risky way of approaching the problem. Taking risks for radical innovation is less risky, we believe.

Based on its scenario 'Dynamics as Usual', Shell tells [1] a tale of a possible energy history in the 21st century:

> In the first two decades of the century, renewable energy grows rapidly in OECD countries, within the framework of established electricity grids and strong government support... Deregulated markets provide opportunities for branded 'green energy', which gain 10% of demand in some regions. . ..

> Governments support a spread of renewable technologies to address public concerns about health, climate and supply security. Renewables experience more than 10% compound growth—with photovoltaic solar and wind growing at over 20% a year. By 2020 a wide variety of renewable sources is supplying a fifth of electricity in many OECD markets and nearly a tenth of global primary energy. Then growth stalls. . . .

> Stagnant electricity demand in OECD limits opportunities for expansion. Although the public supports renewables, most are unwilling to pay premium prices. In spite of significant cost improvements, photovoltaic power gains only niche markets. And with little progress on energy storage, concerns about power grid reliability block further growth of wind and solar energy. . . .

> Since 2025 when the first wave of renewables began to stagnate, biotechnology, materials advances and sophisticated electric network controls have enabled a new generation of renewable technologies to emerge. . . . A range of commercial solutions emerge to store and utilize distributed solar energy. By 2050 renewables reach a third of world primary energy and are supplying most incremental energy.

For Shell, this is not the only scenario. Many others may exist, as they clearly state, but in another one they present, the so called 'The Spirit of the Coming Age', strongly based on the technological revolutions they perceive around fuel cells and H_2 technology, they also give crucial and similar weight to 'second-wave' PVs because, as they state at the beginning of their report, for the energy system,

> Two potentially disruptive energy technologies are solar photovoltaics, which offer abundant direct and widely distributed energy, and hydrogen fuel cells, which offer high performance and clean final energy from a variety of fuels. Both will benefit from manufacturing economies but both presently have fundamental weaknesses.

A similar message is the one implicit in our study. Silicon-based PV technology will permit a tremendous expansion in PV electricity but most probably 'in spite of significant cost improvements, photovoltaic power will gain only niche markets'.

Let us analyse why crystalline silicon cell technology will fail to reach lower prices. From a model point of view it is the low learning factor of only 17.5%, compared to the 32% of semiconductor memories. Why is this factor so low for silicon cell modules? We are going to advance a suggestion. The reason is related to the nature of solar energy. Solar energy reaches the Earth in huge amounts but it comes to us in a relatively dispersed form. For the exploitation of a resource this is very important. Mineral beds are only exploitable if the concentration of ore is above a certain limit. Solar energy is a unique kind of mineral bed. Unlike other mineral beds, it is available everywhere but, as in many others, its resource concentration is modest. Not so modest to make it non-exploitable but modest enough to make its exploitation relatively expensive. Almost every collecting material—including those that are used only for support or auxiliary mechanisms—is already expensive. Making them cheaper is certainly necessary but extracting the ore with more efficiency is as important. However, in the classic single-gap PV cell and, in particular, in the silicon ones, only the energy of the photons that are close to the semiconductor bandgap is extracted effectively. For the remaining photons, the energy is extracted rather ineffectively or totally wasted. Consequently, the possibility of greatly improving the silicon cell behaviour is limited. The top efficiency of a single-gap solar cell is limited to 40% under the very best conditions. In contrast, the PV limit of efficiency when this limitation is removed goes to efficiencies of about 85% under the same very best conditions. Improving the efficiency of multi-junction solar cells is an active area of research today: 32% has been reached and 40% is a medium-term goal. This should be compared with the 15% of most commercial silicon cells or the 25% of the best laboratory silicon cell.

But multi-junction solar cells are not the only way of making better use of the solar spectrum. The authors in this book will present proposals that might realize second-wave PV generators.

From our model viewpoint, innovations based on a better use of this solar spectrum should lead to a faster experience factor since more improvements are possible and more things remain to be learned. As an exercise, let us consider

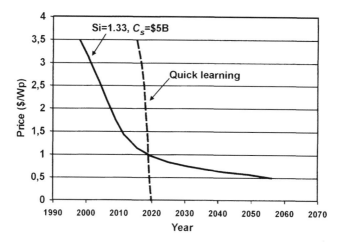

Figure 1.10. Cost evolution of a quick learning technology compared to the baseline case.

a solar converter learning at the rate of semiconductor memories (32%) that has been able to enter into the market and reach a cumulated market of 10 MWp by 2015 with a price in this year as high of 3.5$ Wp^{-1}. If this situation is reached, in four years such modules will be able to compete with incumbent electricity!

It might seem difficult to sell a technology that is so expensive compared with the concurrent silicon technology that, at that time, will cost only about 1$ Wp^{-1}, as shown in figure 1.10. Of course it would be better if the price of the new technology were to be smaller. However, most probably any new technology would start competing at a higher price and, therefore, it will have to have distinctive characteristics to make it attractive. Furthermore, someone will be forced to take an entrepreneurial risk to enable this product to expand. Anyway, from the model viewpoint, if the starting price were to be smaller, the price of competition would be reached slightly faster. We want, however, to stress again the importance of a fast learning curve, which is commonly associated with new technologies.

1.6 Conclusions

We are about to experience a revolution in energy. The social push towards deregulation, concerns about sustainability, the scarcity of oil by the second quarter of the century, the disrupting role of the ever present and modular PV technology, together with that of the hydrogen technology driven by fuel cells will all constitute the driving forces of this revolution.

Present-day silicon PV technology will be at the onset of this revolution. It will grow tremendously in this decade constituting one of the first big new economic activities of the 21st century. But then its growth will stagnate, as the cost-reducing capacity of present commercial PV technology is moderate.

One reason for this moderate cost-reducing capacity is in the poor utilization of the solar resource that is huge but dispersed. Only photons with energy close to the semiconductor energy bandgap will be used effectively. For the rest, their energy will be ineffectively converted or totally wasted.

A new generation (after silicon and single-gap thin films) of technologies making better use of the solar spectrum will constitute the second wave of the energy revolution announced by Shell. Its potential will be based on its stronger capacity for cost reduction by experience due to its higher limiting efficiency and to the fact that it will be based on novel and unexplored concepts.

This technology is not yet ready but, in this book, you will find the germ of many of the solutions which will form this second wave. For Shell, this will ripen in the second quarter of the century. We must do our best to start as soon as possible. The later we go to market, the more difficult it will be to displace the already established and cheap silicon technology. And we will need the support of silicon technology manufacturers, not their opposition, to succeed in marketing the potentially cheaper new technologies. This must be kept well in mind.

May the germ in this book blossom into actions leading to the establishment of the second-wave PV technology necessary to mitigate the adverse climatic changes and to permit higher equity in a world that must be based on enriching everyone.

References

[1] Shell International 2001 *Exploring the Future. Energy Needs, Choices and Possibilities. Scenarios to 2050* Global Business Environment. Shell International Limited, London
[2] Watson R T *et al* 2001 Climatic Change 2001, Synthesis Report *IPCC Plenary XVIII*, Wembley
[3] Johansson T B, Kelly H, Reddy A K N, Williams R H and Burnham L (ed) 1993 *Renewable Energy Sources for Fuel and Electricity* (Washington, DC: Island Press)
[4] Luque A 2001 Photovoltaic markets and costs forecast based on a demand elasticity model *Prog. Photovoltaics: Res. Appl.* **9** 303–12
[5] Yamaguchi M and Luque A 1999 High efficiency and high concentration in photovoltaics *IEEE Trans. Electron. Devices* **46** 2139–44

Chapter 2

Trends in the development of solar photovoltaics

Zh I Alferov and V D Rumyantsev
Ioffe Physico-Technical Institute, 26 Polytechnicheskaya,
194021, St Petersburg, Russia

2.1 Introduction

Current civilization is based on mankind's economic and social experience of the organization of life, accumulated over thousands of years, resulting in increasing material consumption but also providing energy and information benefits. Radical alterations in the material base of civilization started at the end of the 18th century industrial revolution (just after the invention of the steam-engine). Since that time, scientific and technical progress has accelerated. To supply energy to power the various technical inventions, a powerful and gradually growing infrastructure leaning upon fossil fuel resources has been created. As it is easily converted into other types of energy, the consumption of electrical energy increases rapidly.

Nature has localized depositions of the fossil fuels necessary for operating thermal and atomic power stations. For this reason the maintenance and development of the fuel-powered complex has become a global problem and not only a technical one as, in many respects, it is also a political problem. However, mankind does not seriously concern itself with the fact that fuel resources are exhaustible and the ecological damage resulting from their use may, in the future, reduce their usefulness. Meanwhile, both these circumstances have already made themselves evident. The exploitation of new deposits of fossil fuels to replace exhausted ones becomes more and more difficult. The number of natural catastrophes is increasing and this is ascribed to the beginning of the 'greenhouse effect' resulting from the rise in the carbonic gas content in the atmosphere from the combustion of organic fuels. With an increase in the number of atomic power stations, the risk of technological catastrophes with serious consequences is growing.

At the present time, there is a growing conviction that the power industry of the future has to be based on the large-scale use of solar energy, its manifestations being quite different. The Sun is a huge, inexhaustible, absolutely safe energy source which both belongs equally to everyone and is accessible to everyone. To rely on the solar-powered industry must be considered not only a sure choice but also the only alternative for mankind as a long-term prospect. We shall consider the possibilities for converting solar energy into electricity by means of semiconductor photocells both retrospectively and for long-term planning. In both scientific and technological aspects, these devices are ready to be considered as a technical basis for large-scale solar photovoltaics of the future.

2.2   Starting period

Edmond Becquerel first observed the photovoltaic (PV) effect in a liquid–solid interface in 1839. W G Adams and R E Day in London carried out the first experiments with a solid-state photovoltaic cell based on selenium in 1876 [1]. It took more than a half of century for the creation of the first solar photocells with an efficiency barely exceeding 1%. These were thallium sulphide photocells with a rectifying region [2]. The investigations were carried out under the leadership of Academician A F Ioffe, who, in 1938, submitted a programme for the use of solar photovoltaic roofs to supply energy for consideration by the USSR government.

However, for the introduction of photovoltaics (even if we ignore economic considerations) essentially the devices needed to be more efficient. A decisive event in this direction was the creation in the USA in 1954 of silicon-based photocells with p–n junctions that were characterized by an efficiency of about 6% [3]. The first practical use of silicon solar arrays took place not on the Earth but in near-Earth space: in 1958, satellites supplied with such arrays were launched—the Soviet 'Sputnik-3' and the American 'Vanguard-1'.

It should be noted here that the achievements in the theory and technology of semiconductor materials and semiconductor devices with p–n junctions provided the scientific basis for the creation of the first solar cells. At that time semiconductor devices were mainly applied as converters of electric power into electric power of a different kind (alternating currents into direct ones, HF generation, switching, and so on) or in electronic circuits for information processing and translation (radio, communication, and so on). In addition to the 'classical' semiconductor materials—germanium and silicon—materials from the A^3B^5 family group were synthesized. One such material—indium antimonide—was first reported by researchers at the Physico-Technical Institute (PTI) in 1950 [4]. Also at the PTI, at the beginning of 1960s, the first solar photocells with a p–n junction based on another A^3B^5 material, gallium arsenide, were fabricated. Being second in efficiency ($\sim3\%$) only to silicon photocells, gallium arsenide cells were, nevertheless, capable of operating even after being significantly heated. The first practical application of improved gallium arsenide

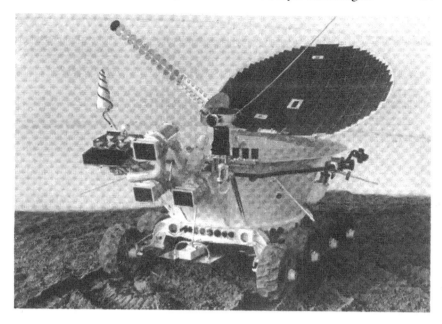

Figure 2.1. At the beginning of 1960s, it was found that p–n homojunction GaAs solar cells had a high temperature stability and were radiation resistant. The first applications of such cells took place on the Russian spacecrafts 'Venera-2' and 'Venera-3', launched in November 1965 to the 'hot' planet Venus and on moon-cars (see photograph), launched in 1970 ('Lunokhod-1') and 1972 ('Lunokhod-2').

solar arrays to supply energy was even more exotic than in the case of silicon ones (figure 2.1). They provided the electricity for the Russian space probes 'Venera-2' and 'Venera-3' operated in the vicinity of Venus (1965) as well as for the moon-cars 'Lunokhod-1' (1970) and 'Lunokhod-2' (1972).

2.3   Simple structures and simple technologies

The practical introduction of A^3B^5 materials opened a new page both in semiconductor science and in electronics. In particular, such properties of gallium arsenide as the comparatively wide forbidden gap, the small effective masses of charge carriers, the sharp edge of optical absorption, the effective radiative recombination of carriers due to the 'direct' band structure as well as the high electron mobility all contributed to the formation of a new field of semiconductor techniques–optoelectronics. Combining different A^3B^5 materials in heterojunctions, one could expect an essential improvement in the parameters of existing semiconductor devices and the creation of new ones. Again, the contribution of the PTI (Ioffe Institute) can be seen to be valuable. Here, in the second half of 1960s and the first half of 1970s, pioneer work on the fabrication and investigation of 'ideal' heterojunctions in the AlAs–GaAs system was performed with the main purpose of making semiconductor injection lasers more perfect. Heterolasers operating in the continuous mode at room temperature were fabricated and these first found application in fibre-optical communication systems. Ways for using multi-component A^3B^5 solid solutions for the creation of light-emitting and photosensitive devices operating in different spectral regions were also pointed out.

One of the results of the study of heterojunctions was the practical realization of a wide-gap window for cells. This idea had been proposed earlier and had the purpose of protecting the photoactive cell region from the effect of surface states. Defectless heterojunctions using p–AlGaAs (wide-gap window) and (p–n)GaAs (photoactive region) were successfully formed; hence, ensuring ideal conditions for the photogeneration of electron-hole pairs and their collection by the p–n junction. The efficiency of such heteroface solar cells for the first time exceeded the efficiency of silicon cells. Since the photocells with a gallium arsenide photoactive region appeared to be more radiation-resistant, they quickly found an application in space techniques, in spite of their essentially higher costs compared with silicon cells. An example (figure 2.2) of a large-scale application of the heteroface solar cells was a solar array with a total area of 70 m^2 installed on the Russian space station 'Mir' (1986).

Silicon and gallium arsenide, to a large extent, satisfy the conditions of 'ideal' semiconductor materials. If one compares these materials from the point of view of their suitability for the fabrication of a solar cell with one p–n junction, then the limiting possible efficiencies of photovoltaic conversion appear to be almost similar, being close to the absolute maximum value for a single-junction photocell (figure 2.3). It is clear that the indubitable advantages of silicon are its wide natural abundance, non-toxicity and relatively low price. All these factors and the intensive development of the industrial production of semiconductor devices for use in the electronics industry have determined an extremely important role for silicon photocells in the formation of solar photovoltaics. Although considerable efforts have been expended and notable advance has been made in the creation of different types of thin-film solar arrays, crystalline silicon (both in single- and poly-crystalline forms) still continues today to make the main contribution to the world production of solar arrays for terrestrial applications.

Until the middle of the 1980s, both silicon and gallium arsenide solar photocells were developed on the basis of relatively simple structures and simple technologies. For silicon photocells, a planar structure with a shallow p–n junction formed by the diffusion technique was used. Technological experience on the diffusion of impurities and wafer treatment from the fabrication of conventional silicon-based diodes and transistors was adopted. The quality of the initial base material in this case could rank below that of the material used for semiconductor electronics devices. For fabricating heteroface AlGaAs/GaAs solar cells, as in growing wide-gap AlGaAs windows, it was necessary to apply

Figure 2.2. The first AlGaAs/GaAs heteroface solar cells were created in 1969–70 [5]. In the following decades their AM0 efficiency was increased up to 18–19% owing to the intensive investigations in the field of physics and technology (liquid-phase epitaxy) of space solar cells. The LPE technology was used in large-scale production of PV arrays for the spacecrafts launched in 1970–80s. For example, AlGaAs/GaAs solar arrays with a total area of 70 m^2 were installed in the Russian space station 'Mir' (see photograph) launched in 1986.

epitaxial techniques. A comparatively simple liquid-phase epitaxy technique developed earlier for the fabrication of the first-generation heterolaser structures was adopted. In the case of heterophotocells (figure 2.4), it was necessary to grow only one wide-gap p–AlGaAs layer, while the p–n junction was obtained by diffusing a p-type impurity from the melt into the n-GaAs base material.

2.4 Nanostructures and 'high technologies'

From the middle of the 1980s, 'high technologies' began to penetrate into the semiconductor solar photovoltaics sphere. Complicated structures for silicon-based photocells, which enabled both optical and recombination losses to be decreased, were proposed. In addition, an effort to improve the quality of the base material was undertaken. The realization of such structures appeared to be possible due to the application of multi-stage technological processing well mastered by that time in the production of silicon-based integrated circuits. These efforts resulted in a steep rise in the photovoltaic conversion efficiency

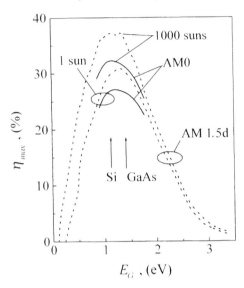

Figure 2.3. The maximum thermodynamically limited photovoltaic conversion efficiencies (η_{max}) of a solar cell with p–n junction in a material with the energy gap E_G as a variable parameter. The full curves correspond to the AM0 sun spectrum, the dotted curves to the AM1.5d spectrum. In both cases, non-concentrated sunlight (1 sun) and sunlight, concentrated up to 1000 suns, were taken into account (from [14]. © John Wiley & Sons Ltd. Reproduced with permission).

of silicon photocells. The efficiency demonstrated by the laboratory cells closely approached the theoretical limit. Unfortunately, the cost of the 'highly efficient' silicon photocells greatly exceeded that of 'conventional' ones.

At the same time, progress in the field of gallium-arsenide-based photocells took place due to the use of new epitaxial techniques for the growth of heterostructures. A metal-organic chemical vapour deposition technique (MOCVD) was mainly used. This technique was elaborated during the development of the second-generation injection lasers based on A^3B^5 compounds. Rapid development in optoelectronics required a reduction in the threshold current density in heterolasers, a rise in their output power, an improvement in their reliability and a wider spectral range for laser action. New epitaxial techniques, to which the molecular beam epitaxy technique (MBE) is also related, essentially allowed the heterolaser structure to be modified and a wider range of semiconductor materials to be used for solving these problems. The MOCVD and MBE methods could provide low rates of epitaxial growth under non-equilibrium conditions. This meant that layers could be as thin as desired as their thickness could be controlled on the monolayer level. A technological basis for the realization of many new projects using heterojunctions in device structures was

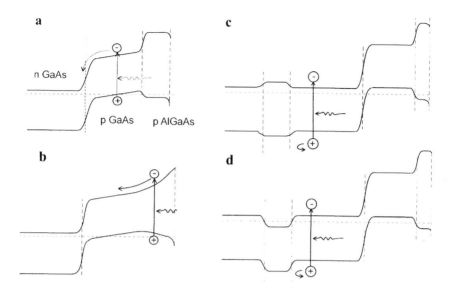

Figure 2.4. Band diagrams of the p–AlGaAs–(p–n)GaAs heteroface solar cells: (*a*) the structure, in which p-GaAs layer with a built-in quasi-electric field is formed by means of Zn diffusion into an n-GaAs base during LPE growth of the wide-gap p–AlGaAs window; (*b*) the structure with a strong built-in quasi-electric field formed during etch-back regrowth of the GaAs base in a non-saturated Ga+Al melt during LPE of the p–AlGaAs window; (*c*) the structure with a back-surface field formed by a wide-gap n-layer which improves minority-hole collection from an n-type part of the photoactive region; and (*d*) the same structure but with the back-surface field formed by a highly doped n^+-GaAs layer.

created. In particular, quantum-size (10–20 nm) active regions (planar quantum wells for the recombination of injected charge carriers) and periodic structures could now be applied in semiconductor lasers. The implementation of quantum wells and short-period (tens of nanometres) superlattices allowed the threshold current density in second-generation heterolasers to be decreased by an order of magnitude (from $\sim$1000–500 A cm^{-2} down to $\sim$100–40 A cm^{-2}) compared with that of first-generation heterolasers with a simple double heterostructure (narrow-gap active region situated between two wide-gap emitters). Built-in reflectors, operating on the principle of interference Bragg mirrors, could be formed in the structures and these were used for the creation of surface-emitting lasers. With careful choice of the composition and thickness for the contacting layers, superlattices could have built-in elastic strains but no growth defects. This property of superlattices was used to solve one of the main problems of heterojunctions—matching the lattice constants of the contacting materials. Pairs of materials with completely matched lattices could only form defectless heterostructures, as is the case in the AlGaAs/GaAs system. By introducing superlattices, this limitation could be moderated essentially. Mastering the multi-component systems of solid solutions, in particular the (Al,Ga)InAsP system, allowed overlapping a wide spectral range for different applications of injection heterolasers. The development and application of MOCVD and MBE techniques, improved not only the parameters of heterolasers but also those of many other devices. Moreover, new high-frequency devices using tunnel effects and high electron mobility effects were created. Together with the progressing technology of silicon integrated circuits, the MOCVD and MBE technologies, developed for growing planar A^3B^5 nano-heterostructures, in fact provided the material base for the 'information revolution', which we are currently witnessing.

Which structural improvements in solar heterophotocells have appeared as a result of the potential of these new technologies? First, the wide-gap AlGaAs window was optimized and its thickness had become comparable with that of the nano-dimensional active regions in heterolasers. The AlGaAs layer also began to serve the function of the third component in the triple-layered interference antireflection coating of a photocell. As in heterostructure lasers, a narrow-gap heavily doped contact layer was grown on top of the wide-gap AlGaAs window, which could be removed during post-growth wafer treatment in the areas between contact fingers. Second, a back (behind the p–n junction) wide-gap layer was introduced, which, together with the front wide-gap layer, ensured double-sided confinement of the photogenerated charge carriers within the light absorption region. The recombination losses of charge carriers before their collection by the p–n junction were reduced. At this stage of optimization of the one-junction AlGaAs/GaAs photocell heterostructures, the newly developed MOCVD technique was still in competition with a modified low-temperature liquid-phase epitaxial technique. In particular, an efficiency value of 27.6%, measured under concentrated sunlight illumination with an AM1.5 spectrum, belongs to photocells grown by the MOCVD technique (this value is an absolute record for photocells with one p–n junction [6]). However, the record efficiency value of 24.6% obtained under AM0 illumination conditions and 100 suns belongs, up to now, to photocells grown by the LPE technique [7].

In the AlGaAs/GaAs photocell structures grown by the MOCVD technique, a single wide-gap AlGaAs layer, forming the back surface field, could be replaced by a system of alternating pairs of AlAs/GaAs layers, making up a Bragg mirror (figure 2.5). The wavelength of the reflection spectrum maximum of such a mirror was chosen in the vicinity of the absorption edge of the photoactive region. The long-wavelength radiation, which was not absorbed in this region during one pass, could be absorbed at the second pass after reflection from the mirror. At the same time, the wide-gap mirror layers continued, as before, to serve the function of the back barrier for photogenerated charge carriers. In these conditions, the thickness of the photoactive region could be decreased twice without loss in current compared with structures without a mirror. As a result, the radiation tolerance of the photocells increased in essence, since the number of lattice defects, generated by irradiation with high-energy particles and, hence, degrading

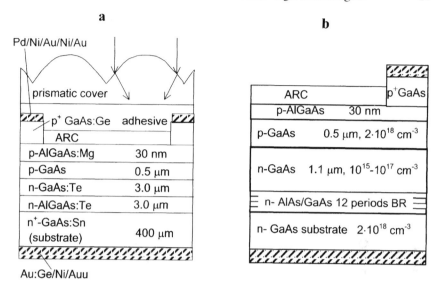

Figure 2.5. Schematic diagrams of single-junction multilayer AlGaAs/GaAs solar cells used in space. (*a*) Solar cell structure with a back-surface field n-GaAs layer and thin p–AlGaAs window grown by low-temperature LPE. An antireflection coating (ARC) and silicone prismatic cover minimized the optical losses caused by contact grid shadowing and reflection from semiconductor surface. A record efficiency of 24.6% under concentrated (100×) AM0 solar spectrum was measured in such SCs [7]. (*b*) Solar cell structure with built-in Bragg reflector (BR) grown by MOCVD. The BR consisted of 12 pairs of AlAs (72 nm) and GaAs (59 nm) layers with a total reflection coefficient of 96% centred at $\lambda = 850$ nm. As a result of this, a two-pass effect for longer-wavelength light was realized allowing a reduction in the thickness of the base n-GaAs region (up to 1–1.5 μm; AM0; 1 sun photocurrent density as high as 32.7 mA cm^{-2}). A high radiation resistance characterized by a remaining power factor of 0.84–0.86 after 1 MeV electron irradiation with fluence of 10^{15} cm^{-2} was realized in these cells [8].

the charge carriers' diffusion length, decreased proportionally to a decrease in the photoactive region thickness [8].

In parallel with the creation of a scientific and technological 'stock' of solar photocell structures created by the development of heterolaser structures, the use of new epitaxial techniques also allowed a number of strictly 'photovoltaic' problems to be solved.

The first problem was to find conditions for the growth of perfect AlGaAs/GaAs heterostructures on a germanium substrate. This became possible by using intermediate superlattices grown under non-equilibrium epitaxy conditions. After this problem was solved, heterophotocells on germanium began to be the main candidates for applications on the majority of spacecrafts. A

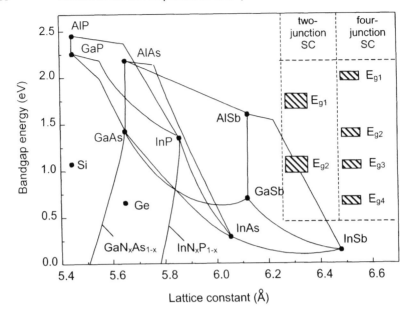

Figure 2.6. Bandgap energies *versus* lattice constant for Si, Ge, III–V compounds and their solid solutions. The boxes correspond to the bandgap intervals for possible materials to obtain the highest PV efficiencies in current-matched two-junction and four-junction solar cells.

decisive factor was the fact that germanium is mechanically stronger than gallium arsenide which had been previously used for the substrates. For this reason, arrays composed of AlGaAs/GaAs photocells on germanium were comparable by weight and strength with the silicon ones but by efficiency and radiation tolerance they outperformed them.

The second problem was of basic importance for solar photovoltaics. The case in point is the creation of cascade multi-junction solar cells (figure 2.6).

2.5   Multi-junction solar cells

The idea of cascade photocells began to be discussed in the early 1960s and was considered to be obvious; however, increasing the efficiency seemed a long way away. The situation started to change in the late 1980s, when many research groups concentrated their efforts on developing different types of double-junction solar cells. At the first stage, the best results on efficiency were obtained with mechanically stacked photocells. However, everyone understood that the really promising cells would be those with a monolithic structure. Researchers from the NREL (USA) were the first to develop such structures. Using germanium

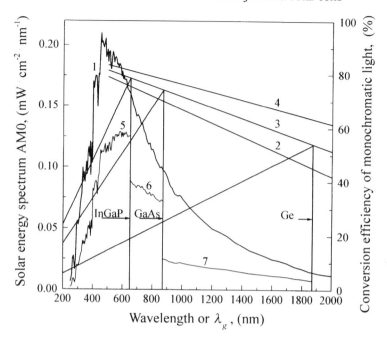

Figure 2.7. Curve 1 is the energy spectrum AM0 for non-concentrated sunlight; curves 2, 3 and 4 are plots of the maximum magnitudes of the 'monochromatic' efficiency of an idealized solar cell for photocurrent densities $I_{SC} = 0.1$, 1.0 and 10 A cm^{-2} respectively. These depend on the boundary wavelength λ_g of the semiconductor material. The sloped lines are the spectral dependences of the conversion efficiency in idealized solar cells based on In$_{0.5}$Ga$_{0.5}$P, GaAs and Ge materials, at $I_{SC} = 1.0$ A cm^{-2}; curves 5, 6 and 7 show the fraction of solar energy converted into electricity in the corresponding cell which is part of a three-junction solar cell (from [14]. © John Wiley & Sons Ltd. Reproduced with permission).

substrates and an MOCVD technique, they grew multilayer structures matched by their lattice constants, in which the upper photocell had a p–n junction in an In$_{0.5}$Ga$_{0.5}$P solid solution and the lower one was in GaAs. The cells were electrically connected in series by means of a tunnel p–n junction specially formed between the cascades. Later a third cascade with a p–n junction in a germanium substrate was included in the photovoltaic conversion process. At the present time, triple-junction photocells are already at the stage of practical application in spacecrafts (figure 2.7).

It may appear that we have paid too much attention to the development of structurally rather complicated and expensive photocells formed from A^{3}B^5 compounds. Being developed for use in a relatively narrow and specific field of energy supply, i.e. space, do they hold any promise for use in the large-scale

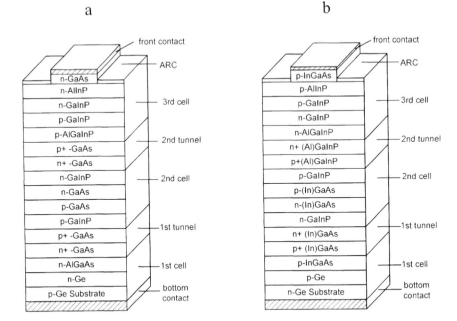

Figure 2.8. A cross section of the triple-junction solar cells of different design: (*a*) (Al)GaInP/GaAs/Ge (n-on-p) in which the second cell, as well as the first and second tunnel junctions are based on GaAs material; and (*b*) (Al)GaInP/(In)GaAs/Ge cell (p-on-n), in which first tunnel junction is based on InGaAs, whereas the second cell and second tunnel junction are based on (Al)GaInP material.

photovoltaics of the future? We believe that the answer is positive and that there exist strong arguments for this conclusion.

The structure of triple-junction heterophotocells is complicated (figure 2.8) and it will be even more complicated after the development, for example, of photocells with four and even five photoactive p–n junctions. However, epitaxial growth of such structures is a totally automatic single-stage process, the success of which depends only on the degree to which the technological base is mastered. The consumption of the initial materials (gases in a MOCVD technique) here has little dependence on the number of cascades. Since all photoactive regions are, as a rule, made of direct-band materials, the total thickness of the epitaxial structure to be grown is only several micrometres.

The substrate material mainly determines the cost of the epitaxial wafer. We have already discussed that the use of a foreign, with respect to A^3B^5 materials, germanium substrate enabled the operational parameters of space solar arrays to be improved. As a matter of fact, this has resulted in the 'second birth' of germanium technology, which was, at one time, the leading 'classical' material in semiconductor electronics and only later superseded by silicon. In principle, the cost of germanium as a substrate material is lower than that of gallium arsenide which is currently used for this purpose. Also, we recall its technological merits (mechanical tolerance to post-growth treatment) and the possibility of being included in photovoltaic conversion in a cascade structure. However, today, turning to look at the successes achieved in nanostructure technologies, one may suppose that germanium, now a substrate material, will probably, once again, be superseded by silicon, which is both cheaper and a more technological material. The work in this direction has already been carried out. Thus, the result of applying such 'high technologies' may not only be a radical increase in efficiency (in multi-junction structures) but also a radical decrease in the cost of heterostructure photocells.

Let us consider now the prospects for an increase in the efficiency of multi-junction photocells. Today, experience in developing single-, double- and triple-junction photocells allows us to hope for the practical realization of increased efficiency in higher-multiple-junction structures (see figure 2.9 and table 2.1). There are no scientific or theoretical doubts that expectations could be realized, if suitable materials for intermediate cascades can be found and if these materials can be grown to an appropriate quality. The search for such materials is on-going and several directions in this regard may be distinguished.

A 'traditional' direction is 'merely' the synthesis of new materials, for instance, semiconductor nitrides. For wide-gap nitrides, considerable technological experience (in growth by the previously mentioned MOCVD technique), motivated by the 'bright prospects of revolution' in the lighting technique, already exists. It is quite likely that we shall witness the replacement of hot mercury and incandescent tungsten in lighting devices by 'cold structures based on A^3B^5 materials of micrometre thickness'. However, narrower-gap materials, which are a closer match to the lattice types and constants of the materials already operating in triple-junction structures, are more favourable for cascade photocells. For example, GaInNAs solid solutions, which are under intensive study at the present time, can be used as such materials. It should be noted that the complicated photocell structure, namely the application of multi-cascade structures, reduces the requirements for the bulk properties of the materials used. In fact, a larger number of cascades gives a thinner photoactive region in each of them and, hence, weakens the effect of such a parameter as the minority charge carrier diffusion length. The method of compensating for insufficiently good bulk properties of the materials by the technological perfection of a cascade structure has also begun to be applied in creating new types of thin-film solar arrays.

Let us consider now some other possibilities for developing cascade solar cells. Once again, we shall refer to earlier experiences in the development of semiconductor electronics (figure 2.9) and, in particular, devices based on A^3B^5 compounds. In the past, one could distinguish two stages in such a development. The first was related to the creation of heterostructures, the second one to the creation of nano-heterostructures. In both cases, the basic initial ideas were

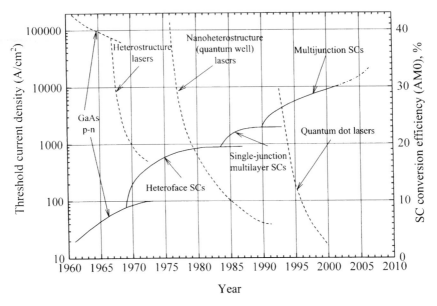

Figure 2.9. Development of A^3B^5-based injection lasers and A^3B^5-based solar cells. Dotted curves (left vertical axis) show the evolution of the threshold current densities in three generations of injection lasers [9] and the full curves (right vertical axis) show the evolution of PV conversion efficiencies in solar cells with different structures (AM0 illumination conditions, no concentration).

directed at improvements in injection lasers and to developing technologies for fabricating such lasers. In the 1970s, there was even a tradition in which the parameters of the injection heterolasers produced by one or other technique always served as the criterion of perfection of this technique. The definition 'material of laser quality' meant that a given heterostructure, merely by high crystallographic perfection, was capable of operating at the super-high excitation densities necessary for realizing a laser effect. That is why we shall analyse the trends to be observed at the present time in the work creating third-generation injection lasers to reveal the trends in the field of solar cell development.

The main feature of third-generation injection lasers is a structure with quantum dots. Quantum dots are narrow-gap semiconductor micro-crystals a few nanometres in size formed in a semiconductor matrix with a wider bandgap. An ideal quantum dot is characterized by an energy spectrum described by a δ-function, i.e. it is similar to an isolated atom. A dense and ordered array of identical quantum dots, free of dislocations and defects, will form an artificially constructed crystal, a new kind of heterosemiconductor. The 'bulk' properties of such a crystal will, to a large extent, be determined not only by the properties of the dot and matrix materials but also by the parameters of the artificial

Table 2.1. Theoretical, realistically expected and experimentally realized PV conversion efficiencies of multi-junction solar cells.

Spectrum	Efficiency	Number of junctions				
		1	2	3	4	5
AM0	Theoretical	28	33	38	42	45
	Expected	23	28	33	36	38
	Realized	21.8 [10]	27.2 [11]	29.3 [11]	—	—
AM1.5	Theoretical	30	36	42	47	49
	Expected	27	33	38	42	44
	Realized [6]	25.1	30.3	31.0	—	—
AM1.5 with concentration	Theoretical	35	42	48	52	54
	Expected	31	38	43	47	49
	Realized [6]	27.6	31.1	34.0	—	—

structure. The use of a heterosemiconductor as the active region in injection lasers, in addition to a decrease in the threshold current compared with that of lasers based on nano-heterostructures, ensures the stability of this parameter with temperature, increases the differential quantum efficiency and improves some other characteristics.

The 'construction' of heterosemiconductors is a new semiconductor technology field. Therefore, new possibilities will also be opened for photovoltaic converters. It may be a 'free design' of materials for photoactive regions, in which the light absorption spectrum is modified in a definite way and the parameters affecting the charge carrier transfer are optimized. To create quantum dot arrays, the most tempting idea is spontaneous self-organization with epitaxial growth of the heterostructures. In recent years this idea has been under active development in a collaboration between research groups from the Ioffe Institute and the Technical University of Berlin. It has been found that ordered arrays of the 'islands' of a monolayer height could be formed spontaneously during the deposition of one material upon another, if the materials are strongly mismatched by the crystal lattice constant. Relaxation of elastic strains and interaction by means of the strains in the bedding layer are the motivating force in the formation of the uniform arrays of strained islands on crystalline surface. In particular, arrays of three-dimensional dots have been grown for lasers with an InAs/GaAs active region under processing controlled up to the precision of one monolayer. The shape of the InAs quantum dots could vary significantly when grown over GaAs layers using specific temperature regimes. A short-period alternate deposition of strained materials splits the dots and results in the formation of superlattices from layers of vertically correlated dots. The observation of ultra-narrow lines of luminescence (< 0.15 meV; $T = 5$–50 K; individual quantum dots) revealing no tendency to broaden with temperature served as evidence for the formation of quantum dots with an energy spectrum described by a δ-function. Much evidence of the quantum dot 'operation' has been obtained in investigating low-threshold heterolasers grown with newly developed techniques [9].

It is evident that heterosemiconductors should be considered as a new object for applications and that these should include solar photovoltaic converters. New approaches involving the use of materials with quantum dots have been proposed. In particular, the case in point is the creation of a photoactive medium with an 'intermediate band' [12]. Also, in the structures of multi-cascade photocells, in addition to the use of newly created materials with the desired absorption spectrum, it would probably be possible to improve the characteristics of commutating tunnel diodes (increase the peak current) by introducing superlattices created from vertically correlated dots between the n^+ and p^+ layers. It should be added that other proposals, including rather old ones, for the enhancement of photovoltaic converter efficiency, the realization of which necessitates 'newly constructed' materials, also exist. Among these is an idea to employ graded heterostructures, in which it is necessary to realize a very large difference in the forbidden gap and, simultaneously, a very high charge carrier mobility. All these proposals result from attempts to enable solar photovoltaic converters to approach (first theoretically and then, maybe, practically) the thermodynamic limit of 93% determined by the Carnot cycle.

2.6   From the 'sky' to the Earth

Previous experience shows that all achievements in solar cell efficiency are first applied to space technology. The same conclusion can be made when estimating the range of problems space techniques have to solve. There are a number of demands, for example the need for a large number of powerful telecommunication satellites, the increasing energy demands by orbital stations and the need to create specialized energy satellites, which could power space transportation facilities. In particular, in energy satellites solar energy would be converted into electrical energy and then into radiation energy for translation by means of a laser or microwave beam. In the more distant future, large quantities of energy could be translated to the Earth for electric power generation. To fulfil such impressive projects, solar cell structures should probably be grown directly in orbit by the MBE technique in the conditions of space vacuum. In any case a substantial enlargement of solar cell production for space applications will simultaneously create a technological base for even larger-scale production of solar cells for terrestrial use.

2.7   Concentration of solar radiation

Up to this point we have not considered one particular possibility for the enhancement of photovoltaic conversion efficiency, i.e. converting preliminary concentrated solar radiation. The limiting calculated concentration ratio at a distance from the Sun corresponding to the Earth's orbit is 46 200×. Just such a concentration ratio is usually set for estimates of thermodynamically limited efficiencies of different types of solar cells. In particular, for multi-junction cells comprising several tens of cascades, the limiting efficiency is rather close to the Carnot cycle efficiency. It is almost 87% and, with respect to this parameter, it is slightly ahead of other types of cells (for example, photocells based on a material with intermediate sub-bands). Thus, multi-junction photocells, in addition to demonstrating the highest efficiency values to date and a good outlook for their rise in the nearest future, also have the best 'fundamental' prospects.

But may one speak about the prospects of the large-scale use of such photocells for terrestrial applications, where the determining factor is the economic one? As a matter of fact, the structure of multi-junction photocells is very complicated. Moreover, their structure is probably the most complicated among all semiconductor devices. In multi-junction cells, the widest variations in forbidden gap values for several photoactive regions with p–n junctions have to be ensured. The doping levels of the layers also have to be varied considerably with sharp changes in conductivity type with the formation of several (in the course of the structure growth) tunnel p–n junctions which commutate the cascades. The designated thickness of the photoactive layers must be adhered to with high accuracy to ensure the absorption of a definite part of the solar spectrum for the generation of the same photocurrent value throughout the cascade. The layers forming the tunnel p^+–n^+ junctions must be extremely thin (in the nanometre range) to minimize light absorption, while the photoactive layers must be approximately two orders of magnitude thicker. The whole set of A^3B^5 materials (in the form of solid solutions) appears to be involved in the formation of multi-junction photocell structures (including the nitrides) grown on a foreign substrate (Ge or Si). However, as has already been mentioned, the economics in this case do not seem to be so dramatic. The growth of the structures is a single-stage automatic process. Its total thickness is only few micrometres and the non-photoactive substrate may be cheap enough. But the deciding factor for an economically justified use of multi-junction cells is that they can operate effectively under highly concentrated solar illumination (figure 2.10).

The fact that heterojunction solar cells based on gallium arsenide can operate effectively at significant concentrations of sunlight (several hundred or even several thousand times) and, in this respect, differ favourably from silicon ones was pointed out as for back as the late 1970s and early 1980s. The first experiments to create solar photovoltaic modules with GaAs cells and concentrators relate to that time. The generated photocurrent increases linearly with light intensity and the output voltage increases, in turn, with the logarithmic

Figure 2.10. Solar PV installation with reflective concentrators. AlGaAs/GaAs solar cells 1.7 cm in diameter are placed on aluminum heat pipes to dissipate waste heat. Automatic sun-tracking is carried out with the help of analogue sun sensors. The mechanical support structure consists of a base platform moving in the azimuth direction by means of three wheels and a suspended platform with PV modules [13].

law. Thus, light intensity following the generated power rises super-linearly with concentrating radiation, so that the efficiency of photovoltaic conversion increases.

This situation can be realized in practice if a higher current does not create a noticeable potential drop across the internal resistance of a photocell. Radical reduction in internal ohmic losses was becoming a key problem in developing concentrator photocells. The prospects for an efficiency rise by operating with concentrators look rather tempting. However, the main motive force in creating concentrator modules was the possibility of reducing the consumption of semiconductor materials for the generation of the required electric power proportional to the sunlight concentration ratio. In this case semiconductor photocells of relatively small area receive power from the sun in the focal plane of the concentrators—mirrors or lenses. These concentrators can be fabricated from relatively cheap materials. The contribution of the photocell cost to the solar module cost begins to be insignificant whereas concentrator module efficiency depends directly on the efficiency of the employed photocells. Thus, the prerequisites for economically justified application of the most effective, even if expensive, photocells based on the A^3B^5 compounds are being created.

The initial period in the development of concentrator photovoltaic modules (the late 1970s) coincided with a period of vigorous growth in the power semiconductor technique. The creation of power semiconductor devices for various applications became one of the main directions of their development.

Solar photocells were not an exception in this respect. In the imagination of the researchers, for unit power increase concentrator photocells had to be promoted at the cost of an increase in both the cell photoactive area and the density of the generated photocurrent. To operate such photocells, large-sized concentrators were required. The easiest way to realize such concentrators was to employ focusing mirrors. The absolute values of the photocurrents generated by unit photocells reached tens of amperes. The design features of the high-current photocells were similar to those of electric power rectifiers. Metallic thermoconductive cases had to be fastened onto a heat dissipation system. Since every cell had to be situated in front of a focusing mirror (along solar beams), compact, however complicated, expensive and non-reliable systems for liquid cooling or cooling by means of special heat pipes were used to diminish the mirror shadowing.

The cell structure was also being optimized in a special way to enable operation in the conditions in which high photocurrent densities and high absolute currents are generated. The sheet resistance of the front p–n junction region in single-junction AlGaAs/GaAs photocells was being minimized by means of heavier doping. To ensure as complete as possible collection of charge carriers in this region, a built-in quasi-electric field at the cost of the doping impurity gradient was being formed. New features in photocell development were proposals to use the luminescent radiation arising in the direct-band part of the cell structure under solar photoexcitation. For instance, to reduce sheet resistance significantly, a direct-band AlGaAs layer was introduced just behind the wide-gap window, which totally absorbed most of the solar spectrum and converted it into comparatively narrow-band photoluminescence with an efficiency close to 100%. This luminescence could penetrate, with low losses, the structure where the p–n junction was situated, to be absorbed and generate a photocurrent. Thus, the front p–n junction region could have a thickness of tens of micrometres and a notably small sheet resistance enabling the photocells to operate at concentration ratios of up to 3000×. More recent proposals on the use of photo- and, also, electroluminescent emission (the latter arises by passing the forward current through the p–n junction of photocells fabricated from direct-band materials) have been concerned with possibilities of estimating the quality and operation capability of photocells in a wide variety of aspects—from the visualization of cell defects to measurements, including contactless ones, of the electrical and photoelectric parameters [14].

Very stringent requirements were placed upon the contact structure of concentrator photocells. The contact grid on the front surface ought to have a transient resistivity similar to that of the contacts for the injection lasers. At the contact grid level, a minimal sheet resistance at minimum darkening of the photosensitive surface had to be realized. These contradicting requirements could be fulfilled by making the contact grid fingers in the form of bulk metallic bars separated by 50–200 μm. Usually the shadowing by contact grid was about 10%. A rise in ohmic losses because of the current-spreading effect in large-

Converters of electric power:	Converters of concentrated sunlight:
- concentrated character of input power transferring by the wires; - a problem to make equal the passing currents; - a small part of heat in comparison with the converted power (< 10%).	- distributed character of initial incident sunlight; - natural equalization of the photogenerated currents; - very large part of heat in comparison with sunlight power (70 - 80%).

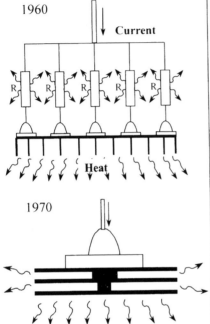

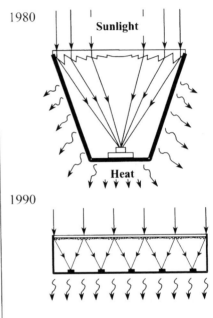

Figure 2.11. Semiconductor converters of electric power and concentrated sunlight: conceptions of the development.

area photocells (several cm^2) and the complexity involved in thermal matching the material of the photocell with that of the thermoconductive base indicated that the use of devices with a smaller area was preferable. In this case acrylic Fresnel lenses could serve as more convenient solar concentrators. A photocell was placed behind a lens (along the beams), so that the back module surface could be used for heat dissipation by means of radiation and surrounding air convection. The area of a heat dissipater corresponded approximately to the area of the lens aperture.

Since the early 1990s, a new approach based on the concept of small-aperture modules appeared in concentrator systems research and development. A significant disparity between the approaches to designing high-power converters in electric power engineering (for example, semiconductor diodes or transistors) and concentrator photocells (figure 2.11) had appeared. At the beginning of the

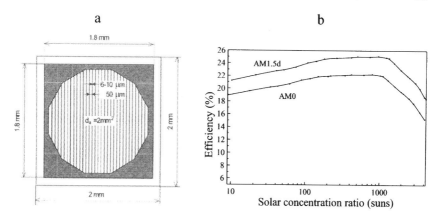

Figure 2.12. (*a*) Top view of a concentrator solar cell with 50 μm spacing between contact fingers; (*b*) dependences of the conversion efficiency on solar concentration ratio for a single-junction AlGaAs/GaAs solar cell under AM0 and AM1.5d illumination conditions [15].

development of the power-converting technique (the late 1950s and the early 1960s), a system's power could rise (by current) by increasing the number of unit devices connected in parallel, so that the use of ballast resistors to equalize currents was necessary. The technology for fabricating larger-area high-current devices, developed by the early 1970s, was successfully solving these problems. In this instance, an important factor is the fact that the converted electric power is 'concentrated' in a natural way in the delivery wires, and a portion of the heat dissipated during the conversion process does not exceed at least 10% of supplied power. A different situation takes place when converting sunlight. Initially solar energy has a low-intensive distributed character, so that the natural equalization of the generated photocurrents is principally ensured in any number of converters of similar area. To convert sunlight in a concentrated form, the light energy must first be focused onto a small-area photocell (figures 2.11 and 2.12) but then a significant portion of this energy, now in the form of waste heat, must be spread by a heat dissipater with dimensions equal to those of the concentrator. The portion of heat during photovoltaic conversion of solar energy appears to be many times greater (70–80%) than in the case of electric power conversion. In a big module, because of the significant linear dimensions of a concentrator and the necessity to dissipate a great amount of waste heat, the heat dissipater has to be made of rather thick metallic material with good thermal conductivity. However, this is not the only disadvantage of large-size modules. For example, in fabricating large Fresnel lenses, a large quantity of polymer material should be used. An additional disadvantage is that assembling rather large concentrator photocells remains a difficult problem bearing in mind that they should operate under thermo-cycling conditions, and so on.

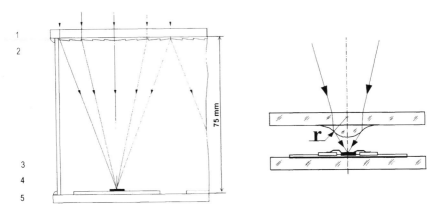

Figure 2.13. (*a*) Cross section of an 'all-glass' solar concentrator module [16]: 1, lens panel base made of glass: 2. Fresnel lens microprisms made of silicone: 3. focused sun rays: 4, solar cell mounted on a thin metallic heat dissipater; and 5, solar cell panel base made of glass. (*b*) Introduction of a secondary smooth-surface mini-lens to increase the sun concentration ratio [18].

Small-aperture concentrator modules have all the prerequisites for ensuring highly efficient and economically justified photovoltaic conversion of solar energy. In decreasing the linear dimensions of a concentrator and in maintaining the predetermined concentration ratio, the focal distance of the lenses decreases. Therefore, in this case, the structural height of a module becomes shorter (figure 2.13). The linear dimensions of the photocells can be reduced down to 1–2 mm, so that they can be mounted by means of automatic equipment used for assembling mass-produced discrete devices in the electronic industry. At the small linear dimensions of the photocells, the substrate for structure growth can be thinner but the proportion of the useful area of wafers can be increased, which results in savings in the semiconductor materials in addition to the fact of sunlight concentration. Cell assembly can be carried out without the thermal expansion coefficients of the cell material and metallic heat-dissipation base being specially matched. Of particular importance is the fact that the heat-dissipater thickness also decreases proportionally with a decrease in the linear dimensions of the concentrator, which results in an impressive reduction in metal consumption during module fabrication. Thus, in small-aperture concentrator modules, the advantages of concentrator systems (rise in efficiency, a saving in semiconductor and structural materials) may be realized with the retention (in the whole) of the distributed character of sunlight conversion and heat dissipation (as in systems without concentration).

It should be noted that when considering the prospects for large-scale production of solar energy, the consumption of any, even the most conventional, structural materials is economically justified only if the converter is very efficient.

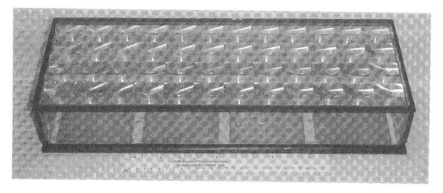

Figure 2.14. Experimental PV module of an 'all-glass' design for the conversion of concentrated sunlight.

This is determined by the necessity to cover considerable areas of the earth's surface to intercept radiation for the generation of high power. Solar power installations should be protected from atmospheric effects to ensure their long-term (20–30 years) operation capability. In this respect, the construction of the 'all-glass' concentrator modules [16] developed in recent years at the Ioffe Institute in cooperation with the Fraunhofer Institute for Solar Energy Systems (Freiburg, Germany) is very promising. Small-aperture Fresnel lenses are united, in this case, into an integrated panel. Concentrator photocells are also mounted on thin (~0.5 mm) metallic heat dissipaters. Both these panels are fastened by sidewalls, so that the inner volume of the module is sealed off. The lens panel has a composite structure. A sheet of conventional silicate glass (which protects the module from the front) serves as a superstrate for a thin silicone Fresnel profile, which focuses the sunlight. The base of the panel of photocells is also a silicate glass sheet, through which heat dissipation occurs. Thus, by its cheap and stable properties, glass becomes the main structural material of a concentrator module. The consumption of optical quality silicone is here reduced to the minimum necessary to form refractive microprisms (mean thickness about 0.2 mm). For fixing and sealing the glass parts of the module, conventional structural silicone serves. Due to the use of materials which the most resistant to radiation and atmosphere effects (glass and silicone), concentrator modules are considered the most promising for long-term operation (figure 2.14).

The efficiency of a concentrator module depends on both the efficiency of the photocell and the optical efficiency of a sunlight concentrator. Fabricating cheap and dimensionally large panels of concentrators characterized by extremely high optical efficiency is a new task for optical technology. It has been solved successfully by using a two-component highly transparent silicone compound for the lens material. The Fresnel lens profile is formed with high accuracy by means of silicone polymerization on the back surface of the protective glass, when it is in contact with a negatively profiled mould. It is possible in this way to fabricate lens panels of up to 1 m × 1 m in total area with the aperture area of the comprising lenses being in the 10–20 cm^2 range. In the experimental 'all-glass' modules, the optical efficiency of the Fresnel lenses achieved 89%. In using double-junction InGaP/InGaAs photocells with an efficiency approximately equal to 30% (AM1.5d), the module efficiency measured under natural sunlight illumination conditions was 24.8% calculated using solar radiation incident on the surface of the lenses [17].

Further improvements in module design are looking at ways of increasing the sunlight concentration ratio. In particular, if every unit lens has dimensions 4 cm × 4 cm, high optical efficiency is ensured by the use of photocells 2 mm in diameter, corresponding to a concentration ratio of about 500×. Further decrease in the dimensions of the photocell and realization of concentration ratios of 1000× and higher is possible by using secondary mini-lenses with a smooth profile placed directly in front of the photocells. In this case the secondary lenses may also be combined into a panel [18]. With highly concentrated solar irradiation, the use of multi-junction photocells has additional advantages. Increase in cascade quantity leads to an increase in output voltage and a decrease in the generated photocurrent. Owing to this circumstance the internal ohmic losses at current collection become lower, so that a high conversion efficiency is realized at higher sunlight concentration ratios.

There are certain misgivings concerning the practical use of concentrator cells, connected with the necessity to ensure sun-tracking by concentrator modules. In fact, in this case, the development of special support and rotating mechanisms supplied with sun-position sensors and electric drives is necessary for the installation of these modules. Compared with flat modules without concentrators, this leads to the additional consumption of structural materials and power for tracking. However, even in a flat module, with continuous sun-tracking during a light day, 30–40% more power is generated than that generated without sun-tracking. Allowing for this gain and for the fact of higher concentrator module efficiency, one may say that this compensates for the additional expense of the materials. As to the consumption of electric power for tracking, it is only 0.2–0.3% of the power generated by concentrator modules arranged on the trackers [19].

Concentrator modules should not be in opposition to conventional (flat) ones when estimating the prospects for solar photovoltaic development. Both of them will be used in the power-generation systems of the future. Apparently, conventional modules comprising crystalline silicon cells or thin-film structures will form the basis of a decentralized power generation system. Whether they are the private property of individual owners or the public property of a broad circle of owners, they will be installed on the roofs and walls of houses and buildings and integrated into a network. Thus, new 'democratic principles' of solar energetics will be embodied compared with the 'dictatorship' of the power giants, which happens at the present time. However, to cover the power demands of power-consuming industries and municipal associations it will be necessary to construct large-scale solar stations to ensure a minimal cost for the generated electric power. Such stations arranged on specially designated territories and served by special personnel will also be part of a decentralized power system. Using concentrator photovoltaic modules in the development of such stations seems to be a reasonable solution. There are numerous economic estimations (see, for example, [20–22]), according to which, in the coming decade, concentrator photovoltaics may not only become economically effective compared with other types of installations for photovoltaic conversion but they may also compete with existing traditional power sources with respect to the cost of the generated electric power. It is also important that solar power stations with a substantial output can be built without the large-scale deployment of new production lines for semiconductor materials. Then, the main efforts can be focussed on fabricating the mechanical parts of the concentrator modules and supporting structures, for which a production basis already exists.

2.8 Concentrators in space

Meanwhile, the semiconductor industry may already need new production capabilities in the nearest future due to the necessity of providing space technology with highly efficient photocells based on A^3B^5 compounds. Again, the concentration approach is considered as a possible way of solving this problem. In space, the most promising way is the use of linear Fresnel lenses as sunlight concentrators. This means that relatively accurate sun-tracking will only be necessary around one axis parallel to the microprisms of the lenses, while the tracking around the second axis can be done much more roughly. The concentration ratios in this case are usually 6–10×. The refractive profile of the lenses is formed from transparent silicone. Lenses for use in space may have a very thin (about 0.1 mm) glass base [23] or even have no base [24]. In the first case, a glass sheet doped with ceria protects the lens from ultraviolet radiation and from the effect of high-energy particles. In the second case, a multilayer coating should protect the front surface of a silicone lens. In both cases, the photocells appear to be better protected from adverse circumstances than in solar arrays without concentrators. In addition, the array's tolerance to radiation rises as it operates at higher photocurrent densities under concentrated sunlight. The current density increase becomes especially important when launching space probes in the direction from the Sun. In this case the light concentration compensates for the efficiency drop with a decrease in sunlight's power density. In concentrator solar arrays in near-Earth space, a specific power per unit weight of about 180 W kg^{-1} and a specific power per unit area of about 300 W m^{-2} are possible [24]. These parameters are expected with the use of triple-junction InGaP/GaAs/Ge photocells and cannot be achieved in any other type of space solar arrays. The Ioffe Institute has been involved in the research and development of space modules with short-focus (23 mm) linear Fresnel lenses. Such modules can be deployed instead of conventional flat arrays without any changes in the design of the transportation containers.

2.9   'Non-solar' photovoltaics

A number of photocell applications which are not connected with sunlight conversion exist. These include, in particular, conversion of the energy generated at nuclear reactions, thermophotovoltaic conversion and laser power conversion. The scale of power generated and the spheres for future applications vary over a very wide range but they are not comparable, which is self-evident, with the scale of the possible use for solar energy conversion. Nevertheless, we shall consider some technical possibilities for the development of such devices.

For some special applications, autonomous low-output electric power sources are required with lifetimes which significantly exceed those of electrochemical ones. The development of such power sources is possible by using radioactive isotopes and semiconductor photocells [25]. Light energy may be delivered from miniature radioluminescent lamps filled with tritium. Tritium is characterized by a half-decay period of 12.3 years and a low β-particle energy ($E_{max} = 18$ keV), which makes the batteries based on tritium ecologically safe. In radioluminescent lamps, the energy of the β-particles is converted into relatively narrow-band light. In this case, photocells based on wide-gap materials with $E_G > 1.9$ eV (GaP, AlGaAs, InGaP) have to be used as photoconverters. Because of the low light intensity, the photocell structure optimization has to be done with the aim of reducing the reverse dark current density down to 10^{-11}–10^{-12} A cm^{-2}. Direct conversion of β-particle energy into electricity is also possible in heterostructures with p–n junctions. In this case, the energy required to create, in a semiconductor material, one electron–hole pair using one fast electron is about 2.8 $E_G + (0.5$–1.0 eV), which is approximately three times higher than the photon energy necessary for the same effect. In connection with this, the theoretical maximum for efficiency for the direct conversion of β-radiation into electricity is only 22-25%. Preference should be given to wide-gap radiation-resistant semiconductor materials. Probably the most suitable materials for this case would be the nitrides A^{3}N. For β-particles to penetrate the p–n junction region, the heterostructures should have a very thin (10–20 nm) wide-gap window serving for passivating the device's front surface. Heterostructure cells may be placed directly in capsules filled with gaseous tritium [26]. In addition, disks of porous titanium saturated with adsorbed tritium can serve as β-particle sources. If converters are fabricated in the form of thin-film structures placed between tritium-saturated foil layers, such a battery will be capable of generating, over 12.3 years, up to 10 W hr^{-1} of electricity from 1 cm^3 volume, which is essentially higher than the capacity of chemical batteries [25].

Let us consider now the possibilities for thermophotovoltaic energy conversion. In this case, the radiation from a heated body (a photon emitter) is converted into electric power by means of semiconductor photocells with p–n junctions. Energy for heating may be obtained from the combustion of an organic fuel, the intensive decay of radioactive isotopes or focused solar radiation. The most important from a practical point of view is the possibility of designing reliable, long-term stable and noiseless thermophotovoltaic (TPV) generators instead of electromechanical ones operating with engines. Like solar photocells, photocells in TPV generators convert photons with energies which exceed the forbidden gap. Since the photon-emitter has to ensure the long-term operation of the whole installation, its temperature is limited, as a rule, to the range 1000–1500 °C. In lowering the temperature, the power and efficiency of a TPV generator is decreased. A temperature rise results in a decrease in system reliability. In addition, an increased amount of toxic nitrogen oxides forms in the combustion products. The emission spectrum of a suitable material for the emitter, such as silicon carbide, is close to the blackbody spectrum. It is assumed that the structure of the TPV generator includes an optical filter to correct the spectrum of radiation incident on photocells. Such a selective filter may be situated between the photon-emitter and photocells reflecting long-wavelength radiation back to the emitter. The photocell itself may play the role of this filter if there is a mirror on its back surface which reflects radiation unabsorbed by the photocell material. Finally, a selective high-temperature multilayer coating, which reflects the long-wavelength radiation, may be deposited directly onto the photon-emitter surface. Optimal values of E_G for photocell materials are: $E_G = 0.5$–0.6 eV for the lower temperatures of the photon-emitter; $E_G = 0.7$–0.75 eV for higher temperatures. The development of the technology for narrow-gap A^3B^5 materials corresponding to these E_G values (GaSb, solid solutions InGaAs and InGaAsSb) has become the impulse which, last decade, stimulated the development of the thermophotovoltaic conversion method.

The high-power density radiated by the photon-emitter in a TPV generator makes the photocell's operating conditions in such a generator similar to those in concentrator solar cells. For example, from a photocell based on GaSb in a limiting regime, one can obtain power densities of up to several Watt per square centimetre. The range of unit outputs of practical TPV generators is quite wide (see figure 2.15)—from about 0.1 W, where a conventional candle plays the role of photon-emitter, up to about 1 kW, where a high-temperature emitter is heated by a gas flame. This power range indicates the advantageous use of TPV generators for power supplies to totally autonomous consumers, especially those located in cold climates, since the dissipated heat of the generator may be effectively utilized for heating.

An analysis shows that in the case of TPV conversion, in contrast to solar energy conversion, employment of cascade photocells does not give such a significant rise in efficiency. This is explained by two circumstances. First, the short-wavelength part of the radiation is smaller in the case of a colder artificial photon-emitter than in the Sun's spectrum. Second, the decrease in current density

Figure 2.15. Different types of TPV generators developed at the Ioffe Institute (from left to right): candle powered TPV mini-generator (output of about 0.1 W) for charging mobile phone batteries; propane/butane fuelled TPV generator with an output of about 7 W [27]; and a fragment of the fuel-powered SiC photon emitter which allows a photocurrent density as high as 4 A cm^{-2} in a GaSb-based TPV cell to be obtained.

on cascading leads, in this case, to a relatively more noticeable decrease in the generated voltage in each cascade, since the p–n junctions are situated in narrow-gap materials. The most significant potential for an increase in TPV-generator efficiency lies in the possibilities for secondary action in photoconverters on a radiation source (photon-emitter). Such a possibility is completely absent in solar power systems. If an idealized TPV generator is considered, one may imagine a situation in which all the long-wavelength radiation which is useless for a photocell keeps returning to the emitter maintaining its temperature at the necessary level for a minimum delivered external power. In addition, all radiation capable of generating electron–hole pairs in the p–n junction material must be transferred without losses from the emitter to the photocell. In this case, the converting radiation spectrum appears to become narrower and the photocell efficiency approaches the efficiency value for the conversion of monochromatic radiation with a wavelength corresponding to the p–n junction material absorption edge. The efficiency with which the thermal energy is transferred from the fuel to the photon-emitter can also be increased at the expense of partial utilization of the exhaust gases heat for predominant heating of the fuel-air mixture. Thus, a TPV generator becomes a more closed and, consequently, a more effective system, if the principles of radiation recirculation and heat recuperation are used.

To conclude this section, let us consider briefly the possibilities for photovoltaic conversion of laser power to supply energy. Such a problem arises with the necessity for a distant power supply for certain objects. To realize the maximum values of conversion efficiency, direct-band semiconductors with a forbidden gap close to the energy of the laser quanta have to be chosen. Since the emission is monochromatic, the problem of photocell structure optimization is much simpler than in solar converters. For example, in practical installations for transferring the energy of a semiconductor laser ($\lambda = 0.85$ μm) with the help of an optical fibre, photovoltaic conversion efficiencies of 60% in single-junction AlGaAs/GaAs heterophotocells have been realized [28]. Allowing for possible high efficiencies in transferring energy by a laser beam, quite large-scale projects are being considered at the present time. In particular, systems for the distant power supply of spacecrafts by laser emission transferred from the Earth, systems for power transfer between spacecrafts and also systems for power transfer from space solar arrays towards the Earth have been proposed. The expected levels of incident power in these cases lie in the range $0.1-100$ W cm^{-2}. For long-distance transfer, a receiver should have a large enough aperture area to intercept the radiation. In this case, it may become worthwhile to use radiation concentrators on the receiver side. The basic features of a laser power conversion system will be similar to those considered for the conversion of concentrated sunlight.

2.10   Conclusions

We are witnesses of and participants in many interesting and important processes which are currently taking place in scientific and technical spheres. Semiconductors, the history of technical development of which is only slightly longer than 50 years, have firmly conquered all fields of electric power conversion techniques and electronics, and now, having prepared and conducted the 'information revolution', are close to 'over-turning' lighting techniques. The scientific and technological successes of recent decades also allow us to hope that a similar 'revolution' based on semiconductors may also take place in the field of electric power generation.

In a broad sense, the cost of developing a new energy basis is never small. It is even hard to imagine how much effort and resources were spent to develop and create the systems for electric power generation, distribution and consumption, when the electric power industry was in its infancy. The further development of this industry required new forms of fuel. Atomic power stations, which seemed to be capable of solving all energy problems, have been built. But could these power stations have been built, if national programmes for creating nuclear weapons had not existed? It is well known that tremendous resources were initially invested into just these programmes, so that power stations appear to be only one result of their realization. The apparent low price of atomic power station engineering has culminated in a paradoxical situation in which it would be necessary to pay a huge price even to refuse to use them.

Solar photovoltaics does not exist in a vacuum. By and large, as a result of the development of electronics, laser techniques and electric power engineering for spacecrafts, a scientific and technological basis has been created, which may serve as the starting point for the development of a terrestrial solar electric power industry based on semiconductors. There comes a time when it should be necessary to provide the much wider investment of resources into this field corresponding to its significance for the future.

References

[1] Adams W G and Day R E 1877 *Phil. Trans. R. Soc.* **167** 313–49
[2] Ioffe A F and Ioffe A V 1935 *Phyz. Z. Sov. Un.* **7** 343–65
[3] Chapin D M, Fueller C S and Pearson G L 1954 *J. Appl. Phys.* **25** 676–7
[4] Goryunova N A 1951 *Thesis* Leningrad State University—Physico-Technical Institute
      Blum A I, Mokrovskii N P and Regel A R 1950 *Presented at VII Conf. on Semiconductors* (Proc. published 1952 *Izv. AN SSSR Ser. Phys* **XVI** 139)
[5] Alferov Zh I, Andreev V M, Kagan M B, Protasov I I and Trofim V G 1970 *Sov. Phys. Semicond.* **4** 12
[6] Green M A, Emery K, King D L, Igary S and Warta W 2002 *Prog. Photovolt: Res. Appl.* **10** 355–60
[7] Andreev V M, Kazantsev A B, Khvostikov V P, Paleeva E V, Rumyantsev V D and Shvarts M Z 1994 *Conf. Record First World Conference on Photovoltaic Energy Conversion* (New York: IEEE) pp 2096–9
[8] Andreev V M, Kochnev I V, Lantratov V M and Shvarts M Z 1998 *Proc. 2nd World Conf. on Photovoltaic Solar Energy Conversion (Vienna)* (Munich: WIP) pp 3757–60
[9] Ledentsov N N, Grundmann M, Heinrichsdorff F, Bimberg D, Ustinov V M, Zhukov A E, Maksimov M V, Alferov Zh I and Lott J A 2000 *IEEE J. Selected Topics Quantum Electron.* **6** 439–51
[10] Bailey S G and Flood D F 1998 *Prog. Photovoltaics* **6** 1–14
[11] King R R *et al* 2000 *Proc. 28th IEEE Photovolt. Spec. Conf.* (New York: IEEE) pp 998–1005
[12] Luque A *et al* 2002 *29th IEEE PVSC (New Orleans)* (New York: IEEE) pp 1190–3
[13] Alferov Zh I, Andreev V M, Aripov Kh K, Larionov V R and Rumyantsev V D 1981 *Solar Eng.* **6** 3–6
[14] Andreev V M, Grilikhes V A and Rumyantsev V D 1997 *Photovoltaic Conversion of Concentrated Sunlight* (New York: Wiley) pp 144–82
[15] Andreev V M, Khvostikov V P, Rumyantsev V D, Paleeva E V, Shvarts M Z and Algora C 1999 5800 Suns AlGaAs/GaAs Concentrator Solar Cells *11th International Photovoltaic Science and Energy Conference (Sapporo)* Technical Digest of the International PVSEC-11 pp 147–8
[16] Rumyantsev V D, Hein M, Andreev V M, Bett A W, Dimroth F, Lange G, Letay G, Shvarts M Z and Sulima O V 2000 *16th European Photovoltaic Solar Energy Conf. (Glasgow)* (Bedford: H S Stephens & Associates) pp 2312–15
[17] Hein M, Meusel M, Baur C, Dimroth F, Lange G, Siefer G, Tibbits T N D, Bett A

W, Andreev V M and Rumyantsev V D 2001 *17th European Photovoltaic Solar Energy Conf.* (Munich: WIP) pp 496–9

[18] Rumyantsev V D, Chosta O I, Grilikhes V A, Sadchikov N A, Soluyanov A A, Shvarts M Z and Andreev V M 2002 *Proc. 29th IEEE Photovoltaic Specialists Conf. (New Orleans)* (New York: IEEE) pp 1596–9

[19] Rumyantsev V D, Andreev V M, Sadchikov N A, Bett A W, Dimroth F and Lange G 2002 *Proc. of the Conf. 'PV in Europe' (Rome)* (Munich: WIP) to be published

[20] Cotal H L *et al* 2000 *Proc. 28th IEEE Photovoltaic Specialists Conf.* (New York: IEEE) pp 955–60

[21] Yamaguchi M 2002 *Workshop Proc. 'The Path to Ultra-High Efficient Photovoltaics'* (Ispra: European Commission) pp 15–28

[22] Yamaguchi M and Luque A 1999 *IEEE Trans. Electron. Devices* **46** 41–6

[23] Andreev V M, Larionov V R, Lantratov V M, Grilikhes V A, Khvostikov V P, Rumyantsev V D, Sorokina S V, Shvarts M Z and Yakimova E V 2000 *Proc. 28th IEEE Photovoltaic Specialists Conf. (Alaska)* (New York: IEEE) pp 1157–60

[24] O'Neill M J, George P J, Piszczor M F, Eskenazi M I, Botke M M, Brandhorst H W, Edwards D L and Jaster P A 2002 *Proc. 29th IEEE Photovoltaic Specialists Conf. (New Orleans)* (New York: IEEE) pp 916–19

[25] Andreev V M 2002 Nuclear batteries based on III–V semiconductors *Polymers, Phosphors, and Voltaics for Radioisotope Microbatteries* (Boca Raton, FL: Chemical Rubber Company) pp 289–363

[26] Andreev V M, Kavetsky A G, Kalinovsky V S, Khvostikov V P, Larionov V R, Rumyantsev V D, Shvarts M Z, Yakimova E V and Ustinov V A 2000 *Proc. 28th IEEE Photovoltaic Specialists Conf. (Alaska)* (New York: IEEE) pp 1253–6

[27] Rumyantsev V D, Khvostikov V P, Sorokina S V, Vasil'ev V I and Andreev V M 1998 *Proc. 4th NREL Conf. on Thermophotovoltaic Generation of Electricity (Denver)* (Woodbury, NY: AIP) pp 384–93

[28] Andreev V M, Egorov B V, Ko'nova A M, Lantratov V M, Rumyantsev V D and Saradzhishvili N M 1986 *Sov. Phys. Semicond.* **20** 3

Chapter 3

Thermodynamics of solar energy converters

Peter Würfel
Institut für Angewandte Physik, Universität Karlsruhe, D-76128
Karlsruhe, Germany

3.1   Introduction

Solar energy is heat from the sun (temperature $T_S = 5800$ K) transported to the earth (temperature $T_0 = 300$ K) in the form of radiation.

When heat is involved, energy conversion is limited by the second principle of thermodynamics which states that entropy must not be destroyed. If an entropy-free form of energy (mechanical, chemical, electrical energy) is to be produced, the entropy which accompanies the solar heat energy has to be given away to a reservoir on earth and with it some energy. This process of energy conversion from heat is realized in a heat engine. Since entropy may also be generated in the conversion process, the highest efficiency for the conversion of heat requiring the smallest amount of energy to remove the entropy is found for reversible processes in which no entropy is produced. A state in which no entropy is produced is a state of equilibrium. However, in equilibrium, no net energy current flows from one body to another. A net energy current requires a deviation from equilibrium.

3.2   Equilibria

Although widely used, a thermodynamic equilibrium between two systems with everything in equilibrium and in which all intensive quantities have the same values in both systems fortunately does not exist because it would be the death of the universe. Instead, there is a possible and independent equilibrium for each intensive quantity, which then has the same value in different systems. There can be pressure equilibrium by the free exchange of volume, but the temperatures will differ. There can be temperature (thermal) equilibrium between two systems by a free exchange of entropy, while the chemical potentials of the particles in each

50

system differ, because their exchange is inhibited. It is important to call equilibria by their proper names to distinguish between them.

For solar energy conversion, there are two possible equilibria between the sun and a converter on earth. One is the well-known temperature equilibrium and the other is a less well-known thermochemical equilibrium. Both equilibria require that the input of the converter exchanges radiation only with the sun and not with other parts of the sky. This restriction is called maximal concentration. Deviations from these equilibria result in net energy currents.

3.2.1   Temperature equilibrium

We consider a black absorber at the input of a heat engine. If it is at temperature T_S, the limiting efficiency for non-entropy-generating processes is the Carnot efficiency and this is given by

$$\eta_{\text{Carnot}} = 1 - \frac{T_0}{T_S} = 1 - \frac{300}{5800} = 0.95. \tag{3.1}$$

If the input is at T_S and if it is in temperature equilibrium with the sun, as much energy is emitted back towards the sun as is absorbed from the sun. A small reduction in the input temperature by dT is necessary for a net energy transfer, resulting in a net energy current per area

$$d j_E = 4\sigma T_S^3 \, dT \tag{3.2}$$

which can be converted with the Carnot efficiency in (3.1) without entropy generation, while the main part of the incident energy current per area

$$j_{E,\text{inc}} = \sigma T_S^4 \tag{3.3}$$

is emitted back towards the sun.

The Carnot efficiency is a good definition only if we have to pay for the net energy current, which is the case for thermal power stations fuelled by fossil or nuclear energy, where only this net energy current is taken out of a reservoir. The problem with solar energy is that it is for free and that we do not get a refund for the energy radiated back towards the sun. Any energy emitted from the absorber at the input of the Carnot engine, therefore, is a loss. In order to account for this loss, we define an absorption efficiency to be the difference in absorbed and emitted energy currents divided by the incident energy current. For the previous example, the absorption efficiency is

$$\eta_{\text{abs}} = \frac{j_{E,\text{abs}} - j_{E,\text{emit}}}{j_{E,\text{inc}}} = \frac{4\sigma T_S^3 \, dT}{\sigma T_S^4} = 4 \frac{dT}{T_S}$$

and this is rather small. Larger deviations between T_S and the temperature T_A of the black absorber at the input of the heat engine result in entropy generation but

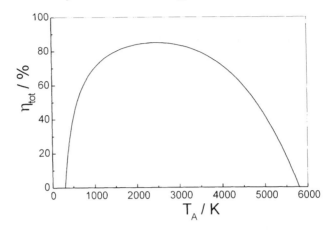

Figure 3.1. The efficiency of an ideal heat engine absorbing fully concentrated solar radiation of temperature $T_S = 5800$ K and working between the temperatures T_A at its input and $T_0 = 300$ K of the environment.

allow for larger absorption efficiencies:

$$\eta_{abs} = \frac{\sigma T_S^4 - \sigma T_A^4}{\sigma T_S^4} = 1 - \frac{T_A^4}{T_S^4}. \qquad (3.4)$$

The overall efficiency of the solar heat engine is

$$\eta_{tot} = \left(1 - \frac{T_A^4}{T_S^4}\right)\left(1 - \frac{T_0}{T_A}\right) \qquad (3.5)$$

and this is shown in figure 3.1. It is zero for $T_A = T_S$, where all the incident energy is emitted back towards the sun and it is zero for $T_A = T_0$, where the efficiency of the Carnot engine is zero. A maximal efficiency of 85% is found for an absorber temperature $T_A = 2478$ K.

A slightly higher efficiency of 86% is obtained if the incident spectrum is divided into many small intervals, each of which is absorbed by a separate Carnot engine with a selective absorber at its input, working at its optimal temperature [1].

For any thermal conversion process employing a material absorber in which entropy is generated, i.e. it is not in equilibrium with the sun, this value is an upper limit for the efficiency of solar energy conversion.

3.2.2 Thermochemical equilibrium

It is not well known that a material may be in equilibrium with the sun even if it is at room temperature if, at the same time, its excitations have a non-zero chemical potential. This equilibrium is limited to an exchange of photons in a photon energy interval $d\hbar\omega$ of the solar spectrum. In semiconductors, the excitations are the electron–hole pairs and they have a well-defined non-zero chemical potential under solar illumination, if their recombination lifetimes are much longer than their energy and momentum relaxation times in their bands. The chemical potential of the electron–hole pairs is also known as the difference in the quasi-Fermi energies $\varepsilon_{FC} - \varepsilon_{FV}$, where ε_{FC} determines the occupation probability of states in the conduction band and ε_{FV} determines the occupation of the valence band.

As a property of the equilibrium between a body on earth and the sun, there must be no net current of anything between the two systems—no net photon current, no net entropy current and no net energy current.

For the photon current balance in the energy interval $d\hbar\omega$, this condition is

$$dj_{\gamma,abs}(\hbar\omega) = a(\hbar\omega)\frac{\Omega}{4\pi^3\hbar^3c_0^2}\frac{(\hbar\omega)^2}{\exp\left[\frac{\hbar\omega}{kT_S}\right] - 1}\ d(\hbar\omega)$$

$$= dj_{\gamma,emit}(\hbar\omega) = a(\hbar\omega)\frac{\Omega}{4\pi^3\hbar^3c_0^2}\frac{(\hbar\omega)^2}{\exp\left[\frac{\hbar\omega - (\varepsilon_{FC} - \varepsilon_{FV})}{kT_0}\right] - 1}\ d(\hbar\omega). \tag{3.6}$$

The incident photon current is given by Planck's formula for radiation of temperature T_S carried by photons of energy $\hbar\omega$, incident from a solid angle Ω. Multiplication by the absorptivity $a(\hbar\omega)$ of the semiconductor results in the absorbed photon current. The emitted photon current is given by a generalization of Planck's formula [2–4] containing the non-zero chemical potential $\mu_\gamma = \varepsilon_{FC} - \varepsilon_{FV}$ of the photons, emitted at temperature T_0 from an excited semiconductor into the same solid angle Ω, from which the incident photons are absorbed. The same solid angle for absorption and emission means that the absorbing and emitting body sees nothing but the sun. Radiation with $\mu_\gamma > 0$ is called luminescence. If the photon currents are balanced, so too are the energy currents which follow from the photon currents in the energy interval $d\hbar\omega$ by multiplying them with the photon energy $\hbar\omega$.

From (3.6) it follows that the occupation probabilities f_γ of photon states in the absorbed and emitted photon currents are equal

$$f_{\gamma,abs} = \frac{1}{\exp\left(\frac{\hbar\omega}{kT_S}\right) - 1} = f_{\gamma,emit} = \frac{1}{\exp\left(\frac{\hbar\omega - \mu_\gamma}{kT_0}\right) - 1}. \tag{3.7}$$

Since the entropy per mode s_γ in either the incident or emitted photon current in

$$s_\gamma = k[(f_\gamma + 1)\ln(f_\gamma + 1) - f_\gamma \ln(f_\gamma)] \tag{3.8}$$

depends only on f_γ, as much entropy is emitted as is absorbed if (3.7) holds, as we would expect for an equilibrium situation.

From (3.7), the chemical potential of the photons and of the electron–hole pairs is given by

$$\mu_\gamma = \hbar\omega \left(1 - \frac{T_0}{T_S}\right) = \varepsilon_{FC} - \varepsilon_{FV}. \tag{3.9}$$

We call this equilibrium between the sun and a semiconductor, in which neither their temperatures nor the chemical potentials of the photons are equal, a thermochemical equilibrium, since it holds for the combination of temperatures and chemical potentials expressed by (3.9).

Again, as in section 3.2.1, the efficiency of producing chemical energy per electron–hole pair $\varepsilon_{FC} - \varepsilon_{FV}$ is

$$\eta_{Carnot} = \frac{\varepsilon_{FC} - \varepsilon_{FV}}{\hbar\omega} = 1 - \frac{T_0}{T_S} \tag{3.10}$$

which is again the Carnot efficiency but since all the absorbed energy is emitted, no net chemical energy current is produced.

The condition (3.6) that absorbed and emitted photon currents must be equal requires the recombination of electron–hole pairs to be entirely radiative, an assumption made throughout these considerations.

As for the thermal conversion in section 3.2.1, a deviation from this equilibrium is necessary for a net chemical energy current. Although it is left to section 3.3 to explain how chemical energy is transformed into electrical energy in a solar cell and withdrawn from it, it is quite clear that the current of electron–hole pairs which is withdrawn is related to the absorbed and emitted photon currents by

$$j_{eh} = j_{\gamma,abs} - j_{\gamma,emit} \tag{3.11}$$

This current of withdrawn electron–hole pairs is positive, if fewer photons are emitted than are absorbed. According to (3.6), fewer photons are emitted if the value of the chemical potential $\varepsilon_{FC} - \varepsilon_{FV}$ of the electron–hole pairs is smaller than the equilibrium value in (3.9). Multiplying the current of electron–hole pairs by their chemical potential $\varepsilon_{FC} - \varepsilon_{FV} = \mu_\gamma$ gives the current of chemical energy converted from solar energy by a semiconductor. Figure 3.2 shows the balance of emitted and absorbed photon currents and figure 3.3 shows the efficiency with which chemical energy is produced from solar energy as a function of the chemical potential of the emitted photons μ_γ. The efficiency is the chemical energy current divided by the incident energy current. No chemical energy can be withdrawn under open-circuit conditions, where $\mu_\gamma = \varepsilon_{FC} - \varepsilon_{FV}$ has the equilibrium value given by (3.9) and all electron–hole pairs recombine radiatively.

The chemical energy current is maximal for an optimal value of its chemical potential, which is found numerically. The efficiency, as this maximum amount of chemical energy current divided by the incident solar energy current, is shown in figure 3.4 as a function of the photon energy, $\hbar\omega$, with which photons are exchanged between the semiconductor and the sun.

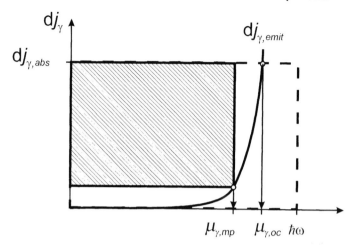

Figure 3.2. The rectangle enclosed by the broken line is the absorbed energy current, given by the absorbed photon current multiplied by the photon energy $\hbar\omega$. The emitted photon current increases as a function of the chemical potential μ_γ of the photons. At $\mu_\gamma = \mu_{\gamma,oc}$, as many photons are emitted as are absorbed and no energy is withdrawn. The maximal current of chemical energy, given by the shaded rectangle, is withdrawn at $\mu_\gamma = \mu_{\gamma,mp}$.

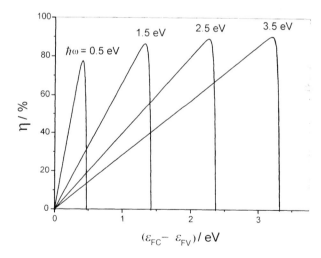

Figure 3.3. The efficiency of a semiconductor of producing chemical energy from incident monochromatic, fully concentrated solar radiation with photon energy $\hbar\omega$ is shown as a function of the chemical potential of the electron–hole pairs $\varepsilon_{FC} - \varepsilon_{FV}$ for different photon energies $\hbar\omega$ (reprinted with permission [19]).

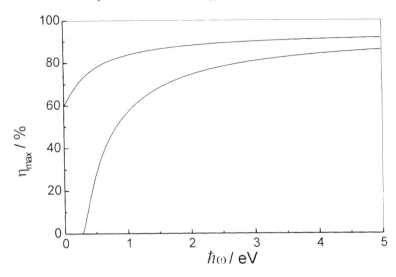

Figure 3.4. The maximal efficiencies of figure 3.3 for monochromatic energy conversion as a function of photon energy $\hbar\omega$, for fully concentrated (upper curve) and non-concentrated (lower curve) solar radiation.

If each photon energy interval $d\hbar\omega$ of the solar spectrum is converted by a separate semiconductor, the overall efficiency is found by integrating over the curve in figure 3.4 after weighting each energy interval by the incident energy current in this interval. For the overall efficiency of converting fully concentrated solar radiation into chemical energy the same value of 86% is found as in the thermal case in section 3.2.1. This is no surprise, since the mathematics are identical if $[\hbar\omega - (\varepsilon_{FC} - \varepsilon_{FV})]/kT_0$ in the chemical energy case is replaced by $\hbar\omega/kT_A$ in the thermal case [1].

The maximum efficiency of 86% must be regarded as the upper limit for any conversion process of solar energy into any entropy-free form of energy. This value is based on the notion that reversible absorption of solar energy occurs only in a situation of thermal or thermochemical equilibrium, where no net energy current flows into the absorber. A net energy current, however, is accompanied by entropy generation, since it requires a deviation from equilibrium.

For comparison, the efficiency of a semiconductor solar cell processing the whole solar spectrum under the condition of only radiative recombination was found by Shockley and Queisser to be 30% for non-concentrated AM0 radiation and 40% for fully concentrated sun light [5]. These efficiencies are smaller than the limiting efficiencies because of the thermalization losses occurring when electrons and holes lose part of their kinetic energy, arising from generation from high-energy photons and scattering with lattice vibrations. These losses are avoided by only allowing the exchange of monochromatic radiation.

The thermo-chemical equilibrium of a semiconductor with the sun is violated if some of the recombination processes are non-radiative, producing phonons instead of photons. Since the chemical potential of the phonons is $\mu_\Gamma = 0$, equilibrium of the electron–hole pairs with the phonons would not allow a non-zero value of the chemical potential of the electron–hole pairs and of the emitted photons. In real materials, the chemical potential of the electron–hole pairs is somewhere in between the value given by (3.9) and zero, depending on the relative contribution of radiative and non-radiative recombination. If the non-radiative recombination dominates by a factor r, the open circuit value of the chemical potential is reduced by $kT_0 \ln(1 + r)$, which still allows a good conversion efficiency, if r is not too large.

3.3   Converting chemical energy into electrical energy: the basic requirements for a solar cell

Converting solar energy into the chemical energy of electron–hole pairs occurs in any semiconductor and most favourably if all the recombination is radiative. One more step is required to make the semiconductor into a solar cell and that is to produce electrical energy and the question is how efficient this conversion of chemical into electrical energy may be.

Electrical energy is transported by a charge current j_Q. Since the electrons and holes generated by the incident radiation are oppositely charged particles, they give rise to a charge current if they move out of the absorber through different ports. To achieve this, we rely on the same mechanisms which would be applied to move the oxygen and hydrogen in a gas mixture out through different ports. For the gas mixture, one would apply semi-permeable membranes in front of the ports, one which conducts hydrogen and blocks oxygen and one which conducts oxygen and blocks hydrogen.

This principle transferred to electrons and holes calls for an electron membrane, an electron-conducting material which blocks holes in front of the electron electrode and for a hole membrane, a hole-conducting material, which blocks electrons in front of the hole electrode. Such materials are well known. A material, which has a large conductivity for electrons and a small conductivity for holes is an n-type semiconductor and a material which conducts holes well and electrons poorly is a p-type semiconductor.

Figure 3.5 shows the basic structure in an energy diagram. For this ideal structure, the electron membrane on the left should have a conduction band edge at a lower energy than the absorber, in order not to impede the electron transport. This requires roughly the same electron affinity χ_e as in the absorber. The bandgap of the electron membrane should be large to reduce the hole concentration further and to transmit the photons to be absorbed in the absorber. The hole membrane on the right should have its valence band edge at the same energy as the absorber and also a large bandgap to reduce the electron

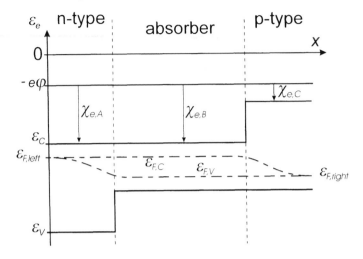

Figure 3.5. In an ideal structure for a solar cell, the absorber, in which chemical energy $\varepsilon_{FC} - \varepsilon_{FV}$ per electron–hole pair is produced, is sandwiched between an n- and p-type semiconductor. The n-type semiconductor is a semi-permeable membrane for electrons, allowing only the exchange of electrons, and the p-type semiconductor is a semi-permeable membrane for holes, allowing only the exchange of holes. A charge current is produced by a very small gradient of ε_{FC}, driving electrons to the left, and a very small gradient of ε_{FV}, driving holes to the right.

concentration and enhance the selectivity of the hole transport. This structure requires three different materials with the appropriate values for the electron affinities and bandgaps. In order to avoid introducing excessive non-radiative recombination, the interfaces with the absorber must be free of interface states within the energy gap of the absorber.

This structure is best realized in the dye cell [6], where the absorber is a dye adsorbed on TiO$_2$ particles which serve as the electron membrane. The dye is also in contact with an electrolyte providing a redox couple as the hole membrane.

An alternative and slightly less ideal structure but one which is much easier to make is based on using the same semiconductor for all functions. An intrinsic or slightly doped absorbing part is placed between highly doped thin n- and p-type layers.

With the road for the electrons and holes properly paved, a driving force is needed to drive the electrons and holes towards their respective membranes. The gradient of the Fermi energy is recognized as the general force, resulting from combining the gradient of the electrical energy per particle (as the force causing a field current, if it is the only force) with the gradient of the chemical energy per particle (as the force causing a diffusion current, if it is the only force). The charge current is given by

$$j_Q = \frac{\sigma_e}{e} \operatorname{grad} \varepsilon_{FC} - \frac{\sigma_h}{e} \operatorname{grad} \varepsilon_{FV} \qquad (3.12)$$

where the first term represents the charge current carried by the electrons and the second term the charge current carried by the holes, both being proportional to the conductivities: σ_e for electrons and σ_h for holes.

Going back to figure 3.5, we see that a small, almost invisible gradient of ε_{FC} drives the electrons from the absorber, where they are generated, through the electron membrane to the electron electrode on the left. There is also a much larger gradient of ε_{FV} in the electron membrane driving holes to the left, i.e. in the wrong direction. This is caused by the very large interface recombination velocity at metal contacts resulting in the Fermi energies ε_{FC} and ε_{FV} coinciding at the interface with a metal. The hole current in the electron membrane, however, is negligibly small, because the electron membrane has a hole conductivity σ_h many orders of magnitude smaller than the electron conductivity σ_e.

On the other side, in the hole membrane, a small gradient of ε_{FV} suffices to drive all the available holes through the hole membrane, while the much larger gradient of ε_{FC} is not large enough for an appreciable electron current due to the very small electron conductivity in the hole membrane.

For an ideal structure such as the one in figure 3.5, a large electron conductivity on the path of the electrons to the left and a large hole conductivity on the path of the holes to the right, allow the gradients of ε_{FC} and ε_{FV} to be negligibly small. Under these conditions, the difference in the Fermi energy $\varepsilon_{F,left}$ in the electron electrode and $\varepsilon_{F,right}$ in the hole electrode is equal to the difference $\varepsilon_{FC} - \varepsilon_{FV}$ in the absorber, the chemical energy per electron–hole pair. Since $\varepsilon_{F,right} - \varepsilon_{F,left} = eV$ (where V is the voltage between the electrodes) is the electrical energy per electron–hole pair of the charge current leaving the solar cell, chemical energy is converted into electrical energy without any loss by the structure in figure 3.5.

3.4 Concepts for solar cells with ultra high efficiencies

From the preceding discussion, it follows that the processes in the absorber by which solar heat is converted into chemical energy are limiting the efficiency of solar cells, whereas the further conversion into electrical energy by separating electrons and holes by means of semi-permeable membranes can, in principle, proceed without any loss. High efficiencies, therefore, require maximal concentration of near monochromatic incident radiation and absorber materials with only radiative recombination in the structure of figure 3.5.

3.4.1   Thermophotovoltaic conversion

In thermophotovoltaic (TPV) conversion [7, 8], the solar radiation is absorbed by an intermediate absorber, which heats it to high temperatures. The absorber is emitting near-monochromatic radiation towards a solar cell, either by virtue of its own selectivity or through a filter. In this arrangement, all thermalization losses in the solar cell are avoided and the unsuitable photons are not lost, since they are either not emitted or they are reflected back onto the intermediate absorber by the filter, which helps to maintain a high absorber temperature. Even losses from radiative recombination in the solar cell are absent, because all photons emitted by the solar cell pass the filter and are also absorbed by the intermediate absorber. Under these conditions, the solar cell can be operated very close to open-circuit conditions, where its efficiency is given by the Carnot efficiency as derived in (3.10). The only losses are due to the emission from the absorber towards the sun as stated in (3.4). The theoretical limit of the overall efficiency of TPV conversion is, therefore, given by (3.5) with a maximal value of 85% if the intermediate absorber is operated at 2478 K, absorbing fully concentrated solar radiation.

3.4.2   Hot carrier cell

Thermalization losses are avoided if electrons and holes leave the absorber through the semi-permeable membranes before they are thermalized by scattering with phonons [9, 10]. Electrons and holes, however, have to cool down to the temperature T_0 of the environment as an important step in the conversion process. If it does not occurr in the absorber, thermalization must happen in the semi-permeable membranes or in the metallic electrodes. In order to avoid thermalization losses there, only monoenergetic electrons (or holes) should be allowed to pass through the membranes, which can be achieved, for example, by a narrow conduction band (or valence band) of the membranes. For the hot carrier cell, the membranes must be energy selective in addition to being carrier selective. Under these conditions, a theoretical efficiency limit of 85% is found, identical to the efficiency of TPV conversion. Realizing this concept seems to be very difficult considering that thermalization times are commonly below 10^{-12} s. It is hoped that materials in which thermalization is much slower as might be expected for quantum dots will be found, which would also provide a monoenergetic exchange of charge carriers [11].

3.4.3   Tandem cells

The most advanced concept for high efficiencies is a tandem arrangement of solar cells with different bandgaps. Placing the cell with the largest bandgap on top, facing the sun, followed by others with decreasing bandgaps, each cell absorbs the photons with energies between its own bandgap and the bandgap of the cell in front of it. With these properties, parallel processing of different photon energy intervals of the solar spectrum in different solar cells is realized and the requirements discussed in section 3.2.2 are fulfilled. For maximal concentration of the incident solar radiation, the limiting efficiency is 86% for infinitely many solar cells with different bandgaps [1]. Even processing the incident solar spectrum with only two or three different solar cells reduces thermalization losses considerably compared to a single material and allows efficiencies substantially above the Shockley–Queisser limit.

3.4.4   Intermediate level cells

An energy level between the valence and conduction bands in the absorber material provides additional transitions at lower energies [12, 13]. In addition to band–band transitions, electron–hole pairs are generated in a two-step process, where an electron is first excited from the valence band to an intermediate level and by a second photon from there to the conduction band. An intermediate-level cell is equivalent to three cells in a tandem, where a series connection of the two cells representing the transitions involving the intermediate level is connected in parallel with the third cell, which represents the band–band transitions. If all transitions are radiative, a maximal efficiency of 63% for full concentration is found for a cell with a bandgap of 1.93 eV and an intermediate level at 0.70 eV above the valence band.

3.4.5   Photon up- and down-conversion

An interesting suggestion is a variation of the intermediate-level cell for up-converting two small energy photons, which would not be absorbed by the solar cell, to one larger energy photon, which is absorbed by the solar cell. This process can also be reversed to down-conversion, where one large energy photon, with an energy larger than twice the bandgap of the solar cell is down-converted into two photons, with an energy just above the bandgap [14, 15]. Figure 3.6 shows an example of the up-conversion process. Up- and down-conversion occur simultaneously in a three-level system. Detailed balance calculations show that for non-concentrated solar radiation, down-conversion is much more probable than up-conversion, except for very large intensities, approaching full concentration, where up-conversion is predominant. Introducing a fourth level slightly below and in electrochemical equilibrium with the upper level of the three-level system, from which the transition to the intermediate level is forbidden but is allowed to the lowest level, also makes up-conversion feasible for non-concentrated radiation. The photon conversion takes place in a material outside the solar cell, which is in good optical but not electrical contact with the solar cell. The photon converter must have the desired optical properties but good transport properties are not required.

For maximal concentration, a maximal efficiency of 63% is found for up-conversion, identical to the intermediate-level cell. For no concentration, the maximal efficiency for up-conversion is 48% and 40% for down-conversion. Up-

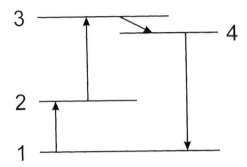

Figure 3.6. In an up-conversion process two photons excite an electron from level 1 to level 3. After a relaxation to level 4, one large energy photon is emitted by the transition to level 1 (reprinted with permission [19]).

and down-conversion can even be applied simultaneously to a single solar cell.

As in all other cases discussed so far, the non-radiative recombination occurring in real materials to varying degrees has been excluded. Intermediate levels, particularly if they are caused by imperfections in a semiconductor, give rise to enhanced non-radiative recombination [16–18]. Having the intermediate level inside the solar cell increases the probability for non-radiative recombination for all electron–hole pairs, whether or not they were generated by band–band transitions or via the intermediate level in a two-step process. Introducing an intermediate level into a real material may, therefore, worsen the efficiency of a solar cell instead of improving it. This is different for photon conversion concepts. An up-converter would be placed behind the solar cell and use the photons with $\hbar\omega < \varepsilon_G$, which are not absorbed by the solar cell and which are wasted in the normal operation of a single solar cell. The up-converter converts two of them into one photon with $\hbar\omega > \varepsilon_G$, which is useful for the solar cell. Non-radiative recombination will also occur in the up-converter but in contrast to the intermediate-level solar cell, it will only affect excitations produced by photons which are not absorbed by the solar cell—it will not affect the recombination in the solar cell. The up-converter can, therefore, only lead to an improvement in the efficiency. A similar argument holds for the down-converter.

3.5 Conclusions

It has been shown that solar cells with only radiative recombination are ideal Carnot engines if operated with monochromatic light. As a result thermophotovoltaic (TPV) conversion or conversion by an infinite tandem configuration, both providing monochromatic illumination, yield large limiting efficiencies of 85% and 86%, respectively. A general model of a solar cell has been presented, demonstrating that all present solar cell concepts adhere to the same physical principles and are subject to the same limitations. Thermalization losses are the most prominent losses and any means of narrowing the spectral width of the photons accepted for an optical transition, without wasting the rejected photons, improves the theoretical efficiency over the Shockley–Queisser limit. Photon up- and down-conversion are particularly attractive, because they transform the incident spectrum only and do not affect any of the processes inside the solar cell and can, therefore, be applied to already existing solar cells.

References

[1] de Vos A 1983 *Proc. 5.E.C. Photovoltaic Solar Energy Conf. (Athens)* (Ispra: European Commission) p 186

[2] Lasher G and Stern F 1964 *Phys. Rev.* A **133** 553

[3] Würfel P 1982 *J. Phys. C: Solid State Phys.* **15** 3967

[4] Würfel P, Finkbeiner S and Daub E 1995 *Appl. Phys.* A **60** 67

[5] Shockley W and Queisser H J 1961 *J. Appl. Phys.* **32** 510

[6] O'Reagan B and Graetzel M 1991 *Nature* **353** 737

[7] Wedlock B D 1963 *Proc. IEEE* **51** 694

[8] Swanson R M 1979 *Proc. IEEE* **67** 446

[9] Werner J H, Brendel R and Queisser H J 1994 *Proc. First World Conf. on Photovoltaic Solar Energy Conversion (Hawaii)* (New York: IEEE)

[10] Würfel P 1997 *Solar Energy Mater. Solar Cells* **46** 43

[11] Nozik A J 2001 *Ann. Rev. Phys. Chem.* **52** 193

[12] Green M A 1995 *Silicon Solar Cells* University of New South Wales, Sydney, Australia

[13] Luque A and Marti A 1997 *Phys. Rev. Lett.* **78** 5014

[14] Trupke T, Green M A and Würfel P 2002 *J. Appl. Phys.* **92** 1668

[15] Trupke T, Green M A and Würfel P 2002 *J. Appl. Phys.* **92** 4117

[16] Shockley W and Read W T 1952 *Phys. Rev.* **87** 835

[17] Hall R N 1951 *Phys. Rev.* **83** 228

[18] Güttler G and Queisser H J 1970 *Energy Conversion* **10** 51

[19] Würfel P 2002 *Physica* E **14** 18–26

Chapter 4

Tandem cells for very high concentration

A W Bett
Fraunhofer Institut für Solar Energie Systeme ISE
Heidenhofstrasse 2, 79110 Freiburg, Germany

4.1 Introduction

Tandem cells for very high concentrations? Why is the Fraunhofer Institute for Solar Energy (ISE) involved in this technology? The answer is simple: in our opinion the tandem cell combined with concentrator technology will play a role in the future energy market. In this chapter, the motivation for high concentration cells, the state-of-the-art of our tandem cell development and the application of dual-junction cells in concentrator modules are discussed.

During the last decade, different laboratories have developed single-junction solar cells with high efficiencies. Crystalline silicon solar cells have exemplified the good progress made with respect to higher efficiencies. So far, an efficiency of 24.7%, which is close to the absolute theoretical limit, has been reported [1,2]. Therefore, one cannot expect higher efficiencies for Si solar cells in the future. The challenge for Si solar cells is achieving these high efficiencies in production. A similar situation is found for other crystalline materials, for example GaAs. If we consider thin-film materials, like CI(G)S, CdTe etc, currently showing efficiencies in the range 16.5–18.5% in the laboratory [1], one could expect some progress in the future. However, even for these materials, the efficiency limits will very soon be reached.

A higher efficiency is desirable to decrease the area-related costs of solar cell modules. Several concepts aiming for efficiencies beyond 30 % are now under discussion [3]. In our opinion, the use of tandem cells is today the only realistic concept. Tandem cells based on different III–V semiconductor materials are already well developed and successfully applied in spacecraft. Recently, a record efficiency of 29.7% was reported for AM0 spectral conditions in space [4]. While these high-efficiency III–V cells gain more and more of the market share for space applications, they are too expensive for standard flat-plate modules on earth. The reason is the high cost of the tandem cell. One factor driving up the costs is the necessity for Ge or GaAs substrates. These substrates cost 50–100 times more than Si substrates. Thus, flat-plate modules made from tandem cells for the terrestrial use are out of the question. There is, however, a solution for this challenge, namely the concentrator technology in which sunlight is concentrated by a cheap optical concentrator. According to the concentration factor, less semiconductor material is necessary. A concentration factor of 100 or higher is mandatory to be cost competitive, as can easily be concluded from the substrate costs. A cost analysis was performed to identify the required concentration level more precisely [56]. Aiming at low-cost photovoltaics, the basis for this study was a goal of 0.1 € kWh^{-1} electricity. A number of assumptions are made for the analysis:

- the cell efficiency increases logarithmically with concentration;
- the operating system efficiency is calculated by considering optical and thermal losses;
- the irradiance is 1826 kWh m^{-2}a (equivalent to the direct normal incidence (DNI) in Madrid);
- a lifetime of 20 years, an interest rate of 4% and an annuity of 7.4% are assumed;
- the power-related system costs are fixed to 350 € kWp^{-1}; and
- the annual operating and maintenance costs are 1.5% of the total system costs.

These assumptions lead to the diagram shown in figure 4.1 [56]. Here one has to emphasize that this result is independent of the applied cell and concentrator technology. Figure 4.1 shows the allowed area-related system costs (including tracking, mounting, frames etc) as a function of cell costs. The dependency on the concentration factor is also shown. The cell efficiencies are values determined in a laboratory measurement under given concentration levels. Therefore, they also reflect the increase in efficiency with high illumination intensity.

The price of current space tandem solar cells is about 10–15 € cm^{-2}. However, if we consider the use of a concentrator solar cell, it seems possible to decrease the cost by a factor of two. One reason is a higher yield in production. The yield of space solar cells is low because large high-quality epitaxial areas are necessary (>16 cm^2). In tiny concentrator solar cells, areas with defects can be sorted out, thus increasing the overall yield. Furthermore, a higher market volume would accelerate the learning curve, thus helping to decrease the cost. Estimating area-related costs, 150–250 € m^{-2} seems realistic. Similar costs are experienced for thermal concentrator systems [5]. Thus, in conclusion, we think that electricity costs of 0.1 € kWh^{-1} are achievable for solar cells with efficiencies above 30% (i.e. tandem cells) and concentration factors higher than 500.

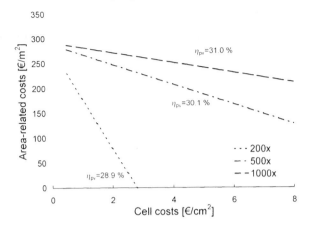

Figure 4.1. Estimation of allowed area-related system costs *versus* cell costs depending on the concentration factor for the assumptions made in the text. The photovoltaic cell efficiencies are realistic values which were measured in the laboratory for the given concentration (from [56]. © 2000 IEEE).

4.2   Tandem solar cells

If we consider tandem solar cells, it is illustrative to look at the benefit of adding further pn-junctions. The theoretical efficiency limits were calculated using the program ETAOPT [6], developed at Fraunhofer ISE and available on the internet [7]. This program considers only radiative recombination, no material properties and is based on the detailed balance formalism suggested by Shockley and Queisser [8]. Figure 4.2 shows a calculation assuming a concentration of 500 and the AM1.5direct spectrum. The gain in efficiency decreases with each additional junction. Thus, one can speculate that only dual-, triple- or quadruple-junctions are of practical interest, especially if, in addition, the complexity of the real device is taken into account. However, for space applications even a quintuple-junction seems of specific interest [9].

Materials with different bandgaps are essential for the processing of tandem cells. Compound semiconductors offer a large variety of different bandgap energies. For example, II–VI or III–V compounds have already been used for tandem cells. The use of III–V semiconductors yields higher efficiencies. Figure 4.3 shows a 'map' of the III–V compounds. Here, the bandgap is plotted as a function of the lattice constant of the materials. The lines between the binary compounds represent the corresponding ternary materials. As a rule, epitaxial growth of high-quality semiconductor material can be achieved if the condition of lattice matching is fulfilled. To give an example, the material $Ga_{0.51}In_{0.49}P$ is lattice matched to GaAs (see figure 4.3).

Pn-junctions using different semiconductor materials are grown by epitaxial

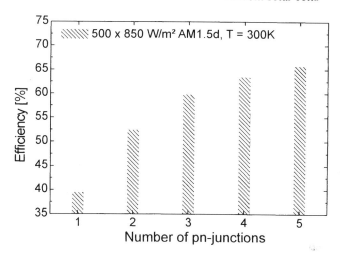

Figure 4.2. Efficiency calculated as a function of the number of pn-junctions using the program ETAOPT. A concentration ratio of 500 is assumed.

methods. A tandem cell which uses only one substrate is called a monolithic tandem cell. The fabrication of monolithic cells is described in section 4.2.2. The different pn-junctions are internally connected by tunnel diodes. Thus, the cells are internally series connected. As a consequence, the total output current of a tandem cell is determined by the minimum current generated in the sub-cells. This is a strong restriction for the monolithic cell, in particular for terrestrial concentrator applications. The tandem cell performance is influenced by the changes in the sun spectrum. This challenge will be discussed in section 4.3.1.

Another possibility for making tandem cells is the so-called mechanically stacked approach. This technology requires pn-junctions fabricated on different substrate materials. The advantage of this approach is the simpler solar cell structure compared with the monolithic tandem cells. Cheaper technologies, like liquid phase epitaxy (LPE) or Zn-vapour diffusion technology, can be deployed to fabricate pn-junctions. The fabrication of mechanically stacked cells is discussed in the following section.

4.2.1 Mechanically stacked tandem cells

The dependence of the efficiency on the choice of the bandgaps for the top and bottom cells in mechanically stacked dual-junction concentrator cells was calculated in [10]. A bandgap combination of 1.4 eV (e.g. GaAs) and 0.7 eV (e.g. GaSb) is close to the optimum combination. Fraas *et al* [11] were the first to report the successful fabrication of mechanically stacked dual junction cells. They originally intended to use these cells for space applications, later the cells were considered for terrestrial concentrator systems [12–14]. As can

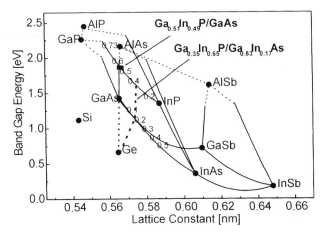

Figure 4.3. Bandgap energy *versus* lattice constant for the III–V semiconductors. The lines between the binary compounds represent the respective ternaries.

be seen from the theoretical calculations [10], another bandgap combination of 1.7 eV/1.1 eV might also be of practical interest [15]. Here, Si can be used as the bottom-cell material. The top-cell can be made of $Al_{0.24}Ga_{0.76}As$. This material is grown epitaxially on a GaAs substrate. However, in our investigations the $Al_{0.24}Ga_{0.76}As$–Si dual-junction cell achieved lower efficiencies (in the range of 25%) compared with GaAs–GaSb (up to 31.1%) [10]. Furthermore, more technological effort is necessary to fabricate the $Al_{0.24}Ga_{0.76}As$–Si cell. For example, the GaAs substrate underneath the $Al_{0.24}Ga_{0.76}As$ cell has to be removed because the bandgap of the top cell is higher than the bandgap of the GaAs substrate. Thus, photons transmitted through the top cell would be absorbed in the GaAs substrate and would be lost for the Si bottom cell. By etching off the GaAs substrate, a high transmissivity is obtained, increasing the current in the Si bottom cell.

Regarding the GaAs–GaSb dual-junction cell, the photoactive top cell has the same bandgap as the substrate. Only absorption due to free carriers in the substrate has to be considered. P-doped GaAs substrates show a high absorption for photons with an energy lower than the bandgap of GaAs. This parasitic absorption does reduce the current in the GaSb bottom cell. In contrast, n-doped GaAs shows a much lower absorption beyond the bandgap [10]. Thus, we investigated p-on-n $Al_{0.8}Ga_{0.2}As$/GaAs heteroface structures for the top solar cell. The GaAs top-cell structure was grown by the isothermal LPE etchback regrowth (LPE-EBR) method. A homemade H_2-purified LPE system was used. A Ga-based melt contains Al (2.4 at%) and Zn (0.7 at%). The melt is saturated with respect to arsenic at a temperature of 850 °C. After homogenization, the melt is in contact with a GaAs substrate at 865 °C for 40 min. Three main processes take place: first, the melt etches the GaAs substrate back due to the As under-saturation. Second, Zn diffuses into the n-doped GaAs substrate material forming the pn-junction; and third, a thin $Al_{0.8}Ga_{0.2}As$ layer is isothermally grown due to the difference in the chemical potentials between substrate and melt. More details about this process can be found elsewhere [16, 17].

It is noteworthy that the simple isothermal LPE process leads to a simple cell structure (see figure 4.4). No highly doped cap-layer nor back-surface field is used. A relatively thick (2 μm) emitter reduces the losses which occur in the base layer of this solar cell [18]. If no back-surface field is used, minority carriers generated in the base layer might diffuse into the substrate and become lost. Furthermore, this thick emitter has the additional advantage of providing a low sheet resistance. This is beneficial for applications under high illumination intensity. The crucial challenge for concentrator solar cells is to achieve a low overall series resistance [19, 20]. One contribution to the series resistance is the sheet resistance of the emitter. Another is the contact resistance. For the simple LPE-EBR structure, the contact resistance was found to limit the series resistance. A contact resistivity of $5 \times 10^{-4} \Omega \ cm^2$ was achieved. This value was obtained for AuMn contacts, a GaAs surface doping level of $2 \times 10^{18} \ cm^{-3}$ and an annealing temperature of 350 °C. The metals are directly evaporated onto the p-doped GaAs emitter. A lower contact resistivity is achieved by increasing the doping level of the semiconductor materials. However, in one-step epitaxy, the Zn surface concentration cannot be increased. Consequently, a post-growth Zn-vapour diffusion process was developed [21]. Here, an additional Zn diffusion process was performed from the vapour phase. The resulting diffusion profile was shallow but steep. The depth was only 30 nm and the surface doping concentration was $1 \times 10^{19} \ cm^{-3}$. Thus, the contact resistivity was successfully lowered to $1 \times 10^{-4} \ \Omega \ cm^2$ [21].

Regarding the GaSb bottom cell, the pn-junction is made by a simple Zn-vapour diffusion process [22, 23] leading to a simple structure, see figure 4.4. The diffusion process is performed in a 'pseudo-closed' box. It is made of graphite and can be closed tightly by screws. The graphite box is put into a quartz tube with an atmosphere of purified hydrogen. Separate pure Zn- and Sb-vapour sources are used in more than sufficient quantities to provide the saturation of the vapour pressures. The antimony is used to prevent non-congruent evaporation of antimony from the unprotected GaSb surface during the diffusion process. Exactly oriented (100) Te-doped GaSb substrates were used. The doping level was in the range $4–7 \times 10^{17} \ cm^{-3}$. Prior to the diffusion process, the surface oxide of the GaSb substrates was etched off in HCl.

The diffusion process and the resulting diffusion profile strongly depend on two parameters: temperature and time. The optimum diffusion regime was found through intensive experiments [22, 24]. Eventually, a diffusion temperature of 480 °C and a diffusion time of 4 hr was used to form the pn-junction in the n-GaSb wafer. However, using these parameters led to a high surface carrier concentration of $2 \times 10^{20} \ cm^{-3}$. The lifetime of the minority carriers in such highly doped layers

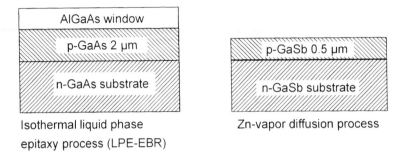

Figure 4.4. Structures of the heteroface $Al_{0.8}Ga_{0.2}As/GaAs$ top cell and GaSb bottom cell. The top cell is fabricated using an isothermal LPE-EBR process and the bottom cell is fabricated by a simple Zn-vapour diffusion process.

is too low to fabricate high-performance PV cells [25]. Consequently, this highly doped layer has to be removed. This can be accomplished by anodic oxidation and selective etching of the anodic oxide layer [26]. Moreover, if the process is properly designed the final anodic oxide layer can be used as an antireflective coating (ARC).

GaAs and GaSb concentrator solar cells with a diameter of 4 mm were fabricated. The applied front contact grids were designed for operating at concentration factors of 160. A prismatic cover was applied reducing the shadowing losses. The prismatic cover consists of micro-lenses reflecting the impinging light away from the grid fingers. This was taken into account when optimizing the ARCs. Here not only the number of absorbed photons (as in single-junction cells) but also the energy output has to be optimized [27]. Successive rays of MgF_2/TiO_x, 115 and 65 nm thick, were applied as top-cell ARCs. Also, at the rear side of the top cell, a spot 4 mm in diameter was evaporated using MgF_2/TiO_x (thicknesses: 225 nm/135 nm). Thus, the transmissivity of the non-absorbed photons is enhanced. Outside this circle Au/Au-Ge serves as the back-contact of the top cell. The bottom cell uses anodic oxide as a single-layer ARC. A prismatic cover was also applied there.

Both cells are mounted on copper plates which serve as heat spreaders. The plates are screwed together. A special thin foil ensures electrical isolation but maintain a good thermal conductivity. Figure 4.5 shows a photograph of a mechanically stacked dual-junction cell. Each cell can be contacted separately, resulting in a four-terminal device. A maximum efficiency of 31.1% for the GaAs–GaSb dual-junction cell under $100\times$ light concentration was confirmed by the Calibration Laboratory of the Fraunhofer Institute. The cells were measured at 25 °C under the AM1.5direct spectrum. The performances of the GaAs and GaSb concentrator cells were also measured at temperatures up to 60 °C. The temperature coefficients of the efficiency at a concentration of ~100 are as follows: GaAs: -0.05 % K^{-1}, GaSb: -0.044 % K^{-1}. For comparison, the

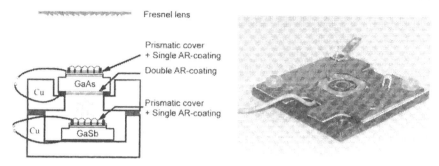

Figure 4.5. Diagram (from [13]. © 1977 IEEE) and photograph of a mechanically stacked GaAs–GaSb dual-junction solar cell. Usually mounted on copper plates which are isolated by a heat-conducting foil. The top cell uses antireflection coatings on the front and rear side. Circular cells with a diameter of 4 mm were fabricated.

coefficient of Si is -0.07 % K^{-1} [28, 29].

As previously mentioned, the series resistance of a solar cell plays a major role under high illumination intensity. Assuming the same solar cell structure and aiming always at the maximum efficiency would require an adapted grid design for each concentration level. The grid design applied for the GaAs–GaSb cell aimed at a concentration factor of 160. In fact, the maximum efficiency was measured for a concentration of 100. A reason for the discrepancy is that the simulation program always assumes full metal coverage of the rear side. The four-terminal mechanically stacked GaAs–GaSb cell has a 4 mm hole without metal on the rear side to allow light transmittance. Consequently, the electrical power losses in the wafer have to be taken into account. A higher doping level for the n-doped substrate would reduce the electrical losses. However, the enhanced parasitic free-carrier absorption increases the optical losses. A trade-off has to be made. Furthermore, for the specific structure under discussion, the wafer doping level corresponds to the doping level of the base (see also figure 4.4). Thus, for our application a good practical trade-off was a 350 μm thick wafer with a doping level of 5×10^{17} cm^{-3} [30]. However, if concentration factors higher than 100 are necessary there are further options, for example a decreased cell size or the application of a grid on the rear side of the solar cell.

Eventually, the developed mechanically stacked GaAs–GaSb dual-junction cells were installed in a concentrator test module. The module uses 24 point-focus Fresnel lenses each with an aperture area of 4.5 cm × 4.5 cm. Thus, the total aperture was 486 cm^2 (see figure 4.6). Twenty-four cells were interconnected as follows: eight top cells are connected in series and three such strings are connected in parallel. All the bottom cells are connected in series and eventually this string is connected in parallel to the top-cell string leading to a two-terminal configuration for the module [31]. A nearly voltage-matched condition was achieved for the top- and bottom-module interconnection. The module was

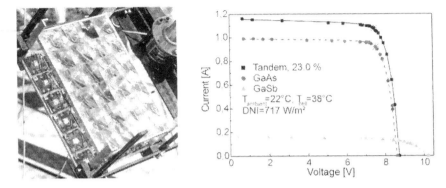

Figure 4.6. Left: photograph of a test module using GaAs–GaSb mechanically stacked dual-junction cells and point-focus Fresnel lenses. The geometrical concentration was 160. Right: measured IV curves for the total module and the two sub-modules consisting of GaAs and GaSb cells. The total module efficiency was 23.0% (from [13]. © 1977 IEEE).

mounted on a tracking system and measured under outdoor conditions in Freiburg, Germany.

The direct radiation at normal incidence was 717 W m^{-2} and an ambient temperature of 22 °C and a solar cell temperature of 38 °C were determined. The operating temperature was 16 K above ambient temperature and comparable to that of flat-plate modules. This low value was expected and is a specific advantage of point-focus concentrators. The photons and, thus, the excess heat are collected within a small spot. Using an efficient heat spreader—like copper—the excess heat can easily be distributed into the two-dimensional area. Consequently, the aperture area (= collection area) nearly equals the heat-spreader area and a similar situation to that for flat-plate modules is obtained. Furthermore, the higher efficiency of the dual-junction cells helps to reduce the excess heat.

The IV curves of the sub-modules consisting of GaAs and GaSb are presented in figure 4.6. The maximum power point of each curve is also indicated by stars. The total module efficiency in two-terminal operation was 23%. A careful analysis of the loss in the module was performed, showing the potential of achieving 25% under operating conditions [31].

4.2.2 Monolithic tandem cells

The advantage of monolithic tandem cells in comparison to mechanically stacked cells is the use of only one substrate. This has a positive impact on the costs of the cell. However, a more complex structure has to be grown. A typical cell structure is shown in figure 4.7.

Obviously, from the point of crystal growth, the structure has a lot of challenges:

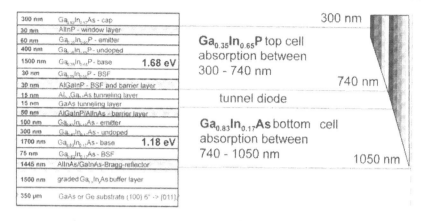

Figure 4.7. Example of a monolithic dual-junction solar cell structure. The high-energy photons are absorbed in the top-cell. The tunnel cell interconnects internally the top-cell with the bottom-cell.

- homogeneity in the composition of the ternary and quaternary III–V compounds,
- thin and thick layers, from 15 nm up to 1700 nm and
- differences in doping levels, from not intentionally doped up to 10^{20} cm^{-3}.

Further demands on the epitaxial growth technique are a high throughput and reproducibility. Fortunately, the metal-organic vapour phase epitaxy (MOVPE) technology fulfils these demands. The Fraunhofer ISE uses an industrial-size MOVPE reactor AIX2600G3. A sketch of the reactor principle is shown in figure 4.8. The metal-organic precursors and the hydrides are mixed in the reactor chamber. The planetary susceptor carries up to eight 4-in or twenty-four 2-in wafers. The homogeneity of the grown layers is ensured by rotating the whole susceptor and each of the substrates separately. Growth rates up to 10 μm hr^{-1} were realized. More details about the growth conditions can be found elsewhere [32, 33].

It is interesting to regard the capacity of such a MOVPE reactor with respect to concentrator solar cells. For example, a grid design for cells with a diameter of 2 mm was developed at the Fraunhofer ISE aiming at concentration ratios of 500. On one 4-inch wafer up to 885 cells can be manufactured. Assuming an efficiency of 30%, this leads to a capacity of 3.2 kWp per epitaxial run. Per year, 2.5 MWp 500×-concentrator solar cells can be produced in one AIX2600-G3 MOVPE reactor. Installing a MOVPE reactor needs only an area of about 100 m². This reduces investment costs which is another advantage of the concentrator cell technology.

Usually lattice-matched conditions are required for the epitaxial growth of high-quality semiconductor materials. Figure 4.3 showed the relationship

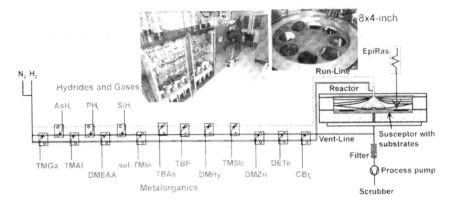

Figure 4.8. Schematic diagram of the AIX2600G3 MOVPE reactor used at the Fraunhofer ISE. The photographs show the gas-mixing cabinet and the susceptor.

between the lattice constant and the bandgap for some III–V compounds. If we considering only the lattice-matching conditions, GaAs and $Ga_{0.51}In_{0.49}P$ would be a suitable choice for dual-junction solar cells. However, calculations using the program ETAOPT [6] reveal this material combination to be far from ideal for dual-junction cells [34]. Higher efficiencies under AM1.5direct spectral conditions can be expected if the bandgaps of the two sub-cells are lowered. This can be achieved by adding indium to GaAs and by increasing the In-content in $Ga_{0.51}In_{0.49}P$. Thus, the absorption of the bottom cell is extended beyond the bandgap of GaAs. However, in this case the lattice-matching condition is not fulfilled (see also figure 4.3). In spite of this disadvantage, the Fraunhofer ISE has fabricated dual-junction solar cells on the basis of $Ga_{0.35}In_{0.65}P/Ga_{0.83}In_{0.17}As$ structures grown on a GaAs substrate (see also figure 4.7). A special buffer layer has to be grown to deal with the difference in the lattice constants between the substrate and epitaxial material [32]. Thus, the dislocation density in the active layers was reduced. Figure 4.9 gives a schematic comparison of the more standard lattice-matched approach using $Ga_{0.51}In_{0.49}P/GaAs$ and the Fraunhofer ISE approach using lattice-mismatched $Ga_{0.35}In_{0.65}P/Ga_{0.83}In_{0.17}As$.

Dual-junction solar cell structures of n-on-p $Ga_{0.35}In_{0.65}P/Ga_{0.83}In_{0.17}As$ were grown. From these, circular solar cells with a diameter of 4 m were fabricated. AuGe and TiPdAg were evaporated as front- and back-contacts, respectively. The contacts were strengthened up to 3 μm by electroplating with Au. This is necessary to reduce the resistance of the metal grid fingers. A prismatic cover was used to reduce the shadowing losses of the front grid design. The Calibration Laboratory of Fraunhofer ISE determined a maximum efficiency of 31.1% at a concentration of 300. The test conditions were AM1.5direct spectrum normalized to 1000 W m^{-2} and a temperature of 25 °C. For comparison, the highest reported efficiency for the lattice-matched

standard

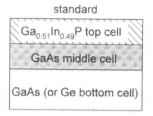

ISE concept

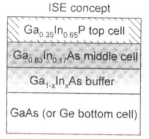

GaInP, GaAs
and
substrate lattice matched
$300 < \lambda < 900$

GaInP and GaInAs
lattice matched,
substrate mismatched
$300 < \lambda < 1050$ nm
higher efficiency potential

Figure 4.9. Comparison of the dual-junction structure used commonly and the structure under investigation at the Fraunhofer ISE. While the standard structure relies on lattice-matched material, the ISE concept is based on lattice-mismatched materials. However, using this approach, higher efficiencies are expected.

$Ga_{0.51}In_{0.49}P/GaAs$ concentrator cell is 30.2% at a concentration level of 180 [35].

Besides the maximum efficiency, it is of practical interest to measure how the efficiency drops at higher concentration levels. For example, if in practical concentrator systems a Fresnel lens is applied as the concentrating optics, the illumination profile is not homogeneous but has a Gaussian profile. Thus, in the centre of a cell, the concentration ratio is much higher compared to the edges. Our experimental results showed an up to five times higher concentration than the average concentration in the centre [36]. In the laboratory, the measurements are performed under homogeneous illumination conditions only. However, the dual-junction cell under investigation showed an efficiency beyond 29% at concentration levels of $\sim$1200$\times$. Efficiencies beyond 30% were achieved for all concentrations between 80 and 1000. Thus, this cell can be applied effectively in real concentrator systems operating at concentration levels of 100 to 200. The results of corresponding experiments are described in section 4.3.1.

The operation of monolithic tandem cells at high concentration is a challenge with respect to the series resistance. As for the mechanically stacked concentrator cells, the series resistance has to be low (in the order of 0.001 Ω). The front grid design has to be optimized with respect to optical (shadowing) and electrical losses. Moreover, one has to take care of the internal series connection in a monolithic dual-junction cell. The current density is typically lower compared to single-junction concentrator cells. This is beneficial because the ohmic drop scales with the second power of the current. However, the internal connection can cause additional series resistance problems. The implemented tunnel diode

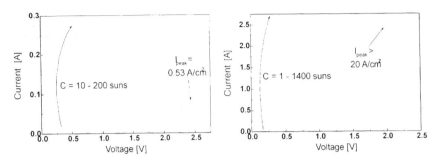

Figure 4.10. *I V* curves taken under different illumination intensities: (*a*) a high bandgap tunnel diode made from n-GaInP/p-AlGaAs was used in a $Ga_{0.51}In_{0.49}P$/GaAs monolithic dual-junction cell; and (*b*) an n-GaInAs/p-AlGaAs tunnel diode was used in a $Ga_{0.35}In_{0.65}P$/$Ga_{0.83}In_{0.17}As$ monolithic dual junction cell.

should perform like a quasi-ohmic series connection. The tunnelling current is influenced by the doping levels and the bandgap of the semiconductor material. As a rule, the p- and n-doping should be as high as possible. Furthermore the tunnelling occurs more easily in materials with a small bandgap compared with materials with a high bandgap [37]. However, for a monolithic tandem cell a material with a high bandgap would be desirable to avoid parasitic absorption. $Al_{0.3}Ga_{0.7}As$ was always used for the p-doped layer in our investigated tunnel structures. A high p-doping level of 1.4×10^{20} cm^{-3} could be achieved in this material by using an intrinsic doping technique. More details about this technique can be found elsewhere [38]. Unfortunately, a high n-doping level is not possible in $Al_{0.3}Ga_{0.7}As$. Instead, Si-doped $Ga_{0.51}In_{0.49}P$ was used for the n-doped layer. The doping level was 5×10^{18} cm^{-3}. Using n-GaInP/p-AlGaAs a high bandgap tunnel diode was realized. Figure 4.10(*a*) shows the *I V* curves taken from a monolithic $Ga_{0.51}In_{0.49}P$/GaAs dual-junction cells under different illumination levels [34].

The *I V* curves show a very good performance for one-sun application. This demonstrates that the high bandgap tunnel diode can be used in space solar cells [39]. Increasing the light intensity reveals the limitation of this tunnel structure. At a higher light-induced current density of 0.53 A cm^{-2}, the tunnelling behaviour breaks down, thus the resistance increases and a voltage drop occurs. Consequently, this tunnel diode can be used for concentration ratios below 10 but is not applicable for high concentration photovoltaics. Figure 4.10(*b*) shows *I V* curves taken from a monolithic dual-junction solar cell using an n-GaInAs/p-AlGaAs tunnel diode [34]. This structure shows a good performance up to current densities of 20 A cm^{-2}. This corresponds to 1400 times the one-sun illumination density. Obviously, this tunnel diode structure is well suited for applications at high concentration and does not limit the series resistance of the monolithic dual-junction concentrator cell. Thus, high-concentration dual-junction solar cells are well developed and ready for use in real concentrator systems.

In a next step, the Fraunhofer ISE is developing monolithic triple-junction cells based on the lattice-mismatched GaInP/GaInAs/Ge structure. Here, the Ge is an active solar cell. This requires an additional tunnel diode between the GaInAs middle cell and the Ge bottom cell. The first devices are being tested. However, other groups have already reported an efficiency of 32.4% at 414 × AM1.5direct [1, 40] using a lattice-matched GaInP/GaAs/Ge monolithic triple cell. This gives confidence that the additional tunnel diode is not a limiting factor for the monolithic triple-junction.

4.2.3 Combined approach: mechanical stacking of monolithic cells

Another possibility for realizing triple-junction solar cells is to combine the approaches described in the previous two sections. Therefore, we investigated a stack based on a monolithic dual-junction cell of $Ga_{0.35}In_{0.65}P/Ga_{0.83}In_{0.17}As$ as the top cell and GaSb as the bottom cell. Using this stack, an efficiency of 33.5% at 308 × AM1.5direct was confirmed by the Calibration Laboratory of the Fraunhofer ISE. However, there is still room for improvement and efficiencies up to 35% should be obtainable. To give reasons for this prediction one has to consider that (1) the ARC was not optimized for this application and (2) the doping level in the substrate was too high, increasing the parasitic absorption losses. A more detailed discussion can be found in [41].

Mechanical stacking of two monolithic dual-junction cells would be a path towards a quadruple cell. Efficiencies beyond 40% can be expected. The monolithic dual-junction top-cell is well developed, as discussed in section 4.2.2. Considering an infrared monolithic dual-junction cell, AlGaAsSb/GaSb would be a suitable choice [42]. However, there are a lot of challenges for the MOVPE growth of Sb-based materials [43, 44]. Anyhow, a first functional infrared monolithic dual-junction cell was recently presented [45]. Still, more efforts are necessary to fabricate high concentration photovoltaic cells with high efficiency.

4.3 Testing and application of monolithic dual-junction concentrator cells

4.3.1 Characterization of monolithic concentrator solar cells

Monolithic dual-junction concentrator cells demonstrated already a high performance under standard test condition measurements. However, no profound knowledge about the performance in real concentrator systems is available. Here, in contrast to single-junction cells, one has to take into account the dependency of the dual-junction cell efficiency on the spectral conditions [46, 47]. The cause is the internal series connection of the sub-cells. Thus, the output current is limited by the minimum of the photo-generated currents of all the sub-cells. The maximum output current can be obtained if the photo-generated current in each sub-cell is the same. This condition is called 'current matched'. Outside this matching point, less output current must be expected, thus reducing the efficiency of the cell. The change in performance of a multi-junction cell due to variations in the spectrum can be measured in the laboratory. At the Fraunhofer ISE, this is only possible for light intensities corresponding to the one-sun illumination level. A three-source simulator is used to adjust the spectral conditions. The intensity of the three different lamps can be set separately, thus generating the desired spectral distribution. For example, using the external quantum efficiency of a dual-junction cell, the spectrum of the simulator can be chosen in such a way that the same photo-current is generated in each sub-cell, as would be under the standardized AM1.5direct spectrum. In so doing, a calibrated efficiency for the AM1.5direct spectral condition can be determined. Furthermore, the spectrum can now be changed to become more red-rich compared with the standard spectrum. The (effective) generated photo-current in the bottom cell (G_{EFF}^{BOT}) increases as well as the ratio $G_{EFF}^{BOT}(E(\lambda))/G_{EFF}^{BOT}(\text{AM1.5d})$ also (see figure 4.11). Regarding the top cell, the (effective) generated photo-current (G_{EFF}^{TOP}) and the ratio $G_{EFF}^{TOP}(E(\lambda))/G_{EFF}^{TOP}(\text{AM1.5d})$ decreases (see figure 4.11). Changing the incident spectrum to more blue-rich, the situation is reversed (see also figure 4.11). The described characterization method is called 'spectrometric characterization' [48, 49]. Figure 4.11 shows an example of a spectrometric characterization of a monolithic dual-junction cell. The AM1.5direct standard spectral conditions were chosen as the reference spectrum. The investigated cell is nearly current matched (broken vertical line) under the AM1.5direct spectral condition. The measured short-circuit current decreases besides this point due to the current limitation of one of the sub-cells. The fill factor shows the opposite behaviour. It has a minimum at the current-matching condition and increases outside this point. The increase in the fill factor can be understood by considering the operating voltage of the sub-cells. Regarding the current-matching point, both cells are operating at the maximum power point leading to the lowest fill factor for the dual-junction cell. Outside this point, one of the sub-cells operates not at its maximum power point but closer to the open-circuit voltage. Consequently, the fill factor of the dual-junction cell increases.

No dependence of the open-circuit voltage on the spectral condition is observed. Regarding the maximum power point (as shown in figure 4.11), the dependency on the spectral conditions is smaller than expected from the short-circuit current. This is due to the compensation effect of the fill factor.

A spectrometric characterization reveals the spectral sensitivity of a monolithic tandem cell under one-sun illumination intensity. Concentrator cell measurements are usually performed using a single-flash light as the illumination source [50]. The flasher technique has the advantage of keeping the cell at constant temperature. No cell heating occurs during the very short measurement time (<1 ms). The obvious problem for the measurement of monolithic tandem concentrator cells is adjusting the flasher spectrum. Multi-source flasher systems comparable with the multi-source one-sun simulator would be a solution. Another

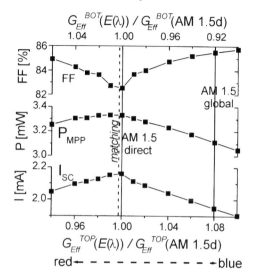

Figure 4.11. Example of a spectrometric characterization of a monolithic dual-junction $Ga_{0.65}In_{0.35}P/Ga_{0.83}In_{0.17}As$ cell. The AM1.5 direct spectrum was chosen as the reference condition. The short-circuit current of the dual-junction cell is given by the minimum of the short-circuit current of each sub-cell. Consequently, one sub-cell limits the current of the dual-junction cell except for the matching point. The voltage is the sum of the voltages of each sub-cell and is not shown here.

option is an iterative process of filtering the flasher light in order to obtain a matched spectrum.

However, in order to demonstrate the effect of changes in the illumination spectrum for concentrator cell measurements, we used flash-bulbs with different aging [51]. The spectra of the different bulbs were measured using a fast diode-spectrometer. Thus, we identified bulbs showing a matched, red-rich or blue spectrum. The matched spectrum generates the same currents as for the standard AM1.5direct spectrum in the respective sub-cells. Red-rich spectra generate more current in the bottom cell, thus the top-cell limits the output current. The opposite is true for blue-rich spectrum. Using different bulbs and, hence, different spectra, IV curves were measured as a function of light intensity. The spectrum-matched measurement led to a calibrated efficiency of 31.1% at a concentration of 300. Interestingly, red- and blue-rich spectra showed higher efficiencies! The highest efficiency of 32% was determined for a red-rich spectrum [51].

The explanation for the increase in efficiency can be understood in the following way. Using a red-rich spectrum for the measurement under concentration, the concentration value calculated by dividing the measured I_{SC} by the one-sun I_{SC} (determined for AM1.5direct, 1000 W m^{-2}) is only correct for the current-limiting top-cell. The bottom-cell is irradiated at a higher concentration.

Thus, this sub-cell does not operate at its maximum power point causing a higher fill factor for the dual-junction cell. The efficiency is over-estimated. This effect increases with increasing current mismatch of the sub-cells. The higher fill factor leads to higher (but wrong!) efficiencies. This was observed for all concentration levels. A more detailed analysis can be found in [51]. These experimental results demonstrate the necessity of careful analysis to obtain calibrated indoor measurements of a monolithic tandem concentrator cell.

Besides the calibrated indoor cell efficiencies, a prediction of the annual energy yield of a real concentrator system is of practical interest. Therefore, it is worth correlating the indoor cell measurement with outdoor module measurements. In this context, it is worth mentioning that the indoor measurement of concentrator modules is very challenging due to the necessity of large parallel light sources and the tendency towards increasingly larger concentrator sub-modules. As a consequence, concentrator modules are usually characterized outdoors, which has the disadvantage that neither the spectrum nor the temperature can be controlled.

Fortunately, a correlation between indoor cell measurements and outdoor concentrator module measurements has been successfully demonstrated [52]. However, there is still a way to go and more data are necessary to obtain the annual energy yield prediction for concentrator modules equipped with monolithic tandem cells.

4.3.2   Fabrication and characterization of a test module

In order to find a correlation between indoor cell measurements and outdoor module measurements, a high-efficiency concentrator module based on Fresnel lenses with an aperture of 4 cm × 4 cm^2 was fabricated. In combination with a cell of 0.13 cm^2 a geometrical concentration of 123 was reached. A 3 × 4 array was processed leading to a module aperture area of 192 cm^2. The Fresnel lenses were fabricated in-house. The technology is based on a stamp process [53]. A thin film of silicon is applied to a 3 mm thick glass sheet. Twelve high-quality matrices with an inverted Fresnel structure are pressed into this film. After drying, the matrices are removed resulting in a composite silicon–glass Fresnel lens panel. The glass sheet is used as superstrate and has an antireflection layer made by the sol–gel technology [52]. The optical efficiency of this Fresnel lens panel was defined as the ratio of the current-dependent concentration and geometrical concentration. The current-dependent concentration was determined by measuring the short-circuit current of a single-junction GaAs cell with and without a Fresnel lens. An overall optical efficiency of 89.1 ± 1.2% was determined for the lens panel. Dual-junction cells made of $Ga_{0.65}In_{0.35}P/Ga_{0.17}In_{0.83}As$ were applied in the module. The cells were characterized indoors using the spectrometric characterization technique. Compared to the standard AM1.5direct spectrum, the cells were top cell limited. Thus, higher efficiencies are obtained if—compared to the standard AM1.5direct

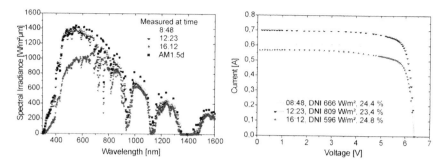

Figure 4.12. Left: measurement of the sun spectrum on 2 October 2001 in Freiburg. Right: simultaneously recorded *IV* curves of the test concentrator module shown in figure 4.13 (from [57]. © 2002 IEEE).

spectrum—more red-rich spectra are applied.

The cells were mounted on 3 cm × 3 cm² copper plates which were glued to a conventional aluminum heat sink. The heat sink acts as the base of the solar cell panel and effectively distributes the waste heat of the solar cells. No active cooling is required. The temperature of the illuminated cells in the module was measured to be less than 10 °C above the ambient temperature.

The experimental set-up for the outdoor measurements of the concentrator module was installed on the roof of the Fraunhofer ISE, Freiburg, Germany. The direct sun spectrum (DNI) was measured using a spectro-radiometer and an additional calibrated GaAs reference cell. The intensity was measured using a pyro-heliometer. The ambient temperature and the cell temperature were recorded. The *IV* curve of the high-efficiency Fresnel lens module was measured during three days in August, September and October 2001. Figure 4.12 shows, as an example, three *IV* curves (right) taken on 2 October 2001 and the simultaneously recorded direct spectra of the sun (left). A maximum module efficiency of 24.8% was obtained in the afternoon at a relatively low DNI of 596 W m^{-2}. At that time the spectrum of the sun was the most red-rich. About noon, the spectrum of the sun was more blue-rich and the DNI showed the highest value of 809 m^{-2}; however, the efficiency, 23.4%, was the lowest. These outdoor module measurements could be directly correlated to the indoor spectrometric characterization of the cell [52].

Figure 4.13 shows the measured efficiencies during three days *versus* the air mass. The air mass was calculated by Smarts2 [54] using longitude, latitude and local time (GMT + 2). The variation in efficiency is less than ±5% relative to the averaged efficiency. This is a promising result, demonstrating that the sensitivity of monolithic dual-junction cells on spectrum variations of the sun are smaller than expected. However, more data taken at different sites and over a longer period are necessary to predict the energy performance of concentrator modules equipped with tandem cells more accurately.

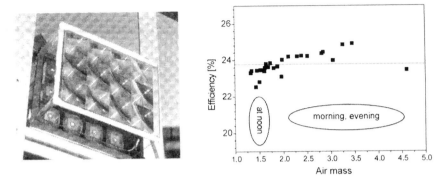

Figure 4.13. Left: photograph of the investigated test module with a geometrical concentration ratio of 123. A maximum efficiency of 24.8% was obtained. Right: the measured efficiencies as a function of the air mass (from [57]. © 2002 IEEE).

4.3.3 FLATCON module

In section 4.3.2 a test module with a concentration ratio of 123 was described. However, from the point of view of costs, this concentration ratio is too low for the expensive III–V semiconductor material (see also figure 4.1). Consequently, we have increased the concentration ratio up to 500. This was achieved by reducing the cell diameter to 2 mm and maintaining the aperture area of the Fresnel lens. The structure of the Fresnel lens was slightly changed. The low-concentration Fresnel lens had two ring sections with different focal distances. The inner ring had a slightly longer focal distance compared to the outer ring [55]. Thus, a more homogeneous light distribution was achieved on the cell. In contrast, the high-concentration Fresnel lens uses only one focal distance. Furthermore, larger modules with an aperture area of 768 cm^2 were fabricated. The modules consist of 48 sub-units. Six cells are connected in parallel and eight of these units are connected in series. The six cells are mounted on one copper plate facilitating good heat spreading. Eight such copper plates are glued directly onto a glass plate. The housing is also made of glass. The Fresnel lens panel is fabricated by the stamp process in silicone as described in section 4.3.2. A focal distance of only 76 mm is used leading to fairly flat modules. This module design is called FLATCON which is an abbreviation of Fresnel Lens All-Glass Tandem Cell Concentrator. The concept has several advantages:

(1) The use of small size cells

   • reduces the ohmic drop in the metal finger at high concentration as the finger length is short;

   • allows better heat dissipation into the two-dimensional area obviating the need for active cooling; and

   • means that the pick and pack technology used for light-emitting diode fabrication can be deployed.

(2) The all-glass design

- reduces the thermal stress which occur if different materials are applied;
- leads to a hermetically sealed housing and, hence, the environment does not degrade either the cell or the Fresnel lens; and
- ensures a high reliability.   Glass is a well-proven material in photovoltaics.

(3) The Fresnel lens structure can be fabricated by a simple stamp process in a 0.2 mm thin silicon film.   Silicone is known to be stable against UV irradiation. For example, it is used for gluing the cover glass on space solar cells.

Figure 4.14 shows two photographs visualizing the in-house module-manufacturing process. On the right-hand photograph, the mounting of the glass housing is shown. A special silicon is used to glue the glass sheets including the Fresnel lens panel. The photograph on the left in figure 4.14 shows the adjustment of copper plates on a glass sheet. The copper sheets carry six monolithic dual-junction cells 2 mm in diameter and a Si by-pass diode in the centre.

The 2 mm concentrator cells mounted in the module were measured using a flash-light simulator. A maximum efficiency greater than 33% at a concentration of 500 was determined.   However, remembering the challenges of testing monolithic dual-junction cells discussed in section 4.3.1, this result does not represent a calibrated value. It demonstrates only that high-efficiency cells are applied in the modules.

Several FLATCON modules were fabricated using these high-efficiency cells. They were mounted on a tracker to perform outdoor characterization. A photograph of the mounted FLATCON modules is shown in figure 4.15. Also, an *IV* curve of one module is shown. The data were taken on 28 September 2001 in Freiburg, Germany. The DNI was 737 W m^{-2} and the ambient temperature was 22.2 °C. A module efficiency of 21.3% was measured. Recently, an even better performance of 22.3 % has been determined.

4.3.4   Concentrator system development

The FLATCON modules described in the previous section showed a promising efficiency under realistic operating conditions.   This type of module is the basis for the efforts at the Fraunhofer ISE to develop a high-concentration photovoltaic system. The size of a concentrator system is a challenge. A large system is desirable to reduce the costs for mounting, land preparation, long-term maintenance etc.   However, larger systems have a greater impact on the mechanical parts, increasing costs and maintenance. For systems which are too small, the cost of the tracking increases and reduces the overall cost advantage of concentrator systems compared to flat-plate systems. As a trade-off, we are

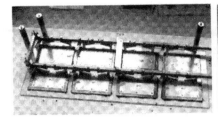

Figure 4.14. Photographs of the in-house manufacture of a FLATCON module. Left: the adjustment of eight copper plates on the glass base is shown. Each of the copper sheets carries six monolithic tandem cells 2 mm in diameter and a Si bypass diode in the centre. Right: the mounting of the all-glass housing is visualized. The Fresnel lens panel and the side walls are glued by silicon.

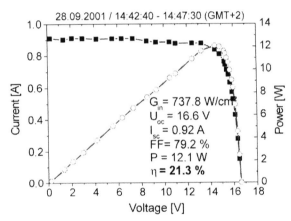

Figure 4.15. Left: photograph of the mounted FLATCON modules. The aperture area of one module is 768 cm^2. Right: the *I V* curve taken on 28 September 2001. An efficiency of 21.3% was determined.

aiming for system sizes of 5 kWp. This corresponds to an aperture area of 30 m^2 for the FLATCON modules.

Today, no low-cost tracking system with sufficient accuracy is commercially available. This partly reflects the problematic situation for the development of high-concentration systems. Consequently, the Fraunhofer ISE and the Ioffe Institute, St Petersburg Russia, developed their own tracking system [56]. A mechanical platform for 5 kWp FLATCON modules has been developed (see figure 4.16). The conceptual design of the tracker is similar to an earlier version built at the Ioffe Institute [57]. The tracker consists of two main moving parts: a base platform moving around a vertical axis and a suspended platform with PV modules moving around the horizontal axis (see figure 4.16).

Geared motor drives powered by 12 V DC are used in this tracker. The

Figure 4.16. Photograph of the 5 kWp tracking system developed at the Fraunhofer ISE in cooperation with the Ioffe Institute, St Petersburg, Russia.

averaged power consumption is less than 10 W. The open-loop control is made fully automatic by the use of analogue sun sensors. The system is equipped with two sensors: a high-precision sensor and a more inaccurate one. The high-precision sensor can align the tracker to the sun within 0.05°. The acceptance angle is ±70° in both the horizontal and vertical directions. The additional and less accurate sensor allows a wider positioning angle for the East–West movement (up to 270°). Thus, for example, the starting position in the morning is found automatically.

The overall accuracy of the system after the FLATCON modules are mounted is better than 0.1°. The installation of the total system can be performed easily. Fifty FLATCON modules are mounted in one row. Eight of these rows are transported to the place of the installation. Each of the rows is fixed and adjusted on the tracker. Regarding the output power of 12.1 W measured at a DNI 737 W m^{-2} in Freiburg, Germany (see figure 4.15), the fully equipped demonstration system will deliver an output power of 4.84 kW. However, Freiburg has a fairly low DNI. In sunny regions with more DNI, like Madrid, the output power of the demonstration high-concentrator system is expected to be even higher than 5 kW.

4.4 Summary and perspective

High-concentration tandem cells are a realistic path to decreasing the cost of photovoltaic-generated energy. Estimates predict costs of 0.1 € kWh^{-1} if the applied cells operate at efficiencies of 30% and concentrations higher than 500. Such high efficiencies are possible with dual-junction concentrator cells achieving values of 31–32% at concentrations of 300. Efficiencies of 30% have also been determined for a very high concentration of 1000. These high-efficiency cells are based on III–V-compound semiconductor materials. The tandem cells are made by two approaches: mechanically stacked cells or monolithic, internally series connected cells. Even a combination of both approaches can be used. With respect to the mechanically stacked approach, the material combination of $Al_{0.24}Ga_{0.76}As$–Si or GaAs–GaSb is of interest: in particular, the latter showed a high performance of 31.4% under concentrated light. The advantage of the mechanically stacked approach is the use of single-junction cells. They can have fairly simple cell structures and each of the cells can be optimized independently. The disadvantage is the need for two substrates, thus increasing the costs.

In the case of monolithic cells, usually lattice-matched materials like $Ga_{0.51}In_{0.49}P$/GaAs are considered. However, as discussed in this chapter, the lattice-mismatched material $Ga_{0.35}In_{0.65}P$/$Ga_{0.17}In_{0.83}As$ shows theoretically and practically better performance. The complex cell structure can be grown by industrially feasible MOVPE technology. This epitaxial growth method has been proven to be highly reproducible, not only for space solar cell fabrication but also for the growth of structures for lasers and LEDs. Furthermore, of particular interest for solar cells, a high throughput is achievable. One industrial-size MOVPE reactor is able to produce 2–3 MWp dual-junction 500×-concentrator solar cells per year. Obviously, the investment costs for the solar cell fabrication are low.

The monolithic high concentration dual-junction cell is well developed and highly reliable. One of the main challenges, to design an internal tunnel diode with low resistive losses, has been successfully met. However, considering their use in different concentrator systems, different cell designs are necessary. Consequently, there is no dual-junction cell on the market which fits all concentrator systems under consideration. This is a remarkable difference between concentrator systems and flat-plate modules. Each concentrator system needs a specific cell design and, therefore, specific development. Anyhow, this is not a fundamental bottleneck but has to be considered when conceiving new concentrator systems.

Looking into the future, monolithic triple-junction cells will be used in concentrator systems. In this case, the dual-junction cell structure is grown on activated Ge. The Ge cell generates more current compared to the two other sub-cells. Thus, this cell does not limit the total output current and acts as a voltage booster. At room temperature, the open-circuit voltage is increased by about 250 mV. Recently, results from the first monolithic triple-junction cell achieving efficiencies of 32.4% at 414 × AM1.5direct and 34.0% at 210 AM1.5global [1] have been reported. Even higher efficiencies can be expected by improving the monolithic triple-junction cell or by using quadruple-junctions. However, besides the issues of semiconductor material quality, there might also be practical limitations for quadruple-junctions. Each additional junction reduces the output current and increases the voltage of the solar cell. The decrease in output current is helpful for concentrator applications because the ohmic drop is drastically reduced. However, the cell becomes more sensitive to changes in the spectrum. As discussed here, the monolithic dual-junction cell is sensitive to changes in the spectrum of the sun. If we consider triple-junction cells, the additional active cell acts only as a voltage booster and does not influence the output current. The sensitivity to variations in the spectrum does not change compared with the dual-junction cell. Regarding a quadruple cell, the output current is reduced compared with the dual- or triple-junction cell. Assuming the same change in the spectrum of the sun, this might cause a higher relative deviation in the output current of the quadruple cell compared with the dual- or triple-junction cell. Consequently, the sensitivity to the spectrum increases. A good trade-off for practical concentrator applications seems to be the use of optimized triple-junction cells rather than quadruple-junction cells.

The dual-junction cell for high concentration is well understood and developed. However, no high-concentration system using tandem cells is on the market. The cells are a fairly new product and have only recently become available. However, system development needs a long time. Moreover, the huge variety in different system designs is a disadvantage compared with flat-plate module technology. Each system requires a specifically adapted type of tandem cell. The long-term reliability of all these possible concentrator systems is an issue. PV systems need a longevity of at least 20 years. Accelerated aging test procedures are necessary. Moreover, no suitable and accurate tracking system is available on the market. Last but not least, the characterization of concentrator modules, in general, and modules equipped with high concentration tandem cells, in particular, is a challenge. The Fraunhofer ISE has initiated developments on all these challenges and has started to build a 5 kWp demonstration concentrator system using the FLATCON concept.

Acknowledgments

I would like to express my deep gratitude to all of the current and former members of the group 'III–V Solar Cell and Epitaxy' at Fraunhofer ISE. The support of the Calibration Laboratory at Fraunhofer ISE headed by Dr W Warta is very important for the progress of this work. Furthermore, I would like to acknowledge the continuous support of Professor W Wettling, Professor J Luther and Dr G Willeke. They always encourage me to go on with the work on III–V tandem cells and concentrator technology. Moreover, they have given valuable input for the ongoing development.

Special thanks go to the Ioffe Institute, St Petersburg, Russia in particular to Professor V Rumyantsev, Professor V Andreev, Dr M Shvarts and Dr N Sadchikov for the good collaboration in the development of the point focus Fresnel modules and trackers. Finally, I would like to acknowledge the continuous financial support of the German Ministry of Economy and Technology (BMWi) in several contracts (0328554). I am solely responsible for the content of this chapter.

References

[1] Green M A, Emery K, King D L, Igari S and Warta W 2002 *Prog. Photovolt.* **10** 355–60

[2] Zhao J, Wang A, Green M and Ferrazza F 1998 *Appl. Phys. Lett.* **73** 1991–3

[3] Green M A 2000 *16th European Photovoltaic Solar Energy Conference (Glasgow)* (London: James & James) pp 51–4

[4] King R R, Fetter M F, Cotter P C, Edmondson K M, Ermer J H, Cotal H L, Yoon H, Stavrides A P, Kinsey G, Krut D D and Karam N H 2002 *29th IEEE PVSC (New Orleans, LA)* (New York: IEEE) pp 776–81

[5] Price H, Lüpfert E, Kearney D, Zarza E, Cohen G, Gee R and Mahoney R 2002 *J. Solar Energy Eng.* **124** 109–25

[6] Letay G and Bett A W 2002 *17th European Photovoltaic Solar Energy Conf. (Munich)* (Munich: WIP) pp 178–80

[7] http://www.ise.fhg.de/english/fields/field2/mb5/index.html:

[8] Shockley W and Queisser H J 1961 *J. Appl. Phys.* **32** 510

[9] Dimroth F, Schubert U, Bett A W, Hilgarth J, Nell M, Strobl G, Bogus K and Signorini C 2001 *17th European Photovoltaic Solar Energy Conf. (Munich)* (Munich: WIP) pp 2150–4

[10] Bett A W, Dimroth F, Stollwerk G and Sulima O V 1999 *Appl. Phys.* A **69** 119–29

[11] Fraas L M, Girard G R, Avery J E and Arau B A 1989 *J. Appl. Phys.* **66** 3866

[12] Fraas L M, Avery J E, Sundaram V S, Dinh V T, Davenport T M, Yerkes J W, Gee J M and Emery K A 1990 *21st IEEE PVSC (Kissimmee, FL)* (New York: IEEE) pp 190–5

[13] Bett A W, Keser S, Stollwerck G, Sulima O V and Wettling W 1997 *26th IEEE PVSC (Anaheim, CA)* (New York: IEEE) pp 931–4

[14] Andreev V M, Karlina L B, Kazantsev A B, Khvostikov V P, Rumyantsev V D, Sorokina S V and Shvarts M Z 1994 *Proc. 1st World Conf. on Photovoltaic Energy Conversion (Waikoloa, HI)* (New York: IEEE) p 1721

[15] Dimroth F and Bett A W 1997 *14th European Photovoltaic Solar Energy Conf. (Barcelona)* (Bedford: H S Stephens & Associates) pp 1759–62

[16] Bett A, Cardona S, Ehrhardt A, Lutz F, Welter H and Wettling W 1991 *22nd IEEE PVSC (Las Vegas, NE)* (New York: IEEE) pp 137–41

[17] Baldus A, Bett A W, Blieske U, Sulima O V and Wettling W 1995 *J. Cryst. Growth* **146** 299

[18] Welter H, Bett A, Ehrhardt A and Wettling W 1991 *10th European Photovoltaic Solar Energy Conf. (Lisbon)* (Dordrect: Kluwer Academic) pp 537–40

[19] Blieske U, Baldus A, Bett A, Lutz F, Nguyen T, Schetter C, Schitterer K, Sulima O V and Wettling W 1993 *23rd IEEE PVSC (Louisville, KY)* (New York: IEEE) pp 735–40

[20] Blieske U, Bett A, Duong T, Schetter C and Sulima O V 1994 *12th European Phtovoltaic Solar Energy Conf. (Amsterdam)* (Bedford: H S Stephens & Associates) pp 1409–12

[21] Blug A, Baldus A, Bett A, Blieske U, Stollwerck G, Sulima O and Wettling W 1995 *13th European Photovoltaic Solar Energy Conf. (Nice)* (Bedford: H S Stephens & Associates) p 910

[22] Bett A W, Keser S and Sulima O V 1997 *J. Cryst. Growth* **181** 9–16

[23] Sulima O V and Bett A W 2001 *Solar Energy Mater. Solar Cells* **66** 533–40

[24] Bett A W, Keser S, Stollwerck G and Sulima O V 1997 *Proc. 3rd NREL Conf. on Thermophotovoltaic Generation of Electricity (Colorado Springs, CO)* (AIP Proceedings) pp 41–53

[25] Stollwerck G, Sulima O V and Bett A W 2000 *IEEE Trans. Electron. Devices* **47** 448–57

[26] Sulima O V, Bett A W and Wagner J 2000 *J. Electrochem. Soc.* **147** 1910–14

[27] Blieske U, Sterk S, Bett A, Schuhmacher J and Wettling W 1994 *1st World Conf. on Photovoltaic Energy Conversion (Waikoloa, HI)* (New York: IEEE) pp 1902–5

[28] Bett A W, Keser S, Stollwerck G and Sulima O V 1997 *14th European Photovoltaic Solar Energy Conf. (Barcelona)* (Bedford: H S Stephens & Associates) p 993

[29] Yoon S and Garboushian V 1994 *Proc. 1st World Conf. on Photovoltaic Energy Conversion (Waikoloa, HI)* (New York: IEEE) p 1500

[30] Bett A W, Dimroth F, Stollwerck G and Sulima O V 1999 *Appl. Phys.* A **69** 119–29

[31] Bett A W, Stollwerck G, Sulima O V and Wettling W 1998 *2nd World Conf. on Photovoltiac Solar Energy Conversion (Vienna)* (Ispra: European Commission) pp 268–72

[32] Dimroth F, Lanyi P, Schubert U and Bett A W 2000 *J. Electron. Mater.* **29** 42–6

[33] Bett A W, Adelhelm R, Agert C, Beckert R, Dimroth F and Schubert U 2001 *Solar Energy Mater. Solar Cells* **66** 541–50

[34] Dimroth F, Beckert R, Meusel M, Schubert U and Bett A W 2001 *Prog. Photovolt.* **9** 165–78

[35] Friedman D J, Kurtz S R, Bertness K A, Kibbler A E, Kramer C, Olson J M, King D L, Hansen B R and Snyder J K 1995 *Prog. Photovolt.* **3** 47–50

[36] Blieske U, Bett A, Duong T, Schetter C and Sulima O V 1994 *12th Photovoltaic Solar Energy Conf. (Amsterdam)* (Bedford: H S Stephens & Associates) pp 1409–12

[37] Yamaguchi M and Luque A 1999 *IEEE Trans. Electron. Devices* **46** 2139–44

[38] Dimroth F, Schubert U, Schienle F and Bett A W 2000 *J. Electron. Mater.* **29** 47–52

[39] Bett A W, Dimroth F, Meusel M, Schubert U and Adelhelm R 2000 *16th European Photovoltaic Solar Energy Conf. (Glasgow)* (London: James & James) pp 951–4

[40] Cotal H L, Lillington D R, Ermer J H, King R R, Karam N H, Kurtz S R, Friedman D J, Olson J M, Ward J S, Duda A, Emery K A and Moriarty T 2000 *28th IEEE PVSC (Anchorage, AK)* (New York: IEEE) pp 955–60

[41] Bett A W, Baur C, Beckert R, Dimroth F, Letay G, Hein M, Meusel M, van Riesen S, Schubert U, Siefer G, Sulima O V and Tibbits T N D 2001 *17th European Photovoltaic Solar Energy Conf. (Munich)* (Munich and Florence: WIP and ETA) pp 84–7

[42] Agert C, Lanyi P, Sulima O V, Stolz W and Bett A W 2000 *IEE Proc. Optoelectron.* **147** 188–92

[43] Biefeld R M 2002 *Mater. Sci. Eng.* R **36** 105–42

[44] Agert C, Lanyi P and Bett A W 2001 *J. Cryst. Growth* **225** 426–30

[45] Agert C, Beckert R, Hinkov V, Sulima O V and Bett A W 2001 *17th European Photovoltaic Solar Energy Conf. (Munich)* (Munich and Florence: WIP and ETA) pp 372–5

[46] Faine P, Kurtz S R, Riordan C and Olson J M 1991 *Solar Cells* **31** 259–78

[47] Kurtz S R, Olson J M and Faine P 1991 *Solar Cells* **30** 501–13

[48] Meusel M, Adelhelm R, Dimroth F, Bett A W and Warta W 2002 *Prog. Photovolt.* **10** 243–55

[49] Adelhelm R and Bücher K 1998 *Solar Energy Mater. Solar Cells* **50** 185–95

[50] Emery K, Meusel M, Beckert R, F. D, Bett A and Warta W 2000 *28th IEEE PVSC (Anchorage, AK)* pp 1126–30

[51] Siefer G, Baur C, Meusel M, Dimroth F, Bett A W and Warta W 2002 *29th IEEE PVSC (New Orleans, LA)* (New York: IEEE) pp 836–9

[52] Hein M, Meusel M, Baur C, Dimroth F, Lange G, Siefer G, Bett A W, Andreev V M and Rumyantsev V D 2001 *17th European Photovoltaic Solar Energy Conf. (Munich)* pp 496–9

[53] Rumyantsev V D, Hein M, Andreev V M, Bett A W, Dimroth F, Lange G, Letay G, Shvarts M Z and Sulima O V 2000 *16th European Photovoltaic Solar Energy Conf. (Glasgow)* pp 2312–15

[54] Gueymarid C 2001 *Solar Eng.* **71** 325–46

[55] Rumyantsev V D, Andreev V M, Bett A W, Dimroth F, Hein M, Lange G, Shvarts M Z and Sulima O V 2000 *28th IEEE PVSC (Anchorage, AK)* pp 1169–72

[56] Rumyantsev V D, Andreev V M, Sadchikov N A, Bett A W, Dimroth F and Lange G 2002 *PV in Europe From PV Technology to Energy Solutions (Rome)* to be published

[57] Alferov Z I, Andreev V M, Aripov K K, Larionov V R and Rumyantsev V D 1981 *Solar Eng.* **6** 3

Chapter 5

Quantum wells in photovoltaic cells

C Rohr, P Abbott, I M Ballard, D B Bushnell, J P Connolly,
N J Ekins-Daukes and K W J Barnham
Imperial College, London, UK

5.1   Introduction

The fundamental efficiency limit of a single bandgap solar cell is about 31% at 1 sun with a bandgap of about $E_G = 1.35$ eV [1], determined by the trade-off of maximizing current with a smaller bandgap and voltage with a larger bandgap. Multiple bandgaps can be introduced to absorb the broad solar spectrum more efficiently. This can be realized in multi-junction cells, for example, where two or more cells are stacked on top of each other either mechanically or monolithically connected by a tunnel junction. An alternative— or complementary (see section 5.4)—approach is the quantum well cell (QWC).

5.2   Quantum well cells

Quantum wells (QWs) are thin layers of lower bandgap material in a host material with a higher bandgap. Early device designs placed the QWs in the doped regions of a p–n device [2] but superior carrier collection is achieved when an electric field is present across the QWs. More recent QWC designs have employed a p–i– n structure [3] with the QWs located in the intrinsic region; a schematic bandgap diagram is shown in figure 5.1. The carriers escape from the QWs thermally and by tunnelling [4–6].

The photocurrent is enhanced in a QWC compared to a cell made without QWs also known as barrier control, and experimentally it is observed that the voltage is enhanced compared with a bulk cell made of the QW material [7]. Hence, QWCs can enhance the efficiency if the photocurrent enhancement is greater than the loss in voltage [8]. The potential for an efficiency enhancement is also discussed in [9, 10]. The number of QWs is limited by the maximum

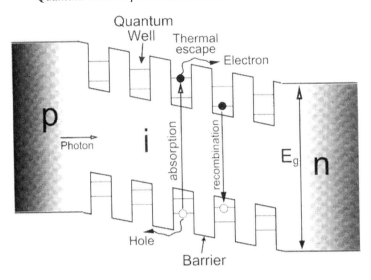

Figure 5.1. Schematic bandgap diagram of a quantum well cell. The absorption threshold is determined by the lowest energy levels in the quantum wells. Carriers escape thermally assisted and by tunnelling.

thickness of the i-region maintaining an electric field across it. QWCs have been investigated quite extensively, both on GaAs as well as on InP substrates, and have been discussed in some detail also in [11, 12].

Historically, the first p–i–n QWCs were in the material system AlGaAs/GaAs (barrier/well) on GaAs [13–17]. AlGaAs is closely lattice-matched to GaAs and the bandgap can be easily varied by changing the Al fraction (see figure 5.2) up to about 0.7 where the bandgap becomes indirect. However, the material quality particularly that of AlGaAs is relatively poor because of contamination during the epitaxial growth, leading to a high number of recombination centres and, hence, a high dark current.

An alternative material to AlGaAs is InGaP which has better material quality and an InGaP/GaAs QWC has been demonstrated [18]. However, the ideal single bandgap for a 1 sun solar spectrum is $E_G = 1.35$ eV, as indicated in figure 5.2, while GaAs has a bandgap of $E_G = 1.42$ eV and that of AlGaAs is still higher. The second material should, therefore, have a smaller bandgap than GaAs to absorb the longer wavelength light, keeping in mind that the quantum confinement raises the effective bandgap of the QWs.

GaAs/InGaAs QWCs fulfil this criterion and they have been studied quite extensively [19–24]. However, because InGaAs has a larger lattice constant than GaAs (see figure 5.2), it is strained. If the strain exceeds a critical value, relaxation occurs at the top and bottom of the MQW stack and the dislocations result in an increase in recombination and, hence, increased dark current [22]. This limitation

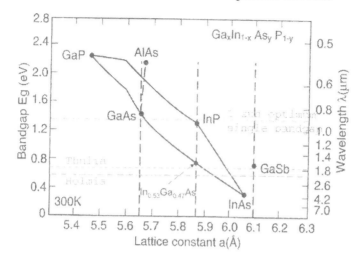

Figure 5.2. Lattice constant *versus* bandgap for $In_{1-x}Ga_xAs_yP_{1-y}$ and AlGaAs compounds. Also indicated is the optimum bandgap for a single-bandgap PV cell under 1 sun, and the emission peaks of selective emitters thulia and holmia.

means that strained GaAs/InGaAs QWCs cannot improve the efficiency compared to GaAs control cells [24].

Strain compensation techniques can be used to overcome this problem (see section 5.3), and QWCs in the material system GaAsP/InGaAs on GaAs have been investigated [24–29]. These devices are also very suitable as bottom cells in a tandem configuration (see section 5.3).

QWCs based on InP are of interest for solar as well as for thermophotovoltaic (TPV) applications. First, material combinations such as InP/InGaAs were investigated [6, 30–32], which was then extended to quaternary material (lattice-matched to InP) $In_{1-x}Ga_xAs_yP_{1-y}$ ($x = 0.47y$) [33–35].

As in the GaAsP/InGaAs system on GaAs, strain-compensation techniques have been employed in $In_{1-x}Ga_xAs/In_{1-z}Ga_zAs$ QWCs on InP [36–41].

QWCs have practical advantages due to both quantized energy levels and the greater flexibility in choice of materials. In particular, this allows engineering of the bandgap for a better match with the incident spectrum. The absorption threshold can be varied by changing the width of the QW and/or by changing its material composition.

This flexibility can be further increased by employing strain-compensation techniques which are explained in more detail in section 5.3. In this way, longer wavelengths for absorption can be achieved than what is possible with lattice-matched bulk material, allowing optimization of the bandgap.

The application of QWCs (based on InP) for (TPV) is discussed in section 5.6. For TPV applications the same concept of strain compensation can be applied to extend the absorption to longer wavelengths. This is important for relatively low temperature sources combined with appropriate selective emitters for example based on holmia or thulia.

Several studies indicate that QWCs have a better temperature dependence of efficiency than bulk cells [16, 42–44].

5.3 Strain compensation

In order to avoid strain relaxation, strain-compensation techniques, first proposed by Matthews and Blakeslee [45], can be used to minimize the stress at the interface between the substrate and a repeat unit of two layers with different natural lattice constants. Layers with larger and smaller natural lattice constant compared to the substrate result in compressive and tensile strain, respectively, as shown in figure 5.3. When these two layers are strained against each other, the strain is compensated and the net force exerted on the adjacent layers is reduced. Therefore, the build-up of strain in a stack can be reduced and, hence, its critical thickness is increased so that more such repeat units can be grown on top of each other without relaxation. If the strain-compensation conditions are optimized to give zero stress at the interfaces between the repeat units, an unlimited number of periods can be grown in principle. In addition, each individual layer has to remain below its critical thickness which means that this concept can only be used for thin layers such as quantum wells.

This technique is very suitable for multi-quantum-well structures; the barriers and wells are made of different materials with larger and smaller bandgaps but they can also have smaller and larger lattice constants (see figure 5.2), i.e. with tensile and compressive strain, respectively (see figures 5.3 and 5.4). This means strained materials can be used for the quantum wells in order to reach lower bandgaps without compromising the quality of the device as dislocations are avoided. This technique, which extends the material range allowing further bandgap engineering, was first applied to photovoltaics in GaAsP/InGaAs QWCs [25].

For highly strained layers, the difference in elastic constants becomes significant and it needs to be taken into account when considering the conditions for zero stress [46]. The strain for each layer i is

$$\epsilon_i = \frac{a_0 - a_i}{a_i} \tag{5.1}$$

where a_0 is the lattice constant of the substrate and a_i the natural lattice constant of layer i. The zero-stress strain-balance conditions are as follows:

$$\epsilon_1 t_1 A_1 a_2 + \epsilon_2 t_2 A_2 a_1 = 0 \tag{5.2}$$

$$a_0 = \frac{t_1 A_1 a_1 a_2^2 + t_2 A_2 a_2 a_1^2}{t_1 A_1 a_2^2 + t_2 A_2 a_1^2} \tag{5.3}$$

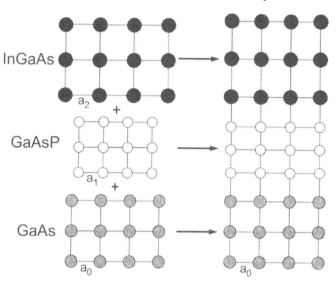

Figure 5.3. Schematic diagram of strain compensation: the natural lattice constant of GaAsP (a_1) is smaller than that of the GaAs substrate (a_0) and GaAsP barriers are therefore tensile strained, while the natural lattice constant of InGaAs (a_2) is larger, and, hence, InGaAs QWs are compressively strained.

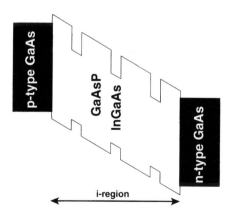

Figure 5.4. Bandgap diagram of a strain compensated QWC with InGaAs QWs and GaAsP barriers.

where

$$A = C_{11} + C_{12} - \frac{2C_{12}^2}{C_{11}}$$

with elastic stiffness constants C_{11} and C_{12}, different for each layer, depending on the material.

The strain energy must be kept below a critical value, however, to avoid the onset of three-dimensional growth [47]. Lateral layer thickness modulations, particularly in the tensile strained material (i.e. barriers), are origins of dislocations and result in isolated highly defected regions if the elastic strain energy density reaches a critical value. In practice, the strain balancing puts stringent requirements on growth.

5.4   QWs in tandem cells

In tandem cells two photovoltaic cells with two different bandgaps are stacked on top of each other. The bandgaps of the two cells can be optimized for the solar spectrum [48] and a contour plot of efficiency as a function of top- and bottom-cell bandgaps is shown in figure 5.5 for an AM0 spectrum. Tandem cells can be grown monolithically on a single substrate, connected with a tunnel junction. The top and bottom cells are connected in series, which means that the lower photocurrent of the two cells determines the current of the tandem device. Monolithic growth requires that the materials are lattice matched in order to avoid relaxation. A tandem with a GaInP top cell lattice matched to a GaAs bottom cell does not have the optimum bandgap combination as one can see in figure 5.5, and there is no good quality material available with a smaller bandgap than GaAs having the same lattice constant.

The standard approach to matching the currents in the GaInP/GaAs tandem is by thinning the top cell so that enough light is transmitted to the bottom cell to generate more current there [49]. This is not an optimum configuration, however, as the quantum efficiency of the top cell is reduced and, in addition, there are losses in the bottom cell due to more high-energy carriers relaxing to the bandedge.

It is possible to grow a virtual substrate, where the lattice constant is relaxed and the misfit dislocations are largely confined to electrically inactive regions of the device. In this way the lattice constant can be changed before growing the tandem. But the bottom and top cell of the tandem should still be lattice matched with respect to each other and, hence, the combination of bandgaps is restricted indicated with a line in figure 5.5; the optimum where the currents in both cells are matched cannot be reached in general. Several groups have grown samples that fall on that line in figure 5.5 [50, 51].

QWs extend the absorption and, therefore, increase the photocurrent compared with a barrier control; hence, a GaAs/InGaAs QWC generates more current than a GaAs cell and can be better matched to a GaInP top cell in a tandem cell under the solar spectrum [17, 23, 52, 53]. However, as mentioned in section 5.2, the voltage deteriorates due to the formation of misfit dislocations with increasing strain when incorporating strained InGaAs QWs [22].

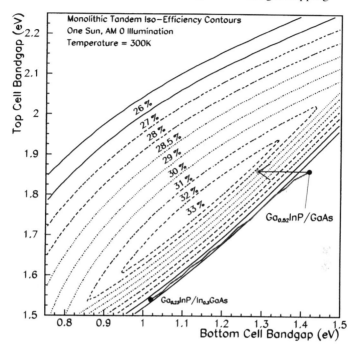

Figure 5.5. Contour plot of tandem cell efficiency as a function of top- and bottom-cell bandgaps. Reprinted from Ekins-Daukes *et al* 2002 Strain-balanced quantum well solar cells *Physica* E **14** 132–5 with permission from Elsevier.

Strain-compensated QWCs offer an attractive solution in that the link between lattice matching and bandgap is decoupled [24]. The bandgap of the bottom cell, for example, can be reduced with a strain-compensated QWC, which means that one can move along a horizontal line in figure 5.5 improving the efficiency of a tandem cell quite rapidly [25]. QWCs for both the top and bottom cell give an extra degree of freedom and optimization of the bandgaps to obtain maximum efficiency becomes possible.

5.5 QWCs with light trapping

Not all the light is absorbed in the QWs because they are optically thin and their number limited. Hence, light-trapping techniques to increase the number of light passes through the MQW is desirable, boosting the QW photo-response significantly. The simplest form is a mirror on the back surface resulting in two light passes. Texturing the front or back surface can further increase the path length of the light, in particular if the light is at a sufficiently large angle with respect to the normal so as to obtain internal reflections. Another option is to incorporate gratings into the structure to diffract the light to large angles [54].

Light trapping is only desirable for the wavelength range where the optically thin QWs absorb. Other parts of the cell are optically thick and, hence, light trapping for photons with energies greater than the bandgap of the bulk material has a minimal effect if any. The energy of photons that are trapped but not absorbed (e.g. below-bandgap photons) and additional high-energy carriers relaxing to the bandedge contribute to cell heating and are, therefore, undesirable.

A solution to this problem is to use a wavelength-specific mirror such as a distributed Bragg reflector (DBR) [29]. A DBR consists of alternating layers of high and low refractive index material, each one-quarter of a wavelength thick. Constructive interference occurs for the reflected light of the design wavelength and adding periods to the DBR causes a higher reflection due to the presence of more in-phase reflections from the added interfaces. DBRs maintain a high reflectance within a region around the design wavelength known as the stop-band. As the refractive index contrast between the two materials in the DBR increases, the peak reflection rises and the stop-band widens.

This technique is particularly attractive for multi-junction cells where the second cell is a QWC with a DBR on an active Ge substrate. In figure 5.6 the quantum efficiency of a typical QWC with 20 QWs is shown with and without a 20.5 period DBR, as well as the reflectance of the DBR; the sample description of the QWC with DBR is given in table 5.1. The quantum efficieny (QE) of the QWC with DBR in figure 5.6 is calculated from the measured QE and the calculated DBR reflectance. The DBR reflects back only the relevant wavelength range that can be absorbed in the MQW, significantly boosting its spectral response.

As the stop-band of the DBR can be made quite abrupt, most longer wavelength light is allowed through, so that an active Ge substrate as a third junction can be employed, which would still produce more than 200 A m^{-2}, more than enough photocurrent compared with the other two junctions (see below). The transmission through to a Ge substrate of a tandem cell with a GaInP top cell and a QWC with a DBR as the bottom cell has been modelled with a multi-layer programme and is shown in figure 5.7 compared with a tandem having a GaAs p-n junction as a bottom cell. As one can see the transmission of photons of energy below the bandgap of the bottom cell is similarly high in both cases.

In table 5.2 the short-circuit current density J_{sc} for AM0 illumination is shown for this QWC (20 QWs) with and without a 20.5 period DBR, compared with a p–n control cell [29]. In a tandem configuration, assuming a cut-off wavelength of 650 nm, corresponding to the bandgap of a GaInP top cell, the p-n control cell is expected to have a typical J_{sc} of 158 A m^{-2}. Introducing a QWC with DBR improves J_{sc} by 16% to 183 A m^2, which is much better current matched to the GaInP top cell. A commercial triple-junction cell with a GaInP top cell, a standard p-n GaAs junction and an active Ge substrate has an AM0 efficiency of 26.0% [55]; if the GaAs junction is replaced with a QWC with DBR the efficiency is calculated to improve by about 3.4 percentage points or by 13% to 29.4%.

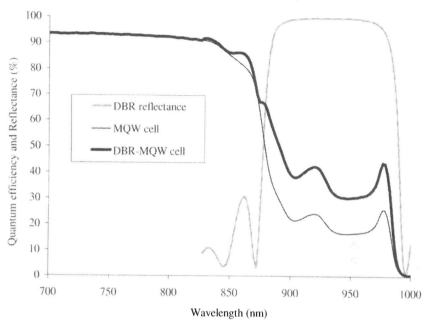

Figure 5.6. Quantum efficiency of a 20 QW device with and without DBR and the reflectance of the DBR.

5.6   QWCs for thermophotovoltaics

Thermophotovoltaics (TPV) is the same principle as photovoltaics but the source is at a lower temperature than the sun, typically around 1500–2000 K instead of 6000 K, and much closer. In TPV applications often a combustion process is used as heat source (e.g. using fossil fuels or biomass) but other heat sources such as nuclear, indirect solar or industrial high-grade waste heat can be used too. Because the source has a lower temperature in TPV, lower bandgap materials are required to absorb the (near-)infrared light more efficiently. Often a selective emitter is introduced between the source and the photovoltaic cells to obtain a narrow band as opposed to a broad black or grey-body spectrum, increasing the cell conversion efficiency. TPV is described in more detail, for example, in [56, 57].

QWCs have advantages for TPV applications, and substantial progress has been made in the development of QWCs for TPV focusing on InP-based materials [58]. QWCs were first introduced for TPV applications independently by Griffin *et al* [32] and Freundlich and Ignatiev [59]. These InGaAs/InP QWCs indicated better performance than InGaAs bulk cells, for example enhancing V_{oc} and improved temperature dependence. These are important parameters in TPV, as the cells are very close to a hot source and the power densities are high giving rise to

Table 5.1. Sample description of QWC with DBR.

Layer	Repeats	Material	Thickness (Å)	Doping	Conc. (cm^{-3})
Cap	1	GaAs	2200	p	2×10^{19}
Window	1	Al$_{0.8}$GaAs	450	p	1×10^{18}
Emitter	1	GaAs	5000	p	2×10^{18}
i-region	1	GaAs	100		
$\frac{1}{2}$ barrier	20	GaAsP$_{0.06}$	210		
QW	20	In$_{0.17}$GaAs	70		
$\frac{1}{2}$ barrier	20	GaAsP$_{0.06}$	210		
i-region	1	GaAs	100		
Base	1	GaAs	24 000	n	1.5×10^{17}
DBR	20	Al$_{0.5}$GaAs	144	n	1×10^{18}
DBR	20	AlAs	538	n	1×10^{18}
DBR	20	Al$_{0.5}$GaAs	144	n	1×10^{18}
DBR	20	Al$_{0.13}$GaAs	520	n	1×10^{18}
DBR	1	Al$_{0.5}$GaAs	144	n	1×10^{18}
DBR	1	AlAs	538	n	1×10^{18}
DBR	1	Al$_{0.5}$GaAs	144	n	1×10^{18}
Buffer	1	GaAs	1000	n	1.5×10^{18}
Substrate		GaAs		n	

Table 5.2. Short-circuit current densities J_{sc} and efficiencies for AM0 illumination. The tandem J_{sc} assumes a cut-off wavelength of 650 nm. The QWC has 20 QWs and the DBR 20.5 periods. The triple-junction efficiencies are based on a device with a GaInP top cell and an active Ge substrate.

Cell type	J_{sc} (A m^{-2})	Tandem J_{sc} (A m^{-2})	Triple-junction efficiency (%)
QWC with DBR	339	183	29.4
QWC without DBR	330	171	
p–n control	320	158	26.0 [55]

higher currents which may pose series resistance problems. Further development of QWCs has been in the quaternary system In$_{1-x}$Ga$_x$As$_y$P$_{1-y}$ lattice-matched to InP ($x = 0.47y$), with very good material quality, and which can be optimized for a rare-earth selective emitter erbia having a peak emission of about 1.5 μm, for example, but which is also attractive for hybrid solar-TPV applications [34].

Many TPV systems operate at temperatures more suitable for longer wavelength selective emitters such as thulia and holmia with peak emissions of

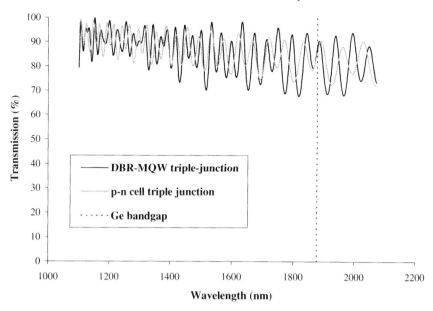

Figure 5.7. Calculated transmission through to a Ge substrate of a tandem cell with a GaInP top cell and, as bottom cell, a QWC with DBR, compared with a tandem with a GaAs p-n bottom cell.

about 1.7 and 1.95 μm respectively. However, the smallest bandgap achievable with lattice-matched material on InP is that of $In_{0.53}Ga_{0.47}As$ with an absorption edge of about 1.7 μm (see figure 5.2). Strain-compensation techniques can be employed to extend the absorption, which is unique to QWCs [36].

$In_{1-x}Ga_xAs/In_{1-z}Ga_zAs$ strain-compensated QWCs have been shown to extend the absorption (see figure 5.8) while retaining a dark current similar to an $In_{0.53}Ga_{0.47}As$ bulk cell and lower than that of a GaSb cell with similar bandgap (see figure 5.9) [37, 39]. A 40 QW $In_{1-x}Ga_xAs/In_{1-z}Ga_zAs$ strain-compensated QWC was designed for a thulia emitter [41], absorbing out to 1.97 μm as shown in figure 5.8. Absorption beyond 2 μm has been achieved with a two-QW cell, although in this case the collection efficiency was incomplete except at high temperatures and reverse bias [38].

A mirror at the back can significantly increase the contribution of the QWs as discussed in section 5.5. Back reflection is particularly important as a form of spectral control in a TPV system, as below-bandgap photons are reflected back to the source increasing the system efficiency. We observe a back reflectivity of about 65% just from a standard gold back contact [39].

As an alternative, Si/SiGe QWCs were grown for TPV applications but the absorption of the SiGe QWs is very low [60].

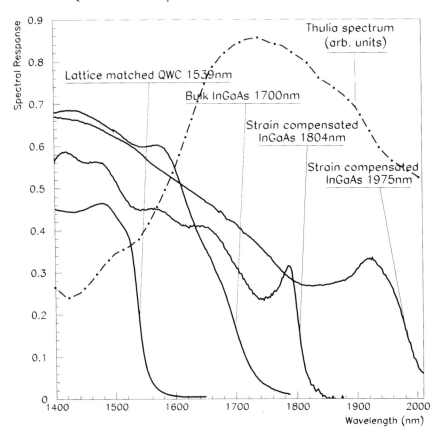

Figure 5.8. Experimental external quantum efficiency of TPV QWC devices (plus a bulk In$_{0.53}$Ga$_{0.47}$As device) with successively longer absorption, including a thulia spectrum (arbitrary units).

5.7   Conclusions

The primary advantage of incorporating quantum wells in photovoltaic cells is the flexibility offered by bandgap engineering by varying QW width and composition. The use of strain compensation further increases this flexibility by extending the range of materials and compositions that can be employed to achieve absorption thresholds at lattice constants that do not exist in bulk material. In a tandem or multi-junction configuration, QWCs allow current matching and optimizing the bandgaps for higher efficiencies.

Light-trapping schemes are an important technique to boost the quantum efficiency in the QWs. DBRs are particularly suited for QWCs in multi-junction devices, allowing light transmission to the lower bandgap junctions underneath.

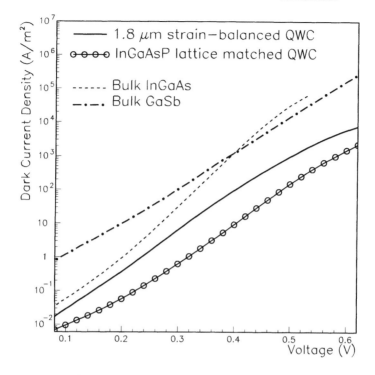

Figure 5.9. Dark current densities of TPV cells.

The flexibility of materials, in particular with strain compensation, combined with the enhanced performance of QWCs make them especially suitable for TPV. QWCs are unique in that the absorption can be extended to wavelengths unattainable for lattice-matched bulk cells, while retaining a similar dark current. Strain-compensated QWCs can be optimized for long-wavelength selective emitters such as thulia.

References

[1] Henry C H 1980 Limiting efficiencies of ideal single and multiple energy gap terrestrial solar cells *J. Appl. Phys.* **51** 4494–500

[2] Chaffin R, Osbourn G, Dawson L and Biefeld R 1984 Strained superlattice, quantum well, multi-junction photovoltaic cells *Proc. 17th IEEE PV Specialists Conf.* (New York: IEEE) pp 743–6

[3] Barnham K W J and Duggan G 1990 A new approach to high-efficiency multi-band-gap solar cells *J. Appl. Phys.* **67** 3490–3

[4] Nelson J, Paxman M, Barnham K W J, Roberts J S and Button C 1993 Steady state carrier escape from single quantum wells *IEEE J. Quantum Electron.* **29** 1460–7

[5] Barnes J, Tsui E S M, Barnham K W J, McFarlane S C, Button C and Roberts J S

1997 Steady state photocurrent and photoluminescence from single quantum wells as a function of temperature and bias *J. Appl. Phys.* **81** 892–900

[6]   Zachariou A, Barnes J, Barnham K W J, Nelson J, Tsui E S M, Epler J and Pate M 1998 A carrier escape study from InP/InGaAs single quantum well solar cells *J. Appl. Phys.* **83** 877–81

[7]   Barnham K W J *et al* 1996 Voltage enhancement in quantum well solar cells *J. Appl. Phys.* **80** 1201–6

[8]   Barnham K, Ballard I, Barnes J, Connolly J, Griffin P, Kluftinger B, Nelson J, Tsui E and Zachariou A 1997 Quantum well solar cells *Appl. Surf. Sci.* **113/114** 722–33

[9]   Anderson N 2002 On quantum well solar cell efficiencies *Physica* E **14** 126–31

[10] Honsberg C, Bremner S and Corkish R 2002 Design trade-offs and rules for multiple energy level solar cells *Physica* E **14** 136–41

[11] Nelson J 1995 Quantum-well structures for photovoltaic energy conversion *Physics of Thin Films* vol 21, ed M Francome and J Vossen (New York: Academic) pp 311–68

[12] Nelson J 2001 Quantum Well Solar Cells *Clean Electricity from Photovoltaics* ed M D Archer and R Hill (London: Imperial College Press) pp 71–87

[13] Barnham K W J, Braun B, Nelson J, Paxman M, Button C, Roberts J S and Foxon C T 1991 Short-circuit current and energy efficiency enhancement in a low-dimensional structure photovoltaic device *Appl. Phys. Lett.* **59** 135–7

[14] Fox A, Ispasoiu R, Foxon C, Cunningham J and Jan W 1993 Carrier escape mechanisms from GaAs/AlGaAs multiple quantum wells in an electric field *Appl. Phys. Lett.* **63** 2917

[15] Paxman M, Nelson J, Braun B, Connolly J, Barnham K W J, Foxon C T and Roberts J S 1993 Modeling the spectral response of the quantum well solar cell *J. Appl. Phys.* **74** 614–21

[16] Aperathitis E, Scott C G, Sands D, Foukaraki V, Hatzopoulos Z and Panayotatos P 1998 Effect of temperature on GaAs/AlGaAs mulitple quantum well solar cells *Mater. Sci. Eng.* B **51** 85–9

[17] Connolly J P, Barnham K W J, Nelson J, Roberts C, Pate M and Roberts J S 1998 Short circuit current enhancement in GaAs/AlGaAs MQW solar cells *Proc. 2nd World PV Energy Conversion Conf. (Vienna)* (Ispra: European Commission) pp 3631–4

[18] Zachariou A, Barnham K W J, Griffin P, Nelson J, Osborne J, Hopkinson M and Pate M 1998 GaInP/GaAs quantum well solar cells *Proc. 2nd World Conf. and Exhibition on Photovoltaic Solar Energy Conversion (Vienna)* (Ispra: European Commission) pp 223–6

[19] Barnes J, Ali T, Barnham K W J, Nelson J and Tsui E S M 1994 Gallium Arsenide/Indium Gallium Arsenide Multi-Quantum Well solar cells *Proc. 12th European PV Solar Energy Conf. (Amsterdam)* (Bedford: H S Stephens & Associates) pp 1374–7

[20] Ragay F, Wolter J, Martí A and Araújo G 1994 Experimental analysis of the efficiency of MQW solar cells *12th European Photovoltaic Solar Energy Conf. (Amsterdam)* (Bedford: H S Stephens & Associates) pp 1429–33

[21] Barnes J, Nelson J, Barnham K, Roberts J, Pate M, Grey R, Dosanjh S, Mazzer M and Ghiraldo F 1996 Characterization of GaAs/InGaAs quantum wells using photocurrent spectroscopy *J. Appl. Phys.* **79** 7775–9

[22] Griffin P R, Barnes J, Barnham K W J, Haarpainter G, Mazzer M, Zanotti-Fregonara

C, Grunbaum E, Olson C, Rohr C, David J P R, Roberts J S, Grey R and Pate M A 1996 Effect of strain relaxation on forward bias dark currents in GaAs/InGaAs multiquantum well p–i–n diodes *J. Appl. Phys.* **80** 5815–20

[23] Freundlich A and Serdiukova I 1998 Multi-quantum well tandem solar cells with efficiencies exceeding 30% AM0 *Proc. 2nd World Conf. and Exhibition on Photovoltaic Solar Energy Conversion (Vienna)* (Ispra: European Commission) p 3707

[24] Ekins-Daukes N, Barnes J, Barnham K, Connolly J, Mazzer M, Clark J, Grey R, Hill G, Pate M and Roberts J 2001 Strained and strain-balanced quantum well devices for high efficiency tandem solar cells *Solar Energy Mater. Solar Cells* **68** 71–87

[25] Ekins-Daukes N, Barnham K, Connolly J, Roberts J, Clark J, Hill G and Mazzer M 1999 Strain-balanced GaAsP/InGaAs quantum well solar cells *Appl. Phys. Lett.* **75** 4195–7

[26] Ekins-Daukes N, Bushnell D, Zhang J, Barnham K and Mazzer M 2000 Strain-balanced materials for high-efficiency solar cells *Proc. 28th IEEE PV Specialists Conf. (Anchorage, AK)* (New York: IEEE) pp 1273–6

[27] Bushnell D, Ekins-Daukes N, Mazzer M, Connolly J, Barnham K, Roberts J, Clark J, Hill G and Forbes I 2000 GaAsP/InGaAs strain-balanced quantum well solar cells *Proc. 16th European Photovoltaic Solar Energy Conf. (Glasgow)* (London: James & James) vA3.5

[28] Ekins-Daukes N, Bushnell D, Barnham K, Connolly J, Nelson J, Mazzer M, Roberts J, Hill G and Airey R 2001 Factors controlling the dark current of strain-balanced GaAsP/InGaAs multi-quantum well p-i-n solar cells *Proc. 17th European Conf. and Exhibition on PV Solar Energy (Munich)* (Munich: WIP) pp 196–9

[29] Bushnell D, Ekins-Daukes N, Barnham K, Connolly J, Roberts J, Hill G, Airey R and Mazzer M 2003 Short-circuit current enhancement in Bragg stack multi-quantum-well solar cells for multi-junction space applications *Solar Energy Mater. Solar Cells* **75** 299–305

[30] Freundlich A, Rossignol V, Vilela M, Renaud P, Bensaoula A and Medelci N 1994 InP-based quantum well solar cells grown by chemical beam epitaxy *Proc. 1st World Conf. and Exhibition on Photovoltaic Solar Energy Conversion (Hawaii)* (New York: IEEE) pp 1886–9

[31] Zachariou A, Barnham K W J, Griffin P, Nelson J, Button C C, Hopkinson M, Pate M and Epler J 1996 A new approach to p-doping and the observation of efficiency enhancement in InP/InGaAs quantum well solar cells *Proc. 25th IEEE PV Specialists Conf. (Washington, DC)* (New York: IEEE) pp 113–16

[32] Griffin P, Ballard I, Barnham K, Nelson J, Zachariou A, Epler J, Hill G, Button C and Pate M 1997 Advantages of quantum well solar cells for TPV *Thermophotovoltaic Generation of Electricity: 3rd NREL Conf. (Colorado Springs) (AIP Conf. Proc. 401)* ed T J Coutts *et al* (Woodbury, NY: AIP) pp 411–22

[33] Griffin P, Ballard I, Barnham K, Nelson J, Zachariou A, Epler J, Hill G, Button C and Pate M 1998 The application quantum well solar cells to thermophotovoltaics *Solar Energy Mater. Solar Cells* **50** 213–19

[34] Rohr C, Connolly J P, Barnham K W J, Griffin P R, Nelson J, Ballard I, Button C and Clark J 1999 Optimisation of InGaAsP quantum well cells for hybrid solar-thermophotovoltaic applications *Thermophotovoltaic Generation of Electricity: 4th NREL Conf., (Denver, CO) (AIP Conf. Proc. 460)* ed T J Coutts *et al* (Woodbury, NY: AIP) pp 83–92

[35] Raisky O Y. Wang W B. Alfano R R. C L Reynolds J. Stampone D V and Focht M W 1998 $In_{1-x}Ga_xAs_{1-y}P_y$/InP multiple quantum well solar cell structures *J. Appl. Phys.* **84** 5790–4

[36] Rohr C, Barnham K W J. Connolly J P, Nelson J, Button C and Clark J 2000 Potential of InGaAsP quantum well cells for thermophotovoltaics *Proc. 26th Int. Symp. on Compound Semiconductors (Berlin) (Institute of Physics Conf. Series 166)* (Bristol: Institute of Physics Publishing) pp 423–6

[37] Rohr C, Connolly J, Barnham K W, Mazzer M, Button C and Clark J 2000 Strain-balanced $In_{0.62}Ga_{0.38}As$/$In_{0.47}Ga_{0.53}As$ (InP) quantum well cell for thermophotovoltaics. *Proc. 28th IEEE Photovoltaic Specialists Conf. (Anchorage, AK)* (New York: IEEE) pp 1234–7

[38] Rohr C, Connolly J P. Ekins-Daukes N, Abbott P, Ballard I, Barnham K W, Mazzer M and Button C 2002 Ingaas/ingaas strain-compensated quantum well cells for thermophotovoltaic applications *Physica E* **14** 158–61

[39] Abbott P, Rohr C, Connolly J P. Ballard I, Barnham K, Ginige R, Corbett B, Clarke G, Bland S W and Mazzer M 2002 A comparative study of bulk InGaAs and InGaAs/InGaAs strain-compensated quantum well cells for thermophotovoltaic applications *Proc. 29th IEEE PV Specialists Conf. (New Orleans, LA)* (New York: IEEE) pp 1058–61

[40] Rohr C, Abbott P. Ballard I. Connolly J P. Barnham K W, Nasi L, Ferrari C, Lazzarini L, Mazzer M and Roberts J 2003 Strain-compensated ingaas/ingaas quantum well cell with 2 μm band-edge *5th Conf. on Thermophotovoltaic Generation of Electricity (Rome) (AIP Conf. Proc. 653)* ed T J Coutts *et al* (Woodbury, NY: AIP) pp 344–53

[41] Abbott P, Rohr C, Connolly J P. Ballard I. Barnham K W, Ginige R, Clarke G, Nasi L and Mazzer M 2003 Characterisation of strain-compensated InGaAs/InGaAs quantum well cells for TPV applications *5th Conf. on Thermophotovoltaic Generation of Electricity (Rome) (AIP Conf. Proc. 653)* ed T J Coutts *et al* (Woodbury, NY: AIP) pp 213–21

[42] Ballard I 1999 Electrical and optical characterisation Of MQW solar cells under elevated temperatures and illumination levels *PhD Thesis* Imperial College, University of London

[43] Rohr C 2000 InGaAsP quantum well cells for thermophotovoltaic applications *PhD Thesis* Imperial College, University of London

[44] Ballard I, Barnham K W J, Connolly J P. Nelson J, Rohr C, Roberts C, Roberts J and Button C 2001 The effect of temperature on the efficiency of multi quantum well cells for solar and thermophotovoltaic applications *Proc. 17th European Conf. and Exhibition on PV Solar Energy (Munich)* (Munich and Florence: WIP and ETA) pp 41–4

[45] Matthews J W and Blakeslee A E 1976 Defects in epitaxial multilayers (III) *J. Crystal Growth* **32** 265–73

[46] Ekins-Daukes N. Kawaguchi K and Zhang J 2002 Strain-balanced criteria for multiple quantum well structures and its signature in x-ray rocking curves *Crystal Growth & Design* **2** 287–92

[47] Nasi L, Ferrari C, Lazzarini L. Salviati G, Tundo S, Mazzer M, Clarke G and Rohr C 2002 Extended defects in InGaAs/InGaAs strain-balanced multiple quantum wells for photovoltaic applications *J. Phys.: Condens. Matter* **14** 13 367–73

[48] Fan J, Tsaur B and Palm B 1982 Optimal design of high-efficiency tandem cells *Proc. 16th PV Specialists Conf. (San Diego, CA)* (New York: IEEE) p 692

[49] Kurtz S, Faine P and Olson J 1990 Modeling of 2-junction, series-connected tandem solar-cells using top-cell thickness as an adjustable-parameter *J. Appl. Phys.* **68** 1890–5

[50] Hoffman R, Fatemi N, Stan M, Jenkins P and Weizer V 1998 High efficiency InGaAs on GaAs devices for monolithic multi-junction solar cell applications *Proc. 2nd World Conf. and Exhibition on Photovoltaic Solar Energy Conversion (Vienna)* (Ispra: European Commission) pp 3604–8

[51] Dimroth F, Beckert R, Meusel M, Schubert U and Bett A 2001 Metamorphic GaInP/GaInAs tandem solar cells for space and for terrestrial concentrator applications at $C > 1000$ suns *Prog. Photovolt. Res. Appl.* **9** 165–78

[52] Freundlich A Multi-quantum well tandem solar cell, US Patent 6,147,296

[53] Freundlich A Multi-quantum well tandem solar cell, continued, US Patent 6,372,980

[54] Bushnell D B, Barnham K W J, Connolly J P, Ekins-Daukes N J, Airey R, Hill G and Roberts J S 2003 Light-trapping structures for multi-quantum well solar cells *Proc. 29th IEEE PV Specialists Conf. (New Orleans, LA)* (New York: IEEE) pp 1035–8

[55] Fatemi N S, Sharps P R, Stan M A, Aiken D J, Clevenger B and Hou H Q 2002 Radiation-hard high-efficiency multi-junction solar cells for commercial space applications *Proc. 17th European Conf. and Exhibition on PV Solar Energy (Munich)* (Munich and Florence: WIP and ETA) pp 2155–8

[56] Coutts T J 1999 A review of progress in thermophotovoltaic generation of electricity *Renewable Sustainable Energy Rev.* **3** 77–184

[57] Barnham K W, Connolly J P and Rohr C (ed) 2003 *Semicond. Sci. Technol.* Special issue: Thermophotovoltaics **18**

[58] Connolly J P and Rohr C 2003 Quantum well cells for thermophotovoltaics *Semicond. Sci. Technol.* Special issue: Thermophotovoltaics **18** S216–20

[59] Freundlich A and Ignatiev A Quantum well thermophotovoltaic convertor, US Patent 6,150,604

[60] Palfinger G, Bitnar B, Sigg H, Müller E, Stutz S and Grützmacher D 2003 Absorption measurement of strained SiGe nanostructures deposited by UHV-CVD *Physica* E **16** 481–8

Chapter 6

The importance of the very high concentration in third-generation solar cells

Carlos Algora
Instituto de Energía Solar, Universidad Politécnica de Madrid
ETSI Telecomunicación, Ciudad Universitaria s/n, 28040
Madrid, Spain

6.1 Introduction

Traditionally, concentration has been proposed as a way of reducing the cost of photovoltaics (PV) [1]. As a rule of thumb, a higher concentration means lower costs. However, high and very high concentrations pose severe technological and economic limitations. For example, at very high concentrations the deleterious effect of series resistance can become important reducing the efficiency of the solar cell. However, suitable optical concentrators become rarer when the concentration increases and such properties as their acceptance angle are seriously affected. As well as this, efficient heat extraction could require complicated solutions. Therefore, the expected cost savings in reducing the solar cell area could be counterbalanced by the expensive cost of other elements such as sophisticated optics, a refined heat removal subsystem, etc or simply by the fact that, for a given concentration level, there is currently no available technology.

Accordingly, an important subject is the determination of the minimum concentration level required for a PV system in order to be competitive in terms of cost. Of the different possibilities provided by third-generation solar cells, only the approaches based on tandem and intermediate-band solar cells will be considered in this chapter. In both cases, only III–V compound semiconductors layers will be assumed.

108

6.2   Theory

In this section theoretical concepts devoted to solar cells when operating under high concentration levels are reviewed. First, solar cell performance under varying concentrations is examined. After this, two of the more important aspects of operating at high concentration, i.e. series resistance and the wide-angle cone of light produced by the optics, are analysed in detail.

6.2.1   How concentration works on solar cell performance

There are two main benefits from the use of concentration: (a) the increase in efficiency and (b) the decrease in cost. Let us begin with a detailed discussion of the first aspect as the second one will be analysed in the section devoted to the economics of PV.

The increase in the efficiency of a well-developed solar cell arises mainly from an increase in the open-circuit voltage. Over a wide range of concentrations, the photocurrent of a solar cell is proportional to the intensity of the impinging radiation. The linearity of the photocurrent with the incident light is a subject which has hardly been studied. In principle, this linearity may not be applicable once the solar cell enters the high injection regime. This may be understandable as at sufficiently high intensities, when the photogenerated carrier density becomes comparable to that of the base doping level (in p-on-n structures), the base lifetime may increase and the photocurrent may increase super-linearly with intensity, provided the series resistance effects are small [2]. In the case of GaAs solar cells, this happens at several thousand suns as some works have observed linearity up to at least 1000 suns [3]. However, other studies have shown a certain super-linearity at concentrations as low as 50 [4], well below the beginning of high injection. And while further study will clarify this topic, the linearity of the short circuit current with concentration (for the range 1–4000 suns) will be assumed throughout this chapter.

If this linearity is assumed, and the photocurrent at 1 sun is denoted by I_{L1}, the photocurrent at concentration X is:

$$I_{LX} = X I_{L1}. \tag{6.1}$$

Assuming a simple one-exponential model, the open-circuit voltage is

$$V_{OC} = m\frac{kT}{e} \ln\left(\frac{I_L}{I_0} + 1\right) \tag{6.2}$$

in which m is the diode's ideality factor and the rest of the parameters have their usual meaning. Therefore, combining (6.1) and (6.2), the open-circuit voltage at X suns is

$$V_{OCX} = V_{OC1} + m\frac{kT}{e} \ln X. \tag{6.3}$$

Equation (6.3) also assumes that m and I_0 do not change significantly when the concentration changes and that 1 in equation (6.2) is negligible with respect to the I_L/I_0 ratio. Equation (6.3) shows that concentration produces a continuous logarithmic increase in V_{OC} weighted by the thermal voltage, kT/e, and the ideality factor, m. Of course, the increase in voltage with concentration is not unlimited because one cannot make X infinitely large. In fact, the highest concentration achievable from the Sun on the Earth is something higher than 46 000 [5].

Equation (6.3) is useful for seeing the basics of the influence of concentration on the open-circuit voltage. However, the majority of the III–V solar cells cannot be conveniently described in terms of a single exponential model; at least a double exponential model is required. In this case, V_{OCX} has to be determined by computing techniques. Figure 6.1 shows the calculated V_{OC} as a function of concentration for a GaAs solar cell. Its kT and $2kT$ recombination current values as well as its short-circuit current have been taken from [6]. As can be seen, a rapid increase in V_{OC} controlled by a $2kT$ recombination appears followed by a slower increase controlled by kT recombination. The influence of $2kT$ recombination ranges from the lowest concentration to about 2 suns while the influence of kT recombination ranges from about 50 suns to the highest concentration. Hence, the shared influence of both kinds of recombination ranges from 2 to 50 suns.

The predominance of kT over $2kT$ recombination when the concentration increases can be explained in terms of carrier concentration variation. Effectively, an increase in light concentration also involves an increase in carrier concentration. Because kT recombination is a band-to-band recombination that depends on the square of carrier concentration while $2kT$ recombination is a defect via recombination that linearly depends on carrier concentration, kT recombination prevails as the concentration increases. This fact does not reveal anything about the existence of non-radiative recombination at a given concentration. In fact, non-radiative recombination could exist in both the depletion and neutral regions. Therefore, non-radiative recombination can be present in both kT and $2kT$ recombinations.

In figure 6.1, the experimental points of V_{OC} are also presented. As can be seen, the agreement with the two-exponential model is fairly good except for high and very high concentrations. At these high light intensities, the experimental V_{OC} decrease with respect to the V_{OC} model based on the two exponentials can be described in terms of series resistance losses. Effectively, although in a first approximation the open-circuit voltage operation involves no current circulation and, therefore, no series resistance losses, a more detailed consideration of the solar cell structure assumes that the current passes through the series resistance and the diode junction located beneath the bus-bar. This situation has been modelled in [7] by means of four diodes plus several distributed series resistances. Its result shows an excellent agreement between theory and experimental data from many concentrator GaAs solar cells.

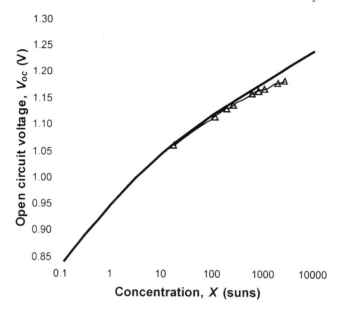

Figure 6.1. Open-circuit voltage as a function of concentration. The full curve represents the theoretical performance of a GaAs solar cell modelled with two exponentials. The values for kT and $2kT$ recombination currents have been taken from the real solar cell of [6]. Symbols are experimental results for the same solar cell.

Once the increase in efficiency with concentration (thanks to the increase in the open-circuit voltage) is understood and once the increase in the open-circuit voltage with concentration has numerically been demonstrated, one can examine the initial reasons for this influence of the concentration on some solar cell parameters. From a qualitative point of view, as concentration increases, the relative weight of recombination decreases. As figure 6.2 shows, as the concentration increases, the total recombination current density (both kT and $2kT$ components) becomes less important with regard to the current density at the maximum power point.

Another complementary point of view is based on the fact that the solar cell equation can be expressed in terms of the excited carrier concentrations since

$$pn = n_i^2 \exp \frac{eV}{kT} \qquad (6.4)$$

Therefore, a given V_{OC} is related to a certain carrier concentration whatever its origin. Therefore, an increase in carrier concentration that produces a V_{OC} increase is achieved by increasing the number of photons impinging on the solar cell (i.e. by increasing concentration). This different perspective opens the door to other alternatives for increasing the carrier concentration such as decreasing

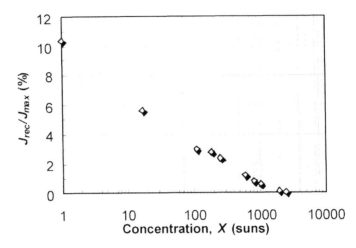

Figure 6.2. The ratio of experimental total recombination current density to current density (at the maximum power point as a function of concentration). The data for the construction of this figure have been taken from [6].

the photons emitted from the semiconductor, for example, by enhancing the self-absorption of the internal generated photons [8].

In addition to the increase in V_{OC}, fill factor (FF) also increases with concentration provided the series resistance is negligible. Therefore, the increase in total efficiency with concentration is due to both factors if the series resistance is low enough. Because the increase in V_{OC} prevails over the increase in fill factor, the increase in efficiency is almost logarithmic with concentration. However, this increase is limited, in practice, by the effects of the series resistance. When series resistance effects become dominant, fill factor begins to decrease rather than increase with concentration. In this situation, the efficiency decreases as a result of fill factor degradation which counterbalances the V_{OC} increase. Therefore, accurate control of the series resistance effects is necessary in order to have very good concentrator solar cells.

6.2.2   Series resistance

The series resistance is a key parameter (but not *the* one) for achieving high efficiencies at high concentrations. Due to its multiple origins, several aspects of the design and manufacture of concentrator cells must be considered carefully. Of course, the lower the series resistance is, the better the performance of the solar cell will be if photocurrent is kept constant. Because of this, one could continuously try to reduce the series resistance but it is a hard work which, in some cases, is even irrelevant. Therefore, it would be better to determine which series resistance value is good enough for a given case and also to detect which technological steps have the biggest influence on the total series resistance value.

Over recent decades, the series resistance has been extensively studied by a great number of authors who have contributed to establishing a coherent theory [9–11] based on a wide variety of experiments [12, 13]. Depending on the focus of the paper, there were two main approaches: (a) the semiconductor structure was fixed and, thus, a given front metal grid was analysed resulting in a grid series resistance component responsible, to a large extent, for the whole series resistance; or (b) particular front metal grid and ohmic contact values were established resulting in the determination of the series resistance of the semiconductor structure (thicknesses and doping levels).

However, neither of these is the best approach. As proposed in [14], the right choice consists of a multi-dimensional optimization considering the whole solar cell simultaneously (i.e. antireflection coatings (ARCs), ohmic contacts, geometry and semiconductor structure). As an example of one of these optimized parameters, the optimum window layer thickness is determined simultaneously by its influence on: (a) the carrier photogeneration in the window (the performance of the semiconductor structure), (b) the transmission of incident light to the photoactive layers (the role of the window layer as an ARC) and (c) the series resistance. The different approach proposed in [14] with regard to previous analysis is that window layer thickness is not optimized in order to minimize the series resistance, to maximize transmissivity or to maximize carrier photogeneration but in order to maximize efficiency.

Let us look at another example. Usually, once the minimum gridline width (derived from given technology) is known, the front grid pattern is designed but this is a mistake. The right way, once the gridline width is known, should be the measurement of the specific front-contact resistance. Only from a combination of both parameters (gridline width and specific front-contact resistance), should the optimum shadowing factor (grid pattern) be calculated by considering the whole structure. Several useful calculations can be found in [15].

In modelling concentrator solar cells, the series resistance, r_S, can be considered as a lumped parameter. The applicability of this concept in 1D modelling has been verified for single-junction GaAs solar cells operating at beyond 1000 suns [7, 14] The total series resistance should include all the contributions from the different parts of the solar cell. In figure 6.3, r_S can be written as

$$r_S = r_{FC} + r_L + r_{BC} + r_V + r_G \qquad (6.5)$$

where

$$\frac{1}{r_L} = \frac{1}{r_E} + \frac{1}{r_W} \qquad r_V = r_B + r_{Su} \qquad r_G = r_F + r_A. \qquad (6.6)$$

Resistance to the lateral flow of current in the semiconductor structure is indicated by r_L while resistance to the vertical flow is indicated by r_V. The grid resistance contribution, r_G, can be expressed as the finger contribution, r_F, plus the annulus contribution, r_A, (only for circular grids). Other symbols are; r_{BC}, the back-contact contribution; r_B, the base contribution; r_{Su}, the substrate contribution; r_{FC},

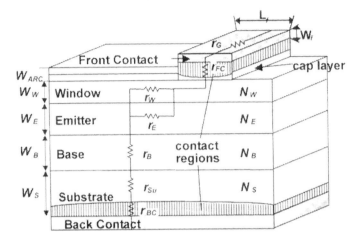

Figure 6.3. Diagram showing the different components of the series resistance of a solar cell (from [14]. © John Wiley & Sons Ltd).

the front-contact contribution; r_W, the window contribution; and r_E, the emitter contribution. Several of these components are very dependent on the front metal grid, so it is necessary to know its geometry accurately. Detailed expressions for these components for circular and square grids can be found in [14].

Although equations (6.5) and (6.6) consider the majority of the origins of the series resistance, there are others. For example, those related to the bus-bar and external connections. In this case, the calculation of the associated series resistance is, of course, dependent on the metals used for the bus-bar and connections as well as their geometries but it is also very dependent on the number of external connections as well as on their positions for extracting the current from the bus-bar. A detailed analysis of these aspects can be found in [16].

Another important origin of the series resistance is related to the tunnel junctions of tandem solar cells. In this case, a careful trade-off between high light transmissivity and high conductivity is stated. High-bandgap materials (such as high-aluminium AlGaAs) are suitable from the point of view of high transmissivity but they are not able to reach high conductivity. In contrast, low-bandgap materials (such as GaAs), which are very suitable for decreasing the series resistance (high conductivity), absorb light deeply. Consequently, models that are as accurate as possible for the whole solar cell and that optimize the efficiency of each third-generation solar cell are necessary.

A final origin of the series resistance in multi-material solar cells is that related to the heterofaces. GaAs–AlGaAs and GaAs–GaInP heterofaces are well known only for specific cases of doping levels and compositions. Thus, careful attention must be paid to finding out the heteroface resistance. This problem will be enlarged when more exotic III–V materials are incorporated into multi-junction structures.

6.2.3 The effect of illuminating the cell with a wide-angle cone of light

One of the challenges of concentration is to match the solar cells and the concentrator system. In this sense, one of the key problems is the optical coupling of the solar cell and the optical concentrator. A main concern comes from the fact that the large area of the concentrator compared to the solar cell size forces the light to impinge on the solar cell in the shape of an inverted cone, pyramid, etc (depending on the shape of the optics).

This situation can be evaluated from basic principles [17]. So, for a loss-less optical concentrator, the conservation of the number of photons leads to

$$A_O f_{ph,S}(\lambda) \sin^2 \theta_S = \hat{n}^2 A_C f_{ph,C}(\lambda) \sin^2 \theta_C \qquad (6.7)$$

where A_O is the optic element area, $f_{ph,S}(\lambda)$ the sun's photon-flux distribution impinging on the optics (expressed in photon flux per unit wavelength interval, λ and unit solid angle), θ_S the sun angle, A_C the solar cell area, $f_{ph,C}(\lambda)$ the photon-flux distribution impinging the solar cell within the angle θ_C and $\hat{n}$ the refractive index surrounding the cell. Defining the geometrical concentration as $X = A_O/A_C$ and because $f_{ph,S} \geq f_{ph,C}$ due to the brightness (or radiance) theorem (an image cannot be brighter than the original object, assuming the refractive indices in the source and object are identical [17]), equation (6.7) transforms into

$$\hat{n}^2 \sin^2 \theta_C \geq X \sin^2 \theta_S. \qquad (6.8)$$

Therefore, considering $\theta_S = 0.267°$ for the sun, $\hat{n} = 1$ and $X = 1000$ suns, the light will impinge on the solar cell within an angle of at least 8.5°. Of course, this is a lower limit because, in fact, practical concentrators reaching such high concentration levels have a θ_C of around 50° or more in order to reach a suitable aspect ratio.

6.2.3.1 Theoretical model

The modelling of a solar cell operating under wide-angle cones of light was first explained in [18] and the performance and optimization of a GaAs solar cell in the ideal case in which the light inside the cone has an isotropic distribution was analysed in [19]. However, practical concentrators do not follow this specific distribution. A generalization of the model for its application to any kind of optical concentrator producing any angular power distribution was carried out in [20]. We reproduce here several equations and figures in order to describe more consistently the effects and conclusions.

The illumination current density of a concentrator solar cell when light impinges on it in the shape of a cone forming an angle θ_C with the normal of

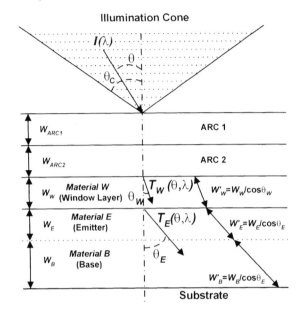

Figure 6.4. The structure of the concentrator solar cell including the ARC together with the light impinging inside a cone forming an angle θ_C with the normal of the cell surface. The meaning of the effective thicknesses as a consequence of refraction is also shown. The position of the substrate suggests a single-junction cell, although the model is also valid for multi-junction cells (from [20]. © 2002 IEEE).

the solar cell (see figure 6.4) is

$$J_L(\theta_C) = \frac{e \times \int_0^{\theta_C} \int_{\lambda_0}^{\lambda_{EG}} [1 - F_s(\theta_C)] N_{ph.C}(\lambda) T(\theta, \lambda) P(\theta) QE(\theta, \lambda) \, d\theta \, d\lambda}{\int_0^{\theta_C} P(\theta) \, d\theta}.$$

(6.9)

Equation (6.9) is slightly different from the one presented in [20] in order to highlight the influence of the angular power distribution, $P(\theta)$, for any kind of concentrator. Other symbols are: T, the transmission; e, the electron charge; $F_s(\theta_C)$, the required shadowing factor for a given cone angle; $N_{ph.C}(\lambda)$ the incident number of photons per wavelength interval on the solar cell (expressed in photons number/time area wavelength); and $QE(\theta\lambda)$ the internal quantum efficiency.

The expression for the internal quantum efficiency must be modified as the light undergoes a change in its direction when it goes from the incident medium with refraction index, $\hat{n}_i$, into the solar cell. In fact, the angle of incidence of the cone of light on the cell, θ_C, changes into the new refracted angles θ_W and θ_E when the light reaches the window layer (of material W) and the emitter (of material E), respectively. Both angles are given by Snell's law:

$$\hat{n}_i(\lambda) \sin \theta_C = \hat{n}_W(\lambda) \sin \theta_W = \hat{n}_E(\lambda) \sin \theta_E. \qquad (6.10)$$

These angles determine the path followed by the light inside the semiconductor layers in such a way that the light does not pass through the 'geometric' thickness of the layers but through longer lengths that can be considered as 'effective' thicknesses.

Therefore, 'effective' thickness must substitute the 'geometric' thickness in the expressions for the quantum efficiency of each layer. However, other expressions in which thickness is involved such as those for the series resistance and the kT-recombination current density are not sensitive to the light path and, consequently, should include the 'geometric' thickness rather than the 'effective' thickness. Consequently, this approach consists of the application of geometrical optics to calculate the length of the rays inside the solar cell. No light-scattering process inside the semiconductor has been included in the model.

Second-order effects such as the additional shadowing produced by the front metal grid when tilted light impinges on the solar cell were also analysed in [20] and a zero influence demonstrated.

6.2.3.2 Results

In order to demonstrate the results, the model must be applied to a given solar cell and to an optical concentrator. Regarding optical concentrators, an interesting case is that of the TIR-R class [21]. This optical concentrator has excellent characteristics. Among the most important ones are: (a) a geometrical concentration higher than 1000×; (b) an acceptance angle (for 90% relative transmission) $\alpha = \pm 1.3°$; (c) an aspect ratio lower than 0.3; and (d) total planarity of the top surface. The angular power distribution of the TIR-R concentrator is shown in figure 6.5.

In [20] the model was applied to a 1000 sun GaAs solar cell operating inside the TIR-R optical concentrator. The results showed that the optimum GaAs solar cell structure for normal incidence was almost the same as in the case of a solar cell operating under a wide-angle cone of light. The main reason for this behaviour is because the critical angle (i.e. the refracted angle when the incident ray is parallel to the surface) of GaAs is only about 17°, so no large variations in the light path inside the semiconductor are expected.

Nevertheless, the conclusion of this case should not be considered of general validity and each combination of optical concentrator and solar cell must be carefully analysed. More specifically, we do not know at present what will happen with tandem cells in which modifications in the light path could affect the current matching of the different pn-junctions. Therefore, studying the optical properties of the concentrators is a key issue in order to predict the final efficiency of the solar cell operating inside the optic element.

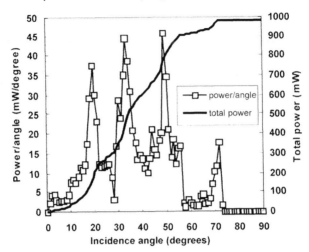

Figure 6.5. Angular light power distribution of the TIR-R concentrator as a function of the incidence angle. The total power impinging on a 1 mm^2 solar cell as a function of the incidence angle is also shown (from [20]. © 2002 IEEE).

6.2.4 Pending issues: modelling under real operation conditions

The principal pending issue in concentrator third-generation solar cells arises from the fact of the almost non-existent experience with complete systems. For example, concentrator solar cell efficiencies are defined under normal incident light with the AM1.5d spectrum when no concentrator solar cell inside an optical concentrator operates at these illumination conditions. This is because the optical concentrator modifies the spectrum received by the solar cell as well as sending the light inside a wide-angle cone.

Therefore, it is necessary to model, to optimize and to characterize concentrator solar cells under real conditions. A first approach is the model of the wide-angle cone of light described earlier. However, other aspects such the non-homogeneity of light, degradation, etc remain.

The non-homogeneity of light is a common characteristic of optical concentrators. When defining the concentration level of an optical concentrator as the current gain experienced by the solar cell regarding 1-sun operation, no supposition on the homogeneity of light on the solar cell is considered. What usually happens is that the concentration level is assumed as the average concentration on the solar cell surface. However, the practice can be very different. Figure 6.6 shows the irradiance pattern produced by an optical concentrator (called RXI) on a solar cell. Although the average concentration is about 1200 suns, there are regions in the centre of the solar cell receiving more than 20 000 suns [22].

Perhaps this example is somewhat extreme but it is real. Other concentrators

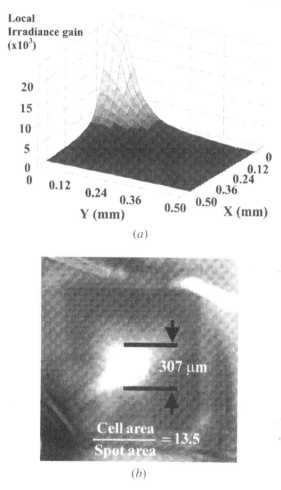

Figure 6.6. (*a*) Irradiance pattern on a quarter of the solar cell surface inside the RXI concentrator and (*b*) photograph of the sunspot produced on the solar cell (its size is 1 mm × 1 mm) inside the RXI concentrator. Average concentration on the whole solar cell area is 1200 suns (from [23]. © James & James).

probably do not produce such extreme differences but concentration ratios ranging from 2 to 10 in different regions of the solar cell can be expected as a rule of thumb. Therefore, specific modelling of the solar cells operating under non-homogeneous light should be carried out. If 1D models are used, pillbox-type calculations should be developed by establishing suitable parameters distribution. However, 2D and 3D models seem to show more potential for this situation. This is not only in order to simulate different illumination conditions in the different regions of the solar cell but also in order to simulate a front metal grid that should have a different shadowing factor in the different regions of the solar cell. Furthermore, 2D and 3D models can include local temperature variations resulting from non-uniform illumination.

Other issues arise from the time evolution of the illumination. For example, while the sun is viewed inside the acceptance angle by the tracking system, the sunspot of maximum efficiency moves across the solar cell surface and even the irradiance pattern can change. Additional concerns arise from the changing spectrum during the day. When a daily average spectrum is well known and stated, this should be a new input into the optimum design of the solar cells, specially for tandems in which current matching is a must.

Finally, extended experiments in order to detect degradation mechanisms and to quantify long-term reliability must be carried out. Very few studies are described in the literature and then only for single-junction GaAs solar cells [23, 24]. Fortunately, preliminary results show that III–V compound semiconductors seem to be resistant to operating under high concentrations.

6.3   Present and future of concentrator third-generation solar cells

Figure 6.7 shows the efficiency as a function of concentration for different third-generation solar cells. The efficiency is obtained by assuming solar cells radiate isotropically in all directions and operate under ideal conditions such as the radiative recombination being much higher than any other losses, unity absorbance, etc. So, assuming the AM1.5d spectrum, the maximum efficiency depends only on the energy gap. In the case of tandems, the efficiency is obtained by assuming that the maximum power of each cell can be extracted independently.

As can be seen, the highest efficiency in the case of two-bandgap systems and the AM1.5d spectrum is 58.5% at full concentration. The intermediate-band solar cell with a main bandgap of 1.93 eV exhibits a slighter higher efficiency for all concentrations [25]. Finally, the efficiency that could be obtained by using an infinite number of gaps is shown. The maximum efficiency is 86.5% and must be considered as the greatest efficiency that could ever be obtained by PV conversion when maximum power is extracted with a fill factor less than unity (for the spectrum considered).

Figure 6.7 also shows the tremendous potential that concentration has in increasing the efficiency of third-generation solar cells. Therefore, it seems recommendable to solve the arising technological problems in order to achieve practical efficiencies close to these limits in the future. Let us look at the present state of the art of these solar cells.

To speak of third-generation concentrator solar cells is to speak of concentrator tandem solar cells because, at present, intermediate-band solar cells are still under demonstration. Two types of tandem have been successfully developed: monolithically grown cells and mechanically stacked cells. The

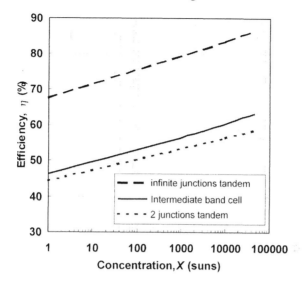

Figure 6.7. Efficiency of different types of third-generation solar cells. Radiative recombination is the only loss considered.

latter have the advantage of tailoring the bandgap materials according to an optimization analysis; they are, therefore, able to achieve very high efficiencies. The first impressive result (in 1989) with mechanically stacked solar cells was achieved by L Fraas and co-workers with a GaAs/GaSb tandem (four terminals). They achieved a 32.6% efficiency at 100 suns [26]. Just recently, this result has been improved to 33.5% at 308 suns by means of a GaInP/GaInAs/GaSb tandem (four terminals) [27].

But despite the high efficiency advantage, mechanically stacked solar cells have two main concerns: (1) more than one semiconductor substrate is necessary, with a subsequent increase in cost; and (2) the manufacturing process is complicated and several processes are repeated for each individual cell (such as ohmic contact formation, ARC deposition, etc) with a subsequent cost increase. Because the high efficiency is not counterbalanced with reasonable cost, it is the opinion of this author that mechanically stacked solar cells are not good candidates for a competitive alternative.

The other type of tandem solar cell (monolithically grown) is limited by semiconductor growth techniques. Semiconductors with a near ideal bandgap for photovoltaic operation, therefore, cannot be grown one on top of each other. The typical example is the impossibility of growing GaAs on GaSb in order to achieve a monolithic solar cell with two terminals. Therefore, monolithic solar cells cannot reach the efficiency limits calculated by putting together optimum bandgap semiconductors and thus, monolithic solar cells will achieve lower efficiencies than mechanically stacked ones with the same number of junctions.

Even so, the best monolithic solar cell with a triple junction of GaInP/GaAs/Ge has an efficiency of 32.4% at 414 suns [28] (i.e. only 1.3% less than the best mechanically stacked solar cell; in addition, its maximum efficiency is achieved at an even higher concentration). So, the present efficiencies of monolithic solar cells are close to those of mechanically stacked ones. This fact has been supported theoretically in [29] which showed that the reduction in efficiency due to constraining the current through a tandem stack by series connection is less than 1.5% (relatively), regardless of the number of cells in the stack. As the number of cells tends towards infinity, the efficiency of a two-terminal stack will tend to the same efficiency as for an unconstrained tandem cell stack of an infinity number of cells. A similar conclusion is obtained in [30] where it is stated that the limiting efficiency of an infinite tandem array is determined by the emission losses through the illuminated face.

The main advantages of monolithic cells are (as opposed to the mechanically ones) the use of a single semiconductor substrate and the similarity of the post-growth process to that of a single-junction one. Consequently, the cost of a monolithic solar cell is relatively similar to that of a single-junction one, being just slightly more expensive as a consequence of the increased number of semiconductor layers. Therefore, the slight efficiency offset is well counterbalanced by the lower cost of monolithic solar cells and, thus, they seem good candidates to be cost competitive if some conditions are complied with (as will be shown in the next section).

6.4   Economics

The definitive driving force that will enable concentration technology to enter the real PV market is ultimately the cost. Accordingly, this section is very important and, thus, considerably long. First, we will analyse the way in which concentration works as a factor in the cost of solar cells and, as a consequence, a minimum concentration level in order to be cost competitive will be recommended. This seems to be a logical path to take: economics should define the concentration level that should be the objective from a technological point of view. Finally, a detailed cost analysis of a complete PV installation based on concentrator third-generation solar cells is presented.

6.4.1   How concentration affects solar cell cost

As we have stated, in addition to increasing the efficiency, concentration also has the advantage of reducing the price, thus making a greater impact on the final product. Even more categorically, the driving force for commercializing concentrator third-generation solar cells would be a suitable cost—not only a high efficiency. This basic rule seems to be forgotten by many people who are continuously looking only for an increase in efficiency without taking any other considerations into account.

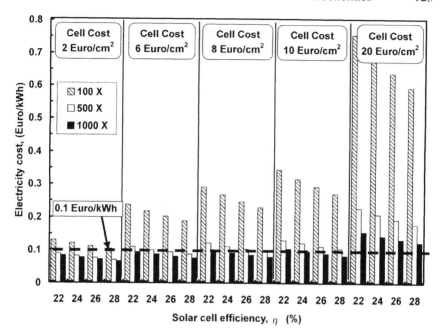

Figure 6.8. Electricity cost in Madrid (Spain) for different scenarios of both cost and efficiency for solar cell operating at concentrations of 100, 500 and 1000 suns. The criterion of PV penetration into the market discussed in the text is indicated with the dotted line at 0.1 € kWh^{-1}. This value could be considered somewhat high.

While efficiency has a clear impact on cost, the concentration level has a far more dramatic impact. A clarifying example is shown is figure 6.8 where the electricity cost in € kWh^{-1} in a place like Madrid is given. The influence on the total PV costs for different prices and efficiencies of the solar cells together with light concentration has been tabulated. The main assumptions for the calculations are described in [31]. For the solar cell cost, five situations have been selected. These are: 2, 6, 8, 10 and 20 € cm^{-2} which correspond to a very low, low, medium, high, very high cost of the solar cell, respectively. Nowadays, we are in a medium-cost situation. The very-low- and low-cost situation would correspond to an improvement in the technological achievement of GaAs substrates or even to its substitution by alternative substrates such as germanium or silicon. The high- and very-high cost situation would correspond to market stresses or to scarcity due to natural catastrophes such as what happened some years ago with Japan's earthquake.

With regard to the commercial efficiency of concentrator solar cells, a standard range 22–28% in industrial production has been considered. Of course, these efficiencies could be surpassed by future third-generation solar cells but this does not restrict the generality of the qualitative reasoning.

As a reference level for the competitiveness of the cost of PV electricity, we have chosen the target of USA utilities for a significant penetration of PV technology into the power plant market [32], i.e. lower than around $0.1 \in kWh^{-1}$. This criterion is even stronger than the achievement of $0.12 \in kWh^{-1}$ proposed by the ALTENER study [33] to comply with the objective suggested at the Madrid Conference which discussed 'An Action Plan for Renewable Energy Sources in Europe' of 16 TWh/year of PV electric energy in Europe by 2010.

As can be seen in figure 6.8, for cheap solar cells the difference between several concentrations is not significant. However, as the price of the solar cell increases from 10 to 20 $\in cm^{-2}$ (as could be the case for very efficient third-generation solar cells), the influence of very high concentration level in dramatically reducing the price is huge.

Therefore, as a rule of thumb, we can summarize that a higher concentration will produce a higher efficiency (if the series resistance is kept low and the heat is effectively extracted) and, hence, a lower price.

In other words, very high concentrations allow the manufacturing of very efficient although expensive third-generation solar cells in order to achieve competitive prices for the whole PV system. More graphically, we could say that if we were able to keep the series resistance low and to extract the heat efficiently, a very high concentration level would allow more and more sophisticated and efficient solar cells to become cost competitive. But what we are calling a very high concentration level (in order to be competitive in cost) should be quantified and this is done in the next section.

6.4.2   Required concentration level

Which concentration level should be used for third-generation solar cells? When, in the 1990s, tandem solar cells started to become widespread the commonly accepted concentration was about 100–200 suns. This level was established because the limiting factors of the solar cells under concentration such as series resistance and heat extraction were effectively controlled only up to 100–200 suns. At higher concentrations, the deleterious effects of the series resistance and heat dramatically reduced the efficiency of the solar cells. A clear example is the concentrator GaAs solar cell with the highest efficiency to date [34]. Its maximum efficiency of 27.6% was achieved at 255 suns while at 1000 suns its efficiency was only about 24%.

Fortunately, solar cell technology is continuously improving so, the required concentration level must appear as a trade-off between the solar cell technology, the available optics and the resulting price. Considering that there is a cross influence between solar cell and optic researchers in the sense that an optical concentrator is usually designed for the concentration level at which there are efficient solar cells and, conversely, the solar cells are designed to operate at the concentrations supplied by the existing optics, we think that the real driving force in determining the concentration level is the price. Accordingly, from economic considerations we think that we should go towards 1000 suns and we will explain why.

Although several years ago to speak of concentration levels of 1000 suns was simply a dream, now it is reality. From the optics point of view, the development has been rapid and now we have concentrators supplying more than 1000 suns with acceptance angles of more than ±1.3 degrees [21].

With regard to solar cells, the achievements have also been impressive. The first very efficient solar cell at 1000 suns was achieved in 1988 at Varian. An efficiency of 27.5% was measured [35] at Sandia National Laboratory. However, this result is not homologated because in 1990 the Sandia reference cell had to be recalibrated. Assuming this new calibration the Varian cell would result in a 26.1% at 1050 suns. There were no measurements at higher concentrations. Higher concentrations but with lower efficiencies were experimented on in Europe in 1995: a GaAs solar cell with an efficiency of 23% at 1300 suns (confirmed by the Fraunhofer Institute) was achieved at the Madrid Solar Energy Institute [36].

In 1999 again the Madrid Solar Energy Institute but now in collaboration with the Ioffe Institute in St Petersburg (Russia) achieved an efficiency of 26.2% at 1004 suns and 25% at 1920 suns again with a single-junction GaAs solar cell [37]. At present, this is the single-junction solar cell with the highest efficiency in this concentration range and for several months it was the highest efficiency for any solar cell in this concentration range until the Fraunhofer Institute of Freiburg (Germany) achieved more than 29% for a GaInP/GaInAs double-junction monolithic solar cell at 1000 suns in 2000 [38]. It is remarkable how in both references [37, 38], the titles do not describe the maximum efficiency of the solar cells but their efficiencies at 1000 suns. Several months later and also in 2000, Spectrolab in collaboration with NREL achieved 31% at 1000 suns with a triple-junction GaInP/GaAs/Ge solar cell [28]. A last important result at very high concentration is the achievement in 2002 at the Madrid Solar Energy Institute of a single-junction GaAs solar cell with an efficiency of 23.8% and 22.5% at concentrations of 2700 and 3600 suns, respectively [39]. The results of these four references [28, 37–39] correspond to the characterization carried out at NREL facilities and always assume linearity of concentration with short-circuit current. In all cases, the cells are not endowed with prismatic covers (see figure 6.9).

Therefore, series resistance is not a big problem in present concentrator third-generation solar cell technology because at 1000 suns the tandem cells have already achieved very good results. In addition, and for future developments, the successful technology of single-junction GaAs solar cells in the range 1000–4000 suns can be transferred to the third generation. Furthermore, and because monolithic solar cells are series connected, as the number of cells increases the photocurrent decreases so the series resistance problems are less important in triple junction than in double junctions and in double junctions they are less important than in single junctions. Similarly, as the number of junctions increases as well as the efficiency, the generated heat is lower, thus its deleterious effects decrease.

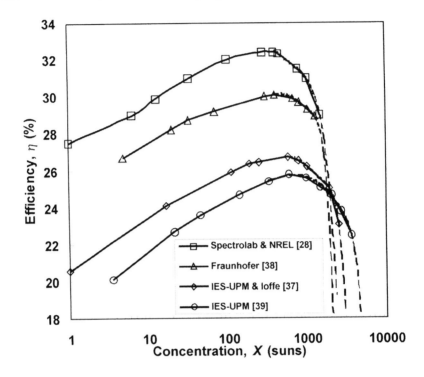

Figure 6.9. Efficiencies of different concentrator solar cells of references [28, 37–39] measured at the NREL. Linearity of the short-circuit current with concentration has been assumed. Dotted curves are extrapolations of the respective full curves (from [39]. © John Wiley & Sons Ltd).

So, once the main technological concerns of earlier solar cells such as series resistance and heat dissipation have been solved for operating at 1000 suns or more and once there are optical concentrators with good properties at this concentration level, it seems that, at present, the concentration level should be at least 1000 suns to achieve a competitive price as it will be shown later.

6.4.3 Cost analysis

Two different production scenarios will be considered for this emerging technology based on third-generation solar cells: 10 and 1000 MW_p cumulated production. First, we start with a calculation based on already existing solar-cell manufacture in the space industry which would be equivalent to 1 MW_p of concentrator terrestrial solar cells. A solar cell size of 1 mm^2 inside the bus-bar in agreement with [6] is considered.

6.4.3.1   Present solar cell cost

The production level for which the majority of costs are currently known is 1 MW$_p$ and this can be considered as a starting point for a learning-based prediction. The information comes from the space solar cell industry.

In order to achieve 1 MW$_p$ with solar cells of 1 mm^2 operating at 1000 suns, a production of about 200 epiwafers (115 mm in diameter, which is about 4.5 inches) of tandem solar cells would be necessary. In this production, the MOVPE growth of the tandem semiconductor structure on germanium substrates including the slight modifications in the semiconductor structure of the space cell for it to become terrestrial results in a cost of 360 € for the epiwafer (substrate + epitaxy). This cost includes all depreciation, labour, gases and other consumables and overheads involved in producing the epiwafers.

Solar-cell processing must be added to this, which involves the cost of contacting both sides involving photolithography for the front side, for antireflecting coatings, for cutting into individual cells and for final measurements of the cell performance.

There could be 5500 cells constructed out of a 115 mm epiwafer (after considering a useful wafer area due to deleterious growth surface effects), each cell having an active area of 1 mm^2 but using a 1.3 mm × 1.3 mm square (1.69 mm^2) epiwafer area including bus-bar and kerf-loss of the dicing saw.

The final quality control step requires the current–voltage-curve semi-automatic control and flash tester to classify the performances of the solar cells diced but not separated from the wafer.

Therefore, post-growth processing together with the quality control increases the cost by 410 € per wafer leading to a total cost per finished wafer of 770 €. This, with a combined electrical and mechanical yield of 80%, results in $770/(0.8 \times 5500) = 0.175$ €/cell. To achieve the final cost, the following must be taken into account:

(a)   the average efficiency is 30% (with a $FF = 0.83$) for solar cells operating at 1000 suns out of the concentrator and under AM1.5d normal incidence, air being the incident medium;
(b)   the non-uniform illumination produced by the optical concentrator reduces the fill factor of the cell to about 0.77;
(c)   optical losses of the solar cell operating inside the concentrator due to tilted incidence of light have been measured to be about 7%; and
(d)   the efficiency of the optics of the concentrator is 90%.

Hence, the peak power of the tandem solar cell inside the concentrator will be 0.231 W$_p$ at 1000 suns. Consequently, the tandem solar cell part of the total cost of the system will be 0.76 €/W$_p$.

6.4.3.2   *The historic learning rate model*

This cost can be reduced by applying the historic learning rate model. The cost reduction is not simply a function of time but is related to cumulative production experience impacted by R&D advances, the amount of technology transfer and the rate of investment in advanced manufacturing process. The latter is closely related to market development, which is influenced by price reduction. The learning model is expressed by

$$\frac{C}{C_0} = \left(\frac{M}{M_0} \right)^{-L.}$$ (6.11)

where C_0 is the cost of a given cumulated production, M_0. So, if the learning rate, L, is known, the new cost, C, of another cumulated production, M, can be calculated. The learning rate of the PV modules is $L = 0.277$ [40]. This rate means a cost reduction of 17.5% each time the production is doubled.

However, if for the manufacturing of the solar cells the processes and guidelines proposed in [41] are applied, the solar cell is closer to both optoelectronics and microelectronics than to classical PV. For these industries a learning rate of $L = 0.56$ (which means a reduction of 32% each time the production is doubled) has been estimated.

Nevertheless, and in order to not to be overly optimistic, we have not assumed any cost reduction by learning when passing from 1 MW_p production to 10 MW_p. This is because the manufacturing costs assumed for the 1 MW_p production are based on the use of techniques closer to those used in optoelectronics such as the use of MOVPE (for the growth of semiconductor structures), photolithography, metal deposition, etc. Because these technologies are very mature and because of the efforts required to adapt the technological processes, no great decrease in cost through learning is expected during the early stages. However, the cost of the optics and encapsulating the cells are again taken from the real prices of optoelectronics and microelectronics for similar devices such as LEDs. Nevertheless, the costs of the small optics and encapsulation will only be valid once the processes are adapted to these new generation solar cells. In other words, we estimate that a cumulated production of 10 MW_p would be necessary in order to validate the costs assumed in the 1 MW_p case. After the transient period from 1 to 10 MW_p production, the learning will operate as usual.

6.4.3.3   *A complete concentrator PV plant: 10 and 1000 MW_p cumulated production*

In spite of the inherent difficulty of a cost study of a currently non-existent product, which is very dependent on assumptions, a rather detailed analysis (including several concepts whose costs are extracted from the experience of related fields) is presented in table 6.1.

The tandem solar cell manufacturing costs from the previous section have been used for the 10 MW_p production (first column of table 6.1) together with an efficiency of 32% at 1000 suns. This is the biggest improvement assumed when passing from 1 to 10 MW_p production and it is well supported by present experimental results.

A solar cell area to total semiconductor area ratio of 60% (1.00/1.69 mm^2) has been applied when cutting. An additional improvement with regard to 1 MW_p production is the assumption of a yield of 85% (with the solar cell ratio complying with the specifications). Therefore, the cost of the individual solar cells with an active area of 1 mm^2 (1.69 mm^2 of the whole dice) will be 14.7 € cm^{-2} (this means 0.147 € for each tandem solar cell which is similar to the cost of a LED of similar dimensions). Finally, the cost of each solar cell must be divided by the concentration level (1000 suns) giving a cost of 0.0147 € cm^{-2} which highlights the dramatic potential of such a high concentration for really reducing the cost.

A problem associated with a small concentrator size is the assembly cost. The formula $1600/N^{0.5} + 0.027N$ (optics cost + encapsulating cost in € m^{-2}) refined after many years of experience [42] has been fitted to real data from optoelectronic products. N is the number of optical concentrators per m^2. The optimization of the formula gives $N = 957$ which means a concentrator unit 32 mm × 32 mm. The corresponding cost is 0.0052 € cm^{-2} for the optics and 0.0026 € cm^{-2} for the encapsulation. To these costs, that of passive cooling based on finned heat exchanger (0.0030 € cm^{-2}) has been added.

In order to obtain the cost of the PV field, the real costs of the EUCLIDES concentrator for assembling the panels, their mounting on metallic structure, tracking and DC wiring (0.014 € cm^{-2}) in the 480 kW_p plant located at Tenerife [43] have been calculated for 10 MW_p production by applying learning. To the cost of the land and its preparation (0.0022 € cm^{-2}), no learning has been applied for the 10 MW_p, it has just has been rounded off to 0.0020 € cm^{-2}. The addition of all these costs produces 0.0415 € cm^{-2} for the PV field.

The power supplied by the PV field (0.0248 W/cm^2) is based on a module efficiency of 24.8% at standard conditions derived by multiplying the tandem solar cell efficiency (32%) by the optics efficiency (80%) and by the module interconnection efficiency (97%).

The cost of the PV plant (0.0465 € cm^{-2}) is obtained by adding the cost of the inverter, transformers and electric equipment (0.0050) to the cost of the PV field (0.0415). This last is obtained by multiplying the figure of 0.2 €/W_p taken directly from the EUCLIDES experience by the power supplied by the PV field (0.0248 W/cm^2). The cost of the PV plant is incremented by 25% commercial profit to reach 0.0581 € cm^{-2}. Finally, the nominal cost of the PV plant is obtained by dividing the price of the PV plant (0.0581 € cm^{-2}) into the power supplied by the PV field (0.0248 W/cm^2) which gives 2.34 €/W_p.

From this price, a great effort should be made to increase both the efficiency and the reliability of tandem solar cells. An increase in standard efficiency of up to 35% will have an impact by itself on the cost (second column) consisting of a decrease in the nominal cost to 2.16 €/W_p (second column of table 6.1).

Table 6.1. Cost analysis based on previously collected data.

Process or concept	Our present technology (10 MW$_p$) Observations	Cost (€ cm^{-2})	Improved solar cell technology Observations	Cost (€ cm^{-2})	Medium term with learning (1000 MW$_p$) Observations	Cost (€ cm^{-2})
Substrate wafer (germanium)		1.2500		1.2500		0.3491
Growth of semicond. structure		2.2500		2.2500		0.6283
Manufacture of the GaAs sol.		4.0000		4.0000		1.1170
Cost of solar cell on subs. wafer		7.5000		7.5000		2.0944
Cost of solar cell as a dice	s.c. area: 60%	12.5000	s.c. area:60%	12.5000	s.c. area:60%	3.4907
Cost of a useful solar cell	Yield: 85%	14.7059	Yield: 85%	14.7059	Yield: 95%	3.6743
Cost of concentrator solar cell	Conc: 1000×	0.0147	Conc: 1000×	0.0147	Conc: 1000×	0.0037
Optics		0.0052		0.0052		0.0027
Encapsulation of cells		0.0026		0.0026		0.0013
Cooling		0.0030		0.0030		0.0015
Structure, tracking, assembling and DC wiring		0.0140		0.0140		0.0039
Land and preparation		0.0020		0.0020		0.0011
Cost of the PV field		0.0415		0.0415		0.0143
GaAs solar cell efficiency	32%		35%		40%	
Optics efficiency	80%		80%		83%	
Module interconnection effic.	97%		97%		97%	
Module effic. at standard condit.	24.8%		27.2%		32.2%	
Power supplied by PV field	0.0248 W/cm^2		0.0272 W cm^{-2}		0.0322 W cm^{-2}	
Inverter, transformers, elect. equi	0.20 €/W$_p$	0.0050	0.20 €/W$_p$	0.0054	0.06 €/W$_p$	0.0034
Cost of the PV plant		0.0465		0.0459		0.0178

Process or concept	Our present technology (10 MW$_p$)		Improved solar cell technology		Medium term with learning (1000 MW$_p$)	
	Observations	Cost (€ cm^{-2})	Observations	Cost (€ cm^{-2})	Observations	Cost (€ cm^{-2})
Price of the PV plant (including commercial profit)	25% of profit	0.0581	25% of profit	0.0587	25% of profit	0.0222
Nominal cost of the PV plant	2.34 €/W$_p$		2.16 €/W$_p$		0.69 €/W$_p$	
PV plant efficiency	Observations	%	Observations	%	Observations	%
Interconnection efficiency (mismatch and wiring)		92.0		92.0		92.0
Average efficiency per year (temperature and irradiance)		87.5		87.5		87.5
Installation efficiency of the photovoltaic field		80.5		80.5		80.5
PV plant and DC efficiency		19.9		21.9		25.9
Conversion efficiency (DC to AC)	91%		91%		91%	
PV plant AC annual efficiency	18.2%		19.9%		23.6%	
Available solar energy & electricity production	Observations	kWh cm^{-2}	Observations	kWh cm^{-2}	Observations	kWh cm^{-2}
Annual direct, 2 axis (MADRID)		0.1826		0.1826		0.1826
Effic. due to tracking 2 axes (%)	100		100		100	

Process or concept	Our present technology (10 MW$_p$)		Improved solar cell technology		Medium term with learning (1000 MW$_p$)	
	Observations	Cost ($€$ cm^{-2})	Observations	Cost ($€$ cm^{-2})	Observations	Cost ($€$ cm^{-2})
Effic. by Shading (GCF = 1/3) (%)	96.9		96.9		96.9	
Annual available energy on the concentrators		0.1769		0.1769		0.1769
Produced annual energy	0.032 18 kWh cm^{-2}		0.035 20 kWh cm^{-2}		0.041 74 kWh cm^{-2}	
PV electricity cost	Observations	$€$ kWh^{-1}	Observations	$€$ kWh^{-1}	Observations	$€$ kWh^{-1}
Investm. by annual kWh ($€$)	1.8048		1.6666		0.5316	
Interest rate	5%		5%		5%	
Inflation rate	2.5%		2.5%		2.5%	
Number of years	30		30		30	
FCR	4.74%		4.74%		4.74%	
Annual payment		0.0855		0.0790		0.0252
Operation and Maintenance	0.0001 $€$ cm^{-2} year^{-1}	0.0031	0.0001 $€$ cm^{-2} year^{-1}	0.0028	0.0001 $€$ cm^{-2} year^{-1}	0.0024
PV electricity cost ($€$ kWh^{-1})	0.089 $€$ kWh^{-1}		0.082 $€$ kWh^{-1}		0.028 $€$ kWh^{-1}	

In the long run, we have applied the historic learning rate of the PV modules to all the process and components for a cumulated production of 1000 MW$_p$ with the exception of the optics, encapsulation, cooling and electric equipment for which a more restricted cumulated production of 100 MW$_p$ has been considered for the learning curve calculation. It must be pointed out that the utilized 17.5% historic learning curve could be pessimistic. This value is that of the PV modules (mostly of silicon) while the concentration technology proposed in [41] is closer to microelectronics and optoelectronics. In this respect, an increase in the yield from the present 85% to 95% (typical in optoelectronics) has been assumed. In addition, an increase in the efficiency of solar cells (from 35 to 40%) and optics (from 80 to 83%) would be expected. All these considerations would drive towards a cost of 0.69 €/W$_p$ (third column) which will completely satisfy the goal of 1 €/W$_p$ for modules by 2010 considered in the Fifth Frame Programme of the European Community (EC). Therefore, we can conclude that, in the long term, the learning rate will have a dominant effect.

By using the data and functions from table 6.1 and by changing the efficiency and concentration level, we have evaluated the nominal cost of a PV plant as a function of concentration and efficiency (figure 6.10). We have included the 1000 sun GaAs solar cells case as a reference because although they do not belong to third generation, they constitute an example of the mature high-concentration technology . The star means the highest efficiency to date (26.2%) of these solar cells [37]. We have assumed a potential efficiency in the future of 30%.

Considering the EC medium-term goal of 3 €/W$_p$, a first conclusion is that GaAs solar cells could already be competitive. Now we go to the cost of tandems at 400 suns and where the star means again the highest efficiency achieved to date by Spectrolab with NREL. So, a second conclusion is that tandems will be competitive (<3 €/W$_p$) when the efficiency becomes higher than 38%; i.e. when tandems have an efficiency corresponding to the real third generation as defined in relation to Shockley's limitations.

With regard to tandems at 1000 suns, where the star indicates the highest efficiency achieved by the Fraunhofer Institute, it can be seen that they are the natural continuation of 1000 sun GaAs solar cells due to their higher potential efficiency. So, the necessity for operating tandem cells at 1000 suns until they achieve the efficiencies of real third-generation cells can be seen. It can also be seen that the concentration level is even more relevant than the efficiency in order to reduce costs. In this sense, from table 6.1 (for 10 MW$_p$), it can be deduced that the solar cell's part of the cost of the PV field is 35% at 1000 suns while at 400 suns it can be calculated as being about 88%.

This cost forecast has been carried out for a cumulated production of 10 MW$_p$, that was, for example, the production of silicon solar cells by the company Isofoton for the year 2000 only. But what is really more amazing is the tremendous potential of this approach for the mid-term after a cumulated production of 1000 MW$_p$ if learning is applied. As can be seen in figure 6.10, costs between 0.5 and 1 €/W$_p$ will be achieved depending on the concentration

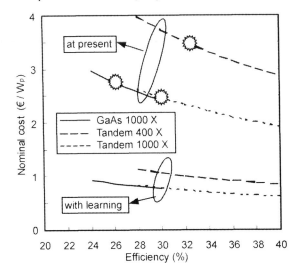

Figure 6.10. Nominal cost of a complete PV plant (based on the approach in [41]) as a function of the efficiency of the solar cells. Two different production scenarios are considered: (*a*) cumulated production of 10 MW$_p$ at present and (*b*) a cumulated production of 1000 MW$_p$ for the medium term when learning is considered. The stars indicate the present highest efficiencies on single-junction GaAs and tandems solar cells described in the text (from [44]. © WIP Munich and ETA Florence).

for PV installations based on third-generation solar cells. Regarding the electricity cost, values of 0.02–0.03 € kWh^{-1} could be achieved with this approach in a place like Madrid (Spain) [44]. These costs would be even lower than those of the classical sources of energy.

6.5   Summary and conclusions

There are two main benefits for the use of concentration: (a) an increase in efficiency and (b) a decrease in cost. Series resistance is a key parameter (but not *the* one) for achieving high efficiencies at high concentrations. Of course, the lower the series resistance (without changing other parameters) is, the better the performance of the third-generation solar cells will be. Because of this, one could try to reduce the series resistance continuously but this is hard work which, in some cases, is even irrelevant. It is better to determine which series resistance value is good enough for a given case and also to detect what technological steps have the biggest influence on the total series resistance value. In addition to the classical origins of series resistance in single-junction solar cells, other important origins are in the tunnel junctions of tandem solar cells. A final cause of series resistance in multi-material solar cells is that related to the heterofaces. This problem will be enlarged when more exotic III–V materials are incorporated into multi-junction structures of third-generation cells.

In addition to series resistance, another relevant effect comes from the large area of the concentrator compared to the solar cell size, which forces the light to impinge on the solar cell in the shape of an inverted cone, pyramid, etc (depending on the shape of the optics). This is called 'operating the cell under a wide-angle cone of light'. The recent existing models allow us to consider specific optical concentrators together with solar cells. Therefore, they would be applied to tandem cells in which modifications to the light path could affect the current matching of the different pn-junctions.

In addition to series resistance and the wide-angle cone of light, there are other important factors but these have not been extensively analysed in the literature. The majority of pending issues of concentrator third-generation solar cells arise from the almost non-existent experience with complete systems. Therefore, it is necessary to model, to optimize and to characterize concentrator solar cells under real conditions. Among the aspects arising in the 'real world' are the non-homogeneity of light on the solar cell surface and its associated temperature gradient, the time evolution of illumination, degradation, etc.

Once these aspects and other concerns are solved, the efficiency limits will be closer. The theoretical predictions show, in the case of two-bandgap systems and the AM1.5d spectrum, a maximum value for the efficiency of 58.5% at full concentration. The intermediate-band solar cell with a main bandgap of 1.93 eV exhibits a slighter higher efficiency for all concentrations. Finally, the efficiency that could be obtained by using an infinite number of gaps is 86.5% at full concentration and must be considered the greatest efficiency that could ever be obtained by PV conversion (for the spectrum considered).

The path to these unbelievable efficiencies has already started. The present highest efficiency is 33.5% at 308 suns achieved by mechanically stacked GaInP/GaInAs/GaSb cells (four terminals). However, the best monolithic solar cell with a triple junction of GaInP/GaAs/Ge has an efficiency of 32.4% at 414 suns (i.e. only 1.3% less than that of the best mechanically stacked solar cell; in addition, its maximum efficiency is achieved at an even higher concentration). The slight efficiency offset is well counterbalanced by the lower cost of monolithic solar cells and so it is the opinion of this author that they are good candidates (as opposed to mechanically stacked ones) to be cost competitive if some conditions are complied with.

This introduces an important idea: although the increase in efficiency by means of concentration is and will be impressive, the definitive driving force allowing the concentration technology to enter the real PV market is finally the cost. Although efficiency has a clear impact on cost, the concentration level has an even more dramatic impact. So, once the main technological concerns of solar cells in the past such as series resistance and heat dissipation have been solved for operation at 1000 suns or more and once there are optical concentrators with good properties at this concentration level, it seems that, at present, the concentration level should be at least 1000 suns to achieve a competitive price.

This threshold concentration level (1000 suns) is proposed by analysing several economic scenarios. For example considering the EC medium term goal of 3 €/W$_p$, a first conclusion is that a PV product based on 1000 sun GaAs solar cells with efficiencies higher than 24% could already be competitive. However, a similar product based on tandems operating at 400 suns will be competitive (<3 €/W$_p$) only when the efficiency of solar cells will be higher than 38%; i.e. when tandems have an efficiency corresponding to real third generation as defined in relation to Shockley's limitations. Of course, a product based on a tandem cell operating at 1000 suns would already be competitive.

In this sense, it has also been shown that the concentration level is even more relevant than the efficiency in order to reduce costs. For example, the solar cell's part of the cost of the complete installation of a PV field is 35% at 1000 suns while at 400 suns, it is about 88%.

Finally, the tremendous potential of third-generation concentrator solar cells for the mid-term after a cumulated production of 1000 MW$_p$ if learning is applied has been detailed. Costs of between 0.5 and 1 €/W$_p$ will be achieved depending on the concentration for complete PV installations. Regarding the electricity cost, values of 0.02–0.03 € kWh^{-1} could be achieved in a place like Madrid (Spain). These costs would be even lower than those of classical sources of energy.

Note added in press

From the time when this chapter was first written, an important result regarding efficiency increase in tandem solar cells has appeared. It is the achievement of a 35.2% of efficiency for a GaInP/GaAs/Ge, two terminals, tandem solar cell [45].

References

[1] Yamaguchi M and Luque A 1999 High efficiency and high concentration in photovoltaics *IEEE Trans. Electron. Devices* **46** 2139–44
[2] Hovel H J 1975 *Solar Cells Vol 11 Semiconductor and Semimetals* (New York: Academic) p 175
[3] Davis R and Knight J R 1975 *Solar Energy* **17** 145
[4] Stryi-Hipp G, Schoenecker A, Schitterer K, Bücher K and Heider K 1993 Precision spectral response and I–V characterisation of concentrator cells *Proc. 23rd IEEE PVSC* (New York: IEEE) pp 303–8
[5] De Vos A 1992 *Endoreversible Thermodynamics of Solar Energy Conversion* pp 76–7 (Oxford: Oxford University Press)
[6] Algora C, Rey-Stolle I, Ortiz E, Peña R, Díaz V, Khvostikov V, Andreev V, Smekens G and de Villers T 2002 26% Efficient 1000 Sun GaAs solar cells for cost competitive terrestrial PV electricity *Proc. 16th European Photov. Solar Energy Conference (Glasgow)* (London: James & James) pp 22–7

[7] Rey-Stolle I 2001 Desarrollo de células solares de GaAs para concentraciones luminosas muy elevadas *PhD Dissertation* Universidad Politécnica de Madrid

[8] Araújo G L 1990 *Physical Limitations to Photovoltaic Energy Conversion* ed A Luque and G L Araújo (Bristol: Adam Hilger)

[9] Handy R J 1967 Theoretical analysis of the series resistance of a solar cell *Solid-State Electron.* **10** 765–75

[10] de Vos A 1984 The distributed series resistance problem in solar cells *Solar Cells* **12** 311–27

[11] Andreev V M, Romero R and Sulima O V 1984 An efficient circular contact grid for concentrator solar cells *Solar Cells* **11** 197–210

[12] Araújo G L, Cuevas A and Ruiz J M 1986 The effect of distributed series resistance on the dark and illuminated current-voltage characteristics of solar cells *IEEE Trans. Electron. Devices* **33** 391–401

[13] Rumyantsev V D and Rodríguez J A 1990 Method of calculating the distributed and lumped components of the resistance in solar cells *Solar Cells* **28** 241–52

[14] Algora C and Díaz V 2000 Influence of series resistance on guidelines of manufacture of concentrator p-on-n GaAs solar cells *Prog. Photovolt. Res. Appl.* **8** 211–25

[15] Algora C and Díaz V 2001 Manufacturing tolerances of terrestrial concentrator p-on-n GaAs solar cells *Prog. Photovolt. Res. Appl.* **9** 27–39

[16] Rey-Stolle I and Algora C 2002 Modeling of the resistive losses due to the bus-bar and external connections in III–V high-concentration solar cells *IEEE Trans. Electron. Devices* **49** 1709–14

[17] Weldford W T and Winston R 1989 *High Collection Nonimaging Optics* (San Diego, CA: Academic)

[18] Algora C and Díaz V 1999 Modelling of GaAs solar cells under wide angle cone of homogeneous light *Prog. Photovolt. Res. Appl* **7** 379–86

[19] Díaz V, Algora C and Rey-Stolle I 1998 Performance and optimization of very high concentrator GaAs solar cells under wide angle isotropic incident light *Proc. 2nd World Conf. of Photovoltaic Solar Energy Conversion (Vienna)* (Ispra: European Commission) pp 8–13

[20] Algora C, Díaz V and Rey-Stolle I 2002 Concentrator III–V solar cells: the influence of the wide-angle cone of light *Proc. 29th IEEE PV Spec. Conf. (New Orleans, LA)* (New York: IEEE) pp 848–51

[21] Alvarez J L, Hernández M, Benítez P and Miñano J C 2001 TIR-R concentrator: a new compact high-gain SMS design *Nonimaging Optics: Maximum Efficiency Light Transfer VI, Proc.* (SPIE) pp 32–42

[22] Algora C *et al* 2000 Ultra compact high flux GaAs cell photovoltaic concentrator *Proc. 16th European Photov. Solar Energy Conf. (Glasgow)* (London: James & James) pp 2241–4

[23] Rey-Stolle I and Algora C 2001 Reliability and degradation of high concentrator GaAs solar cells *Proc. 17th European Photovoltaic Solar Energy Conf. (Munich)* (Munich and Florence: WIP and ETA) pp 2223–6

[24] Rey-Stolle I and Algora C 2003 High irradiance degradation tests on concentrator GaAs solar cells *Prog. Photovolt.* **11** 249–54

[25] Luque A and Martí A 1997 Increasing the efficiency of ideal solar cells by photon induced transitions at intermediate levels *Phys. Rev. Lett.* **78** 5014–17

[26] Fraas L M, Avery J E, Sundaram V S, Kinh V T, Davenport T M, Yerkes J W, Gee J M and Emery K A 1990 Over 35% efficient GaAs/GaSb stacked concentrator cell

assemblies for terrestrial applications *Proc. of the 21st IEEE Photovoltaic Spec. Conf. (Kissimmee, FL)* (New York: IEEE) pp 190–5

[27] Bett A, Baur C, Bekkert R, Dimroth F, Letay G, Hein M, Meusel M, van Riesen S, Schubert U, Siefer G and Sulima O 2001 Development of high-efficiency mechanically stacked GaInP/GaInAs-GaSb triple-junction concentrator solar cells *Proc. 17th European Solar En. Conf. (Munich)* (Munich and Florence: WIP and ETA) pp 84–7

[28] Cotal H, Lillington D, Elmer J, King R, Karam N, Kurtz S, Friedman D, Olson J, Ward J, Duda D, Emery K and Moriarty T 2000 Triple junction solar cell efficiencies above 32%: the promise and challenges of their application in high concentration-ratio PV systems *Proc. 28th IEEE PV Spec. Conf. (Anchorage, AK)* (New York: IEEE) pp 955–60

[29] Brown A S and Green M A 2002 Limiting efficiency for current-constrained two-terminal tandem cell stacks *Prog. Photovolt. Res. Appl.* **10** 299–307

[30] Tobías I and Luque A 2002 Ideal efficiency of monolithic, series-connected multi-junction solar cells *Prog. Photovolt. Res. Appl.* **10** 323–9

[31] Algora C and Díaz V 1997 Guidance for reducing PV costs using very high concentrator GaAs solar cells *Proc. 14th European Photov. Solar Energy Conf. (Barcelona)* (Bedford: H S Stephens & Associates) pp 1724–7

[32] de Meo E A, Goodman F R, Peterson T M and Schaefer J C 1990 *Proc. 21st IEEE PVSC* (New York: IEEE) p 16

[33] Photovoltaics in 2010, European Commission, Directorate-General for Energy, Luxembourg 1996

[34] Vernon S, Tobin S, Haven V, Geoffroy L and Sanfacon M 1991 High efficiency concentrator cells from GaAs on Si *Proc. 22nd IEEE PV Spec. Conf. (Las Vegas, NM)* (New York: IEEE) pp 353–7

[35] McMillan H F, Hamaker H C, Kaminar N R, Kuryla M S, Ladle Ristow M, Liu D D, Virshup G F and Gee J M 1998 28% efficient GaAs concentrator solar cells *Proc. 20th IEEE PV Spec. Conf. (Las Vegas, NM)* (New York: IEEE) pp 462–8

[36] Maroto J C, Martí A, Algora C and Araujo G L 1995 1300 suns GaAs concentrator solar cell with efficiency over 23% *Proc. 13th European PVSEC* (Bedford: H S Stephens & Associates) pp 343–8

[37] Algora C, Ortiz E, Rey-Stolle I, Díaz V, Peña R, Andreev V, Khvostikov V and Rumyantsev V 2001 A GaAs solar cell with an efficiency of 26.2% at 1000 suns and 25.0% at 2000 suns *IEEE Trans. Electron. Devices* **48** 840–4

[38] Bett A, Dimroth F, Lange G, Meusel M, Beckert R, Hein M, Riesen S V and Schubert U 2000 30% monolithic tandem concentrator solar cells for concentrations exceeding 1000 suns *Proc. 28th IEEE PVSC* (New York: IEEE) pp 961–4

[39] Ortiz E and Algora C A 2003 high-efficiency LPE GaAs solar cell at concentrations ranging from 2000 to 4000 suns *Prog. Photovolt.* **11** 155–63

[40] Wallace W 1995 Government acceleration terrestrial programs *Solar Cells and their Applications* ed L Partain (New York: Wiley) p 495

[41] Algora C Universidad Politécnica de Madrid High efficiency photovoltaic converter for high light intensities manufactured with optoelectronic technology, Patent no. PCT/ES0100167 and US-2002-0170592-A1

[42] Miñano J C Personal communication

[43] Sala G Personal communication

[44] Algora C, Rey-Stolle I and Ortiz E 2001 Towards 30% efficient single junction GaAs

solar cells *Proc. 7th European Photovoltaic Solar Energy Conf. (Munich)* (Munich and Florence: WIP and ETA) pp 88–91

[45] King R R *et al* 2003 Lattice matched and methamorphic GaInP/GaInAs/Ge concentrator solar cells *Proc. 3rd WCPEC (Osaka)* at press

Chapter 7

Intermediate-band solar cells

A Martí, L Cuadra and A Luque
Instituto de Energía Solar, Universidad Politécnica de Madrid
ETSI Telecomunicación, Ciudad Universitaria s/n, 28040
Madrid, Spain

7.1 Introduction

An *intermediate-band solar cell* (IBSC) is a photovoltaic device conceived to exceed the limiting efficiency of single-gap solar cells thanks to the exploitation of the electrical and optical properties of *intermediate-band* (IB) *materials*. This type of material takes its name from the existence of an extra electronic band located in between what in ordinary semiconductors constitutes its bandgap, E_G. The IB divides the bandgap E_G into two forbidden energy intervals (sub-bandgaps), E_L and E_H as drawn in figure 7.1. Although in the figure, the lowest of the intervals involved, E_L, has been drawn above the IB, to have located it below the IB would have not make any conceptual difference. For reasons that will become clearer shortly, it will also be required for this IB to be half-filled with electrons. This is the reason why, in the diagram in figure 7.1, we have drawn the Fermi level, corresponding to the structure in equilibrium, crossing the IB. Due to this feature, we also refer sometimes to this IB as possessing a metallic character.

When light in this IB material is absorbed, it has the potential to cause electronic transitions from the valence band (VB) to the IB, from the IB to the conduction band (CB) and, as in conventional semiconductors, also from the VB to the CB. These transitions are labelled as generation processes g_{IV}, g_{CI} and g_{CV}, respectively, in the figure. The inverse or recombination processes are also labelled r_{IV}, r_{CI} and r_{CV}, respectively.

The absorption of photons takes the IB material out of equilibrium and causes the electron occupation probability in each of the bands to be described by Fermi–Dirac statistics with its own quasi-Fermi level. The temperature that corresponds to this statistic will be the lattice temperature, T_C, and the quasi-

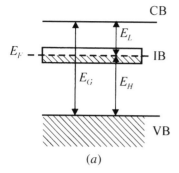
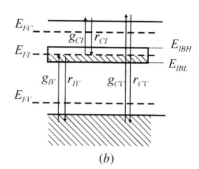

(a)   *(b)*

Figure 7.1. Representation of the simplified band structure of an intermediate-band material *(a)* in equilibrium, with the IB half-filled with electrons and *(b)* out of equilibrium. In the latter case, the different generation and recombination processes involved between bands are also shown as well as the quasi-Fermi levels corresponding to each of the bands. The hatched regions indicate energy states predominantly filled with electrons.

Fermi level will be E_{FC}, E_{FV} and E_{FI} depending on the band we are referring to (CB, VB or IB). We admit three different quasi-Fermi levels exist, one for each of the bands, because we assume that the carrier recombination lifetimes between bands are much longer than the carrier relaxation times within the bands. In particular, for proper IBSC operation, the quasi-Fermi level related to the IB, E_{FI}, will be considered clamped at its equilibrium position. The conditions for this hypothesis to hold are related to the excitation level and the density of states in the IB. Hence, a low excitation and a high density of IB states are parameters assisting this goal. Some figures for this will be given later when the implementation of the IBSC by means of quantum dots is discussed.

When compared with conventional single-gap solar cells, an IBSC bases its superior potential, from one side, on its capability for absorbing and converting into electricity below-bandgap photons and, from the other, on its potential for doing so without voltage loss. Both factors, current and voltage, are important. If the reader is not sufficiently convinced of this, s/he should just consider how easy is to 'invent' a cell with the potential for a superb output current: it is sufficient to manufacture the cell with a low-bandgap semiconductor.

The model describing the basic operation of the IBSC will be described in the next two sections, following the essentials of our work in [1,2]. Sections 7.4 and 7.5 will describe a possible implementation of the IBSC on the basis of quantum dot technology.

Wolf [3] first pointed out the possibility of using the energy levels within the forbidden gap to achieve efficiencies higher than that of single pn-junction devices (multi-transition solar cells). He was also concerned that the recombination properties of these centres could well deteriorate the efficiency of the cell rather than improve it. As it will be reviewed later, our thesis is that these intermediate levels should be grouped together, leading to an electronic (intermediate) band, rather than to a mere collection of energy levels. This might possibly require that the spatial periodicity of the electron potential be preserved whatever the means are used to induce the IB formation within a semiconductor bandgap. The electron properties in this IB would then be similar to those in the more conventional semiconductor conduction and valence bands. An electron in a *band* is characterized because its wavefunction is *delocalized* whereas when it belongs to one of the *levels* of what we have called *collection* of energy levels, its wavefunction is *localized*. Actually, these energy levels are often referred to as localized states.

Hence, the band-to-band recombination processes involving the IB could be ruled by large lifetimes, even radiatively limited just as is the case, for example, in good quality GaAs. The electrons in each band would also have their own quasi-Fermi level in order to describe the electron occupation probability as previously mentioned. In this respect, Kettemann and Guillemoles [4] have also approached the study of the performance of solar cells using several quasi-Fermi levels.

Finally, Trupke, Green and Würfel have proposed the utilization of IB materials as up- or down-converters [4, 5] (see also chapter 3 in this book). This is a clever idea since, as they claim, it is compatible with the use of existing conventional solar cells (with, in principle, some minor modifications) and their effect on the efficiency of the cell is null in the worst cases or only beneficial.

7.2 Preliminary concepts and definitions

The IBSC is manufactured by sandwiching the IB material between two ordinary single-gap semiconductors, called emitters, of n- and p-type. The simplified band diagram of this structure, both in equilibrium and when biased, is plotted in figure 7.2.

Ideally, the gap of the semiconductor that constitutes the emitters should be higher than the total gap of the IB material to provide a good contact selectivity for electrons and holes. This selectivity means that the p-emitter should only allow holes to go through and the n-emitter, only electrons. The emitters will be assumed not to play any optical role (for example, to absorb light) in this idealized model but their existence will be essential to enable a high output voltage to be obtained from the cell. In this respect, the electron quasi-Fermi level in the emitter, E_{Fn}, fixes the electron's quasi-Fermi level in the CB of the IB material, E_{FC}, and the hole's quasi-Fermi level in the p^+-emitter, E_{Fp}, fixes the hole quasi-Fermi level at the VB of the IB material E_{FV}. The cell output voltage is determined by the split between these two levels:

$$eV = E_{Fn} - E_{Fp} = E_{FC} - E_{FV}. \qquad (7.1)$$

The emitters also serve the purpose of isolating the IB from the external contacts so that no current is extracted from the IB. This isolation is what allows

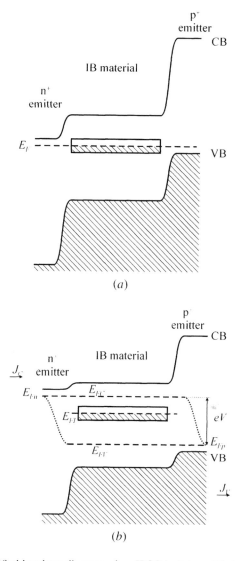

Figure 7.2. Simplified bandgap diagram of an IBSC in (*a*) equilibrium and (*b*) forward biased under illumination (from [2]. © John Wiley & Sons Ltd. Reproduced with permission).

the effective splitting of the quasi-Fermi levels between the CB and the IB and between the VB and the IB and, therefore, according to equation (7.1), the possibility of extracting a high output voltage from the cell.

The quasi-Fermi levels will assumed to be flat along the IB material. This is the situation that causes a null contribution from the carrier transport to the irreversible entropy creation rate [6, 7] and, therefore, is one of the necessary assumptions in order to approach the limiting efficiency of the device. It is interesting in this respect to recall [8] that the carrier current density, J, is given by $J \propto \mu \nabla E_{FX}$, where E_{FX} is the quasi-Fermi level related to the concentration of the carrier being considered. Then, in the CB and VB, the quasi-Fermi levels are assumed to be flat because the mobility is supposed to be infinite. This can also be the case in the IB although, if the optical excitation is homogeneous, this requirement is not necessary because the quasi-Fermi level becomes flat due to the absence of current in this band.

Different absorption coefficients associated with each of the absorption processes that can take place in the IB material can be defined. These coefficients are labelled α_{IV}, α_{CI} and α_{CV} so as to describe the strength of the absorption of photons that cause electron transitions from the VB to the IB, from the IB to the CB and from the CB to the VB, respectively.

The absorption of a photon that, for example, causes a transition from the VB to the CB is symbolically represented by a chemical reaction such as

$$h\nu_{CV} \rightarrow e_C + h_V. \tag{7.2}$$

Aiming for the maximum efficiency, this process has to be reversible, i.e.

$$e_C + h_V \rightarrow h\nu_{CV} \tag{7.3}$$

which requires, from one side, that the recombination of an electron from the CB with a hole in the VB to be exclusively radiative in nature (emitting one photon) and, from the other, that the absorption coefficients do not overlap in energy. Figures 7.3 and 7.4 illustrate these two reversibility related concepts better. It must be mentioned, nevertheless, that a more rigorous thermodynamic treatment of the performance of the IBSC and, in particular, its accomplishment with the second law of thermodynamics can be found in [9]. There, it was also shown that the true zero entropy production rate (full reversibility) can only be achieved in open-circuit conditions (a pity, because no electrical power is extracted from the cell under this mode of operation).

That the absorption coefficients do not overlap in energy means that for a given photon energy, E, only one of the three absorption coefficients is zero or, in other words, that only one of the transitions illustrated in figure 7.1 is possible (either from the VB to the IB or from the IB to the CB or from the VB to the CB). According to the diagram in figure 7.1, the energies E_L, E_H and E_G are the natural energies to define the intervals where the corresponding absorption coefficient is ideally different from zero. Hence, we shall assume that

$$\alpha_{CI} \neq 0 \qquad \text{if } E_L < E < E_H \tag{7.4}$$

$$\alpha_{IV} \neq 0 \qquad \text{if } E_H < E < E_G \tag{7.5}$$

$$\alpha_{CV} \neq 0 \qquad \text{if } E_G < E. \tag{7.6}$$

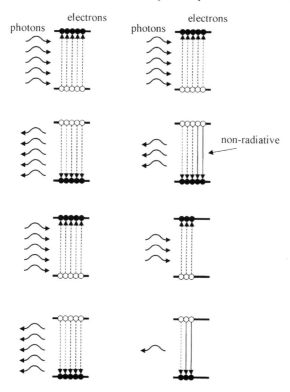

Figure 7.3. Qualitative illustration of the reversibility concept associated with the existence (or not) of non-radiative recombination processes. Only two energy levels are involved in the description. In the left-hand column, non-radiative recombination processes are assumed not to exist while in the right-hand column, they are assumed to be present. The figure illustrates the mental experiment in which some photons are absorbed, re-emitted as a product of radiative recombination and re-absorbed again in a to-and-fro process. After some iterations, the phenomenon is extinguished in the right-hand column, because some recombination processes do not produce a photon, while this does not happen in the left-hand column. Note also that the process in the left-hand column does not distinguish the arrow of time while the process on the right-hand does. In this case, the arrow of time is the same as the direction in which the emission of photons is extinguished.

Now it is also understood why we initially fixed the Fermi level within the IB. It is both to provide enough empty states in the IB to accommodate the electrons pumped from the VB and also to supply enough electrons to be pumped to the CB. In non-equilibrium conditions, this Fermi level turns into three quasi-Fermi levels for describing the concentration of electrons in the three bands. Moreover, the quasi-Fermi level in the IB is assumed to be clamped at its equilibrium position when the cell becomes excited. This is in order to prevent the IB from becoming

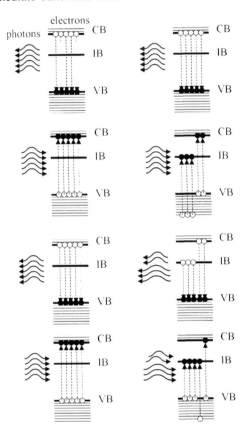

Figure 7.4. Illustration of the reversibility concept associated (or not) with the overlap in absorption coefficients. In the left-hand column, the absorption coefficients are assumed not to overlap while in the right-hand column an overlap is assumed. Assuming again a mental experiment of several to-and-fro absorption and emission processes, the picture in the right-hand column does not repeat itself because, due to the absorption coefficient overlap, some of the emitted photons are reabsorbed in a band other than the one in which they were generated. In the illustrated case, photons emitted from a recombination process between the CB and the VB are absorbed through a transition between the VB and the IB. In addition, because the energy of the photons emitted as a consequence of a recombination from the IB to the VB is lower than the energy of the photons emitted as a recombination from the CB to the VB, they cannot be reabsorbed as a transition between the VB and the CB. In this case, the arrow of time would be determined by the direction in which the number of up-transitions between the VB and the CB decreases.

fully filled or emptied of electrons, which would change the absorption properties of the material as a function of the biasing. In principle, this is not necessarily bad within certain limits, but allowing the IB quasi-Fermi level to move up and down complicates the analysis and obscures the essentials of the description.

Similarly, for the other transitions, we state that

$$h\nu_{IV} \leftrightarrow e_I + h_V \qquad (7.7)$$

which means that the absorption of a photon with energy $E_H < E < E_G$ generates an electron in the IB and a hole in the VB, the reverse process also being exclusively radiative, and

$$h\nu_{CI} \leftrightarrow e_C + h_I \qquad (7.8)$$

which represents that a photon with energy $E_L < E < E_H$ generates an electron in the CB and a hole in the IB; and that when these electrons and holes recombine, they do so by emitting one photon.

Equalling the chemical potentials of the elements involved in equation (7.2) [together with (7.3)] and (7.7) and (7.8), we state that

$$\mu_{CV} = E_{FC} - E_{FV} \qquad (7.9)$$
$$\mu_{IV} = E_{FI} - E_{FV} \qquad (7.10)$$
$$\mu_{CI} = E_{FC} - E_{FI} \qquad (7.11)$$

where μ_{XY} is the chemical potential of the photons that are in equilibrium with the creation and annihilation of electron-hole pairs between band X and Y. The spectral density (number of photons per unit of energy and solid angle fluxing across a unit surface whose normal is oriented with the same angles that define the differential solid angle) of these photons is given by $b(\varepsilon, \mu_{XY}, T_C)$ where

$$b(\varepsilon, \mu, T) = \frac{2}{h^3 c^2} \frac{\varepsilon^2}{\exp\left(\frac{\varepsilon - \mu}{kT}\right) - 1}. \qquad (7.12)$$

In equation (7.12), ε is the photon energy and T the temperature. The parameters h, c and k have their usual meaning: Planck's constant, speed of light in vacuum and Boltzmann constant respectively. In equations (7.9)–(7.11), it has also been taken into account that the electrochemical potential of the holes in a given band reverses the sign of the electrons in the same band.

There is a limitation to the quasi-Fermi level split that can be tolerated and this is related to the IB width. If the upper and lower energy levels of the IB are designated by E_{IBH} and E_{IBL}, then we must have $\mu_{CI} < E_C - E_{IBH}$ and $\mu_{IV} < E_{IBL} - E_V$, where E_C and E_V are the energy limits of the CB and VB respectively, otherwise stimulated emission would take place [10]. In fact, this is the same principle by which, in single-gap solar cells, the bandgap limits the open-circuit voltage. For maximum sunlight concentration (46 050 suns), it has been estimated that this bandwidth limit is in the range of 100 meV while, for 1 sun operation, this limit enlarges up to 700 meV. Nevertheless, in order to prevent this conflict in the model, the IB width will be assumed to approach zero as much as necessary. Conversely, there is a potential limitation for the narrowness of the IB bandwidth. In this respect, it is known that as a band narrows, the carrier mobility tends to decrease. This is because the carrier effective mass is related to the inverse of the curvature of the band at its extrema. This curvature, or second derivative, is close to zero as the band narrows so that the carrier effective mass tends to infinity and, therefore, the carrier mobility approaches zero. However, because no current is extracted from the IB, no carrier transport in the IB is required either as long as we keep in the model idealizations that assume that the generation and recombination processes are not dependent on position all over the IB material [11].

With respect to equation (7.12), it will be useful to define the following auxiliary functions:

$$N_S(\varepsilon_L, \varepsilon_H) = \pi \int_{\varepsilon_L}^{\varepsilon_H} b(\varepsilon, 0, T_S) \, d\varepsilon \qquad (7.13)$$

$$N(\varepsilon_L, \varepsilon_H, \mu_{XY}, T) = \pi \int_{\varepsilon_L}^{\varepsilon_H} b(\varepsilon, \mu_{XY}, T) \, d\varepsilon. \qquad (7.14)$$

The purpose of equation (7.13) is to enable the number of photons that a solar cell absorbs, when the photon spectral density from the sun is assumed to be that of a black body at temperature T_S and the maximum concentration is used, to be counted more easily later on. Similarly, equation (7.14) is intended to be a tool to describe the number of photons emitted from the cell.

7.3   Intermediate-band solar cell: model

Armed with these hypotheses, we shall now work towards the determination of the current-voltage characteristic of the IBSC. In this respect, we shall use detailed balance arguments like Shockley and Queisser when they originally computed the limiting efficiency of single-gap solar cells [12].

The electron current density, J_C, extracted from the n$^+$ contact can be obtained from using detailed balance arguments with respect to the conduction band. Hence, this current is stated to be given as the difference between the number of photons from the sun that are absorbed per unit of area causing transitions ending in the CB (J_{LC}/e), and those photons that are emitted outside of the cell ($J_{0C}(V)/e$) and originated in recombination mechanisms that started in the CB. For this purpose, the sun will be taken as a black body at $T_S = 6000$ K, the cell at a temperature $T_C = 300$ K and maximum light concentration will be assumed. Therefore, in mathematical terms, we write

$$J_C = J_{LC} - J_{0C}(V) \qquad (7.15)$$

where

$$J_{LC} = e N_S(E_G, \infty) + e N_S(E_L, E_H) \qquad (7.16)$$

$$J_{0C}(V) = eN(E_G, \infty, eV, T_C) + eN(E_L, E_H, \mu_{CI}, T_C). \qquad (7.17)$$

A similar equation holds for holes in the VB. In this case, the current extracted through the p contact, J_V, is the difference between the photons absorbed, J_{LV}, and emitted, $J_{0V}(V)$, among those that are related with generation and recombination process involving holes in the VB. That is,

$$J_V = J_{LV} - J_{0V}(V) \qquad (7.18)$$

where

$$J_{LV} = eN_S(E_G, \infty) + eN_S(E_H, E_G) \qquad (7.19)$$
$$J_{0V}(V) = eN(E_G, \infty, eV, T_C) + eN(E_H, E_G, eV - \mu_{CI}, T_C). \quad (7.20)$$

Note also that equation (7.1) has been used in both equations (7.17) and (7.20). It is also worthwhile mentioning that from the way in which the emission of photons has been formulated, it already counts for photon recycling effects [13].

Since no current is extracted from the IB,

$$J_V = J_C \qquad (7.21)$$

which enables μ_{CI} to be obtained as a function of the voltage V by inserting its value back into (7.15) or into (7.18), the current–voltage characteristic of the solar cell. The efficiency of the IBSC, as a function of the gaps E_L and E_G (or E_H), is then trivially obtained by maximizing the output power from the cell, $P = J(V)V$, and dividing the result by the input power, $P_{in} = \sigma T_S^4$. In a further optimization process, E_G can be optimized as a function of E_L. The results have been plotted in figure 7.5. The maximum efficiency is 63.2% and is obtained for $E_L = 0.71$ eV and $E_H = 1.24$ eV (i.e. $E_G = 1.95$ eV). This limiting efficiency is superior to that of single-gap solar cells (40.7%) and that of a tandem of two solar cells connected in series (55.4%) as also shown in figure 7.5.

Green [14] has extended the study to the case in which an infinite number of IBs would exist reaching the conclusion that its limiting efficiency would, in this case, be the same as that of a stack with an infinite number of single-gap solar cells. Würfel has pointed out that there is no significant advantage in going from three bands (CB, IB and VB) to more bands given the difficulty in having them all half-filled with electrons [15] in order to provide the sufficient strength to all the absorption coefficients involved. Bremner *et al* have realized that, conversely to single-gap solar cells, Auger recombination might be beneficial for some non-ideal devices [16].

The fact that the absorption coefficients have been assumed not to overlap requires admittting that some tailoring of the selection rules is possible. Instead of this specific tailoring, Brown *et al* [17] assumed that the absorption coefficients could be prevented from overlapping by considering the bands have finite widths. The result, with these assumptions, is again that there is no significant advantage in going from one IB to more.

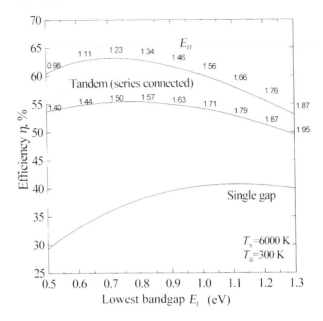

Figure 7.5. Maximum efficiency of the IBSC as a function of the lowest of the bandgaps E_L and for the optimum bandgap E_H. The limiting efficiency of single-gap solar cells and of a tandem of two solar cells connected in series is also shown for comparative purposes. In the tandem case, the figures on the curve label the optimum value of the highest of the bandgaps involved (from [1]. © 1997 APS).

Nevertheless, although the absorption coefficients have been assumed not to overlap in the previous discussion, even if they do, that does not necessarily mean the total destruction of the potential efficiency of the device. Actually, it has been shown [18] that if they overlap so that

$$\alpha_{CI} \neq 0 \qquad \text{if } E > E_L \tag{7.22}$$

$$\alpha_{IV} \neq 0 \qquad \text{if } E > E_H \tag{7.23}$$

$$\alpha_{CV} \neq 0 \qquad \text{if } E > E_G \tag{7.24}$$

the limiting efficiency of 63.2% is recovered as far as $\alpha_{CI} \ll \alpha_{IV} \ll \alpha_{CV}$, i.e. one of the absorption coefficients becomes dominant over the rest and light-trapping techniques are used.

7.4   The quantum-dot intermediate-band solar cell

We have proposed the use of quantum dots (QD) as one of the possible means for implementing the IBSC [19, 20]. We refer to this particular implementation of the IBSC as the quantum-dot intermediate-band solar cell (QD-IBSC). The IB would, in this case, arise from the energy of the confined electrons in the dots.

The way the problem of finding the energy spectrum of an electron in a quantum dot is usually addressed and simplified is by means of the use of the single-band effective mass equation. Focusing on the conduction band, under this approximation, the time-independent wavefunction of an electron in this band, ψ_0, is approximately given by

$$\psi_0 \approx u_C(\mathbf{r})\psi(\mathbf{r}) \tag{7.25}$$

where u_C is a periodic function, with the periodicity of the lattice of the host material, and ψ, the so called envelope wavefunction, is obtained from solving the effective mass equation,

$$-\frac{\hbar^2}{2m_C^*}\nabla^2\psi + E_C(\mathbf{r})\psi = \varepsilon\psi \tag{7.26}$$

where m_C^* is the effective mass of electrons in the conduction band, $\hbar$ is the reduced Planck constant and $E_C(\mathbf{r})$ is the three-dimensional confinement potential well induced for electrons as a consequence of the conduction band-offset between the dot and barrier materials (figure 7.6). The function $u_C(\mathbf{r})$ is also assumed to be independent of the wavevector $\mathbf{k}$ of the reciprocal crystal lattice. The theory underlying the effective mass equation as well as, for example, the problems of using it for the valence band, particularly when multiple bands exist, are well discussed in books like the one by S Data [21]. The short opening discussion here has been mainly made with the purpose of introducing the notation that will be used later.

Let us consider first the electron energy spectrum of an isolated QD as illustrated in figure 7.6. From one side, we have discrete energy levels appearing with energies in-between the 3D potential well minimum, E_{C0}, and $E_{C+} = E_{C-}$. In our description, we also shall assume that very few levels of this kind appear per QD, perhaps only one. This can be achieved by controlling the size of the dot. Above these discrete levels, we have a continuum of energy levels that we shall refer to as the CB continuum. It is not often recognized that these spectral properties of the electron effective mass Hamiltonian are just a mathematical consequence derived from the fact that the potential well, $E_C(\mathbf{r})$, has a minimum at E_{C0} and $E_C(\mathbf{r})$ following E_{C+} and E_{C-} (constants) when $r \rightarrow \pm\infty$ [22]. This spectral result is, therefore, valid whatever the shape of the potential well (spherical, pyramidal, etc).

Similarly, from the hole's effective mass equation, given the shape of the potential created from the valence band offset, it is concluded that a discrete energy spectrum exists for energies ε such that $E_{V+} = E_{V-} > \varepsilon > E_{V0}$ and a continuum for energies below $E_{V+} = E_{V-}$. The diagram in figure 7.6 also corresponds to the so called single type I superlattice [23]. Approaches to the QD-IBSC based on type II superlattices have also been discussed [24].

However, because the hole's effective mass is much higher than that of the electron, there are much more allowed discrete energy levels for holes in the VB than there were for electrons in the CB [25]. Since, as will be seen, dots will be ideally grouped into a closely spaced array, we also expect the levels originating from the hole confinement to merge with the levels in the VB continuum and, together, lead to a new equivalent VB as represented in figure 7.7.

Some things can also be added to describe the electron wavefunction in these levels in terms of electron wavefunction *localization*. Generally speaking, we say that the electron wavefunction in the discrete spectrum is *localized* while the electron wavefunction in the continuum spectrum is *delocalized*. *Localized* means that the modulus of the wavefunction is mainly different from zero only in some *localized* region of the space, which in our case is the dot region. In terms of probability, it is equivalently stated that the probability of finding the electron is concentrated in a given region of the space. 'Delocalized' means the opposite, i.e. that the wavefunction cannot be considered particularly concentrated in some region of the space. Furthermore, in this last case, the wavefunction is usually extended all over the space. It must be mentioned, however, that while for the treatment of a specific physical problem, a given wavefunction can be considered as localized, for the treatment of another different problem, the same wavefunction might be better described as delocalized. They are, therefore, statements that are better assured in relative terms and, in this sense, we say that the electron wavefunction in a discrete level is localized when compared to the electron wavefunction in the CB continuum. However, even in this case, this affirmation is not completely true for states in the continuum characterized by an energy close to the potential well edge, $E_{C+} = E_{C-}$. Conversely, electrons with an energy well above the conduction band edge are delocalized and, in fact, their envelope wavefunction approaches that of a free electron.

From the perspective of manufacturing an IBSC, we are interested in the possibility of absorbing a photon by promoting an electron from one state to another. Symmetry considerations and selection rules [26] will often be able to tell us, for a given geometry, which of these transitions are forbidden in a rigorous framework. However, in a simplistic approach prior to any symmetry consideration, we can expect, as a necessary (although not sufficient) condition, that in order for the probability of such transition not to be very small, there has to be a significant overlap in the wavefunctions of the electrons for the states involved. This is because the probability of this transition is proportional, according to the dipole approximation [27], to the matrix element M_{ij} given by

$$M_{ij} = \langle \psi_i | - e\mathbf{r} \cdot \mathbf{E} | \psi_j \rangle \qquad (7.27)$$

where ψ_i and ψ_j are the wavefunctions of the electrons in the states involved, $\mathbf{r}$ is the position and $\mathbf{E}$ is the electric field related to the radiation and which has the direction of its polarization. Hence, trivially, this matrix element approaches zero when the electron wavefunctions are localized in different regions of the space while we cannot say anything about its value when they overlap (unless we can apply the selection rules previously mentioned). In fact, often, ψ_i and ψ_j have a very significant overlap and the matrix element M_{ij} is strictly zero because ψ_i and $r E \psi_j$ are orthogonal. That is why we said that the condition was necessary but not sufficient.

Therefore, in the case of the isolated QD, we can expect the optical transition from CB discrete levels (those corresponding to confined electrons) to high-energy states in the CB continuum to be very weak because they would take place between delocalized (continuum) and localized (discrete) states. In contrast, photon absorption via a transition between discrete energy levels in the CB and VB can be significant because, selection rules apart, although they involve two localized states, the overlap between the wavefunctions takes place in the same region of the space, the dot region. This is also the reason why the recombination between a confined electron and a confined hole within a QD can be radiative in nature, thus making the existence of QD lasers possible.

As has just been described, there is some risk that the photon absorption involving states in the CB continuum and discrete energy levels of being too weak. It is clear that the absorption of such photons is essential for the success of the theory because they provide the second step in the collection of sub-bandgap photons in the two-step process that is characteristic of IBSC theory. Furthermore, due to the van Roosbroeck–Shockley relation [28], to speak about a weak absorption coefficient is synonymous with saying that the probability of radiative recombination is small. As mentioned in the previous section, the fact that this type of transition becomes predominately radiative is important since it allows, as is also the case for single-gap solar cells, the high efficiency limits predicted by the theory to be approached.

It is in order to solve this potential problem, particularly for the case when absorption between low-energy localized states in the CB continuum and intermediate states is not sufficient, that we suggest that QDs should be periodically distributed in space, leading to the formation of a proper band rather than to a multiplicity of discrete levels. By doing this, the electron wavefunctions in the intermediate energy states become delocalized so that the electrons in both the intermediate states and the CB continuum are now delocalized. It is then that we also say that the CB discrete levels have turned into a proper IB. In fact, the envelope wavefunction of the electrons in this IB can be approached, according to the tight-binding approximation [29], by the Bloch sums:

$$\psi_k = \frac{1}{N^{1/2}} \sum_i \psi_i (r - R_i) \exp(i k \cdot R_i) \qquad (7.28)$$

where k is the wavevector of the reciprocal lattice of quantum dot array, R are vectors labelling the position of the dots in the real space, N is the number of dots in the array and ψ_i are the normalized wavefunctions for the electrons in the isolated dots.

In spite of this, we ignore, however, whether a non-periodic distribution could also lead to some manifestation of the principles proposed for the operation

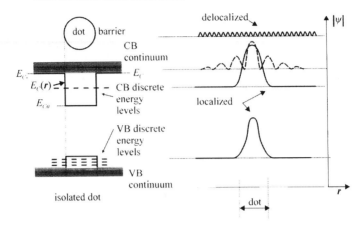

Figure 7.6. Representation of the simplified bandgap diagram of an isolated quantum dot and its energy spectrum. The qualitative shape of the modulus of the electron wavefunction, associated with the energy levels in the quantum dots, is also shown to illustrate the localization and delocalization concepts.

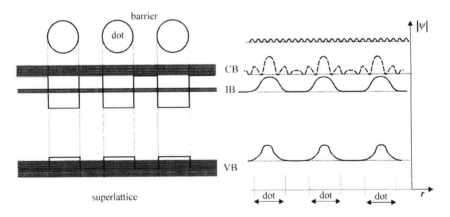

Figure 7.7. Illustration of the formation of an IB through the formation of an array of quantum dots. If the periodicity of the array is not perturbed, confined electrons in the dot become delocalized.

of the IBSC so as to prove that they are physically possible and not merely a purely theoretical conception. This should perhaps be the first experimental issue to aim for and some experiments in this respect are underway [30].

Another point to emphasize is why we prefer to use QDs rather than other low dimensional structures like quantum wells (QWs) or quantum wires that also, apparently, produce electron quantum confinement. The first reason is that only QDs cause confinement in the three spatial dimensions which, in the energy band diagram, implies that a true zero density of states separates the energy of the

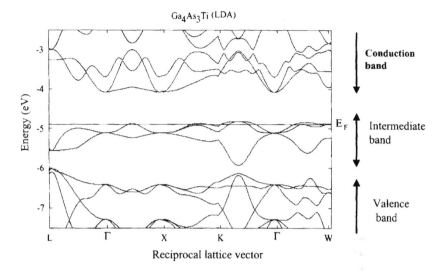

Figure 7.8. Theoretical band diagram of Ga_4As_3Ti compound exhibiting an intermediate band [36]. Note that the Fermi level, E_F crosses the IB conferring on it a metallic character (from [36]).

confined electrons from the energy of the electrons in the conduction and valence bands. The existence of this true zero density of states opens the possibility for the phonon-bottleneck effect [31], thus preventing the electrons in the CB from relaxing to the IB and, therefore, an effective split between the quasi-Fermi level of electrons in the CB and in the IB. A second reason is motivated by the fact that, for example, when using QWs, the QW layers are parallel to the plane of growth. Then, applying symmetry selection rules, since solar cells are illuminated perpendicularly (or almost) to its surface, photon absorption causing a transition from the IB to the CB would be forbidden [32–35] so that $g_{CI} = 0$ (figure 7.1).

QDs have not been the only means that have been proposed to engineer the IB material. Remarkably, Wahnón and Tablero [36–39] have determined that some GaAsTi compounds, if synthesizable, could also lead to the formation of an IB as shown in figure 7.8. In the illustrated case, Ti substitutes for some proper As atoms in the original GaAs lattice in a periodical pattern. In this sense, this approach could be regarded as a limiting case of the QD approach in which the QDs would have reduced their size to that of an atom.

7.5 Considerations for the practical implementation of the QD-IBSC

Several practical considerations concerning the implementation of the IBSC by means of the QD approach deserve to be discussed. Among them, maybe the first one is related to the size of the dots.

The appropriate size for the dots to provide the required quantum confinement depends on their geometry and on the materials being used. Considering the potential use of III–V semiconductors for manufacturing the array, assuming spherical dots and using the effective mass approximation [40], we have estimated that the radius of the dots should be in the range of 39 Å [20, 41]. Suitable materials could be ternary compounds in the InGaAs family for the dot material and in the AlGaAs family for the barrier material. In principle, such dot sizes are achievable by using MBE self-assembly growth techniques such as the Stranski–Krastanow (SK) mode [42]. In addition, one of the remarkable aspects of this technique is that it is able to manufacture almost defect-free QDs avoiding the potential threat to the performance of the cell that the existence of a high surface recombination between the dot and barrier material implies. This is because these dots appear spontaneously by elastically accommodating the strain that arises from the different lattice constants that exist between the dot and barrier materials. For example, the relative lattice mismatch between InAs (dot material) and GaAs (barrier material), one of the better known and more widely used material systems, is about 7%. Hence, the strain is indeed the driving force leading to the formation of high-quality arrays of QDs. In addition, under certain growth conditions [43], it is possible to obtain a very dense array of InAs QDs embedded in GaAs, these dots being laterally ordered in a square lattice with a primitive unit vector in the [100] and [010] directions. As we shall explain later, stacking several layers of such ordered QDs [42], [44] could be a method to manufacture a 3D array of dots leading to the formation of an IB. Note that, nevertheless, the existence of this strain makes it difficult to predict where the energy of the confined electrons in the dots will be located and, therefore, the precise engineering of the QD-IBSC [45]. Another difficulty comes from the fact that while, in the theory, an abrupt transition from the dot material to the barrier material is commonly assumed, in practice, these transition regions tend to be graded [46].

Often inherent to the growth of the QDs by means of self-assembly techniques is the formation of a wetting layer (WL) beneath the dots (figure 7.9). This is a thin layer of dot material that grows prior to the dot formation itself. It is characteristic of the Stranski–Krastanov (SK) mode but its formation is prevented in the Volmer–Weber (VW) growth mode [47]. This second mode, however, requires the existence of a higher lattice mismatch between the barrier and the dot materials than in the SK mode. A suitable material system for the VW growth mode is that constituted by InAs dots embedded in GaP [48]. In principle, the VW growth method would be the one preferred for manufacturing the QD-IBSC because no WL layer appears, unlike dots manufactured using the SK technique. This is because, unless the WL is very thin, the WL would lead to the formation of an undesired QW structure. This QW, as discussed in previous sections, could open a path for the quick relaxation of the electrons from the CB continuum towards the discrete levels in the QW since these levels would not be isolated from the continuum by a zero density of states.

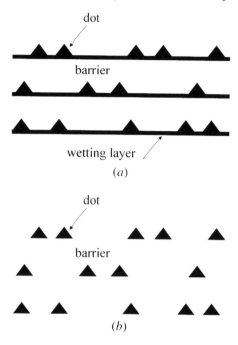

Figure 7.9. (*a*) Illustration of the appearance of a wetting layer in the Stranski–Krastanov growth mode. (*b*) The formation of such a layer is prevented in the Volmer–Weber method which is characterized by the existence of a higher strain between the barrier and the dot materials.

In addition, it is worth mentioning that the shape of the dots grown under these techniques is typically pyramidal [49], although different shapes have been reported [25].

Another topic of interest is the density of the dots (number of dots per unit of volume). In this respect, several aspects point to the fact that this should be as high as possible with some limitations briefly discussed later.

In theory, although the dots are very well spaced, even infinitely spaced, the IB structure would still hold as long as the dots are identical and located in a perfect periodic scheme. However, as the dots become more and more spaced, there is more risk for a minimal perturbation in the periodicity (Anderson and Mott localization effects [50]) causing the electron wavefunctions to transit from being delocalized (as would be the case when they form part of a proper band structure) to being localized. It is in this case that we would again say that the IB has lost its band structure to return to coming a mere collection of energy levels. Therefore, from this perspective, the density of dots should be as high as possible.

As mentioned earlier, in self-assembly growth, the dots are grown by layers. A high density of dots requires that the next layer to be located close to the

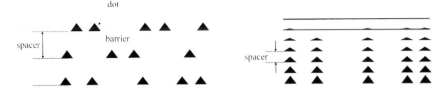

Figure 7.10. Illustration of the trend of the patterns produced in quantum dot formation through self-assembly methods. On the left, the spacer is long and the dots on the new layer have no memory of where the dots in the former layers grew. On the right, the spacer is short and dots in the new layers tend to be formed over preceding layers. However, in this case, as new layers grow, strain is reduced and the proper formation of new dots becomes more difficult.

previous one, with a reduced spacing between dot layers. This has one more potential benefit consisting in the fact that the growth of the dots in the next layer is induced at the location where a dot in the preceding underlying layers has already been formed (figure 7.10). In this way, some periodicity in the growth direction is also induced which can also add to the spatial periodicity in the lateral directions if, for example, substrates are patterned before growth [47]. As discussed earlier, this periodicity is considered to be beneficial for the formation of a proper IB structure. The problem which arises is that as the strain is accommodated from one layer to the next, there is a total structure thickness from which no further QD growth seems to be possible. Another problem, also related to the reduction in the strain, is that the size of the dots slightly increases and the density of dots reduces [42] from one layer to another so that we cannot be completely rid of the perturbation in the periodicity of the potential. Nevertheless it is expected that as the dot density is sufficiently high, the band structure will be prevented from disappearing.

Another benefit derived from a high density of dots is simply related to the increment in the strength of absorption of light by the dots. If the density of dots is very low, light absorption by the dots will be weak and, in order to compensate for it, the use of a high volume of QD material or, alternatively, the use of light-trapping techniques, is necessary.

A third benefit derived from a high dot density is related to the clamping of the IB quasi-Fermi level at its equilibrium position. In order for this clamping to occur, the density of states in the IB has to be sufficiently high. The specific number depends on the injection level at which the cell is planned to be operated. When using QDs, the density of states is given by twice the density of dots (if, as mentioned, dots with one confined level are manufactured), the factor two arising from taking the electron spin degeneracy into account. In this respect, it has been determined that a density of dots in the order of 10^{17} cm^{-3} would be necessary to clamp the quasi-Fermi level within kT for cell operation in the range of 1000 suns [51].

A fourth advantage comes from the fact that as the dots become more closely spaced, the IB broadens. This fact tends to increase carrier mobility in the IB and, although this mobility is not required in an idealized structure in which the IB would be uniformly excited by the generation–recombination mechanisms, it could be required, in practice, to facilitate carrier redistribution through the IB from the places where they are more intensely generated to the regions where they are more intensely recombined [51].

There is, however, a limit to this IB broadening, the most obvious of which being to prevent the IB from mixing with the CB or VB. Another limit comes from the prevention of the appearance of stimulated emission effects as mentioned when the general theory of the IBSC was described in previous sections. In spite of this, it seems that the centres of the dots could be safely located at a distance of 100 Å [52,53] without any of these problems appearing.

But, still, the picture of the QD-IBSC is not complete. As discussed in the first section, the theory of the IBSC required the IB to be half-filled with electrons. In our picture of the QD-IBSC, we have described the creation of the IB but this band would still be 'empty' of electrons. To half-fill this band, it is necessary to introduce some n-type doping. The donors, when ionized, would then supply the necessary electrons to half-fill the IB. The n-doping concentration should approach the volumetric dot density and, ideally, should also be periodically distributed. Nevertheless, it must be said that our understanding of the role and potential of this doping has evolved with time. In this sense, in [54] we thought that the idea would not work but in [55] we realized that it could lead to the required effect if the impurity was located in the barrier region, in essence, as a consequence of modulating doping effects. At present, we think that it could even also work with the impurity located within the dot. In former works we thought that an impurity located within the dot would not half-fill the IB because it would be difficult for it to become ionized and, therefore, supply the required electron. This was because we also thought that the ionization energy of the impurity in relation to the conduction band minima of the bulk dot material was the same regardless of whether the bulk material was forming a dot or not. We have realized that this does not necessarily have to hold as seems to be the case in QW structures [56].

With all these considerations, figure 7.11(a) represents an artistic view of the structure of a QD-IBSC.

If we now draw a simplified bandgap diagram of this structure (figure 7.12), we shall realize that due to band bending near the junctions, dots located close to the n-emitter interface are empty while those located close to the p-emitter interface are full of electrons. According to the previous discussion, the IB will not work properly in these regions. Some QD layers would be wasted in providing the necessary electric charge to sustain the electric field at the junctions but not in playing the IB role expected from them according to IBSC theory [57]. We have estimated that the number of layers that could be wasted in this way is about three per interface. This means that if, for example, ten is the number of layers that can

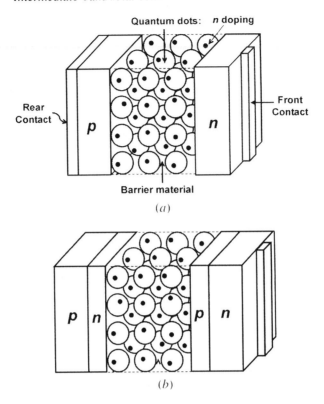

Figure 7.11. (*a*) Artistic view of the structure of a QD-IBSC. (*b*) An evolved structure to prevent dots close to the n and p interface from becoming completely full or emptied with electrons respectively.

be successfully grown using self-assembly techniques, only four serve their true purpose. This suggests that, instead of wasting such QD layers, a conventional p-layer could be inserted at the interface with the n-emitter to provide the required negative charge density and, similarly, an n-type layer at the interface with the p-emitter to provide the necessary positive charge density (figure 7.11(*b*)). In this way, a higher number of QD layers could be saved for their devised purpose.

Hence, with all these considerations, the QD approach reveals itself as a challenging workhorse for testing the operating principles of the IBSC.

7.6   Summary

The IBSC is a type of solar cell that has been conceived to exceed the limiting efficiency of single-gap solar cells thanks to the two-step absorption of sub-bandgap photons via an IB and also thanks to the subsequent extraction of such

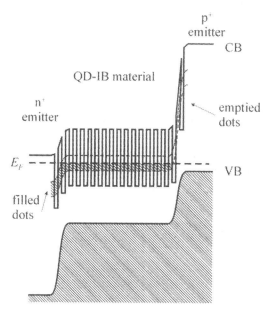

Figure 7.12. Simplified band diagram of a QD-IBSC with n-doping introduced to half-fill the IB. Dots near the n$^+$-emitter would be filled with electrons while those close to the p$^+$-emitter would be emptied due to band bending.

an enhanced photocurrent without voltage degradation. The performance of this solar cell relies on a material exhibiting an IB that ideally should be half-filled with electrons. In the solar cell structure, this IB is isolated from the external contacts by means of two conventional p- and n-emitters. The carrier population in each band is described by its own quasi-Fermi level so that a total of three quasi-Fermi levels are required to describe the operation of this cell. Furthermore, it is the existence of these quasi-Fermi levels which prevents the output voltage from being limited by the lowest of the bandgaps involved in the structure. The limiting efficiency that has been stated for this cell is 63.2%. To approach this limit, the cell should operate in the radiative limit and the quasi-Fermi levels should be flat. In the conduction and valence bands, where transport is required, this condition is approached by demanding that the carrier mobility is infinite. Absorption coefficients related to electron transition between bands should not overlap.

To put this concept into practice, the use of QDs has been described. Under this approach, the IB would arise from the confined electrons in the dots. The introduction of n-doping is suggested as the means of supplying the necessary electrons for half-filling the IB. The use of self-assembled QD technologies has been identified as a suitable one to begin the experimentation. Different trade-offs, particularly concerning electron wavefunction localization and dot distribution, have also been discussed.

Acknowledgments

This work has been supported by the European Commission under Contract ENK6 CT200 00310 and by the Spanish Plan Nacional de I+D TIC2000-1399-C02-01. L. Cuadra is indebted to the Comunidad de Madrid for their financial support.

References

[1] Luque A and Martí A 1997 Increasing the efficiency of ideal solar cells by photon induced transitions at intermediate levels *Phys. Rev. Lett.* **78** 5014–17
[2] Luque A and Martí A 2001 A metallic intermediate band high efficiency solar cell *Prog. Photovolt. Res. Appl.* **9** 73–86
[3] Wolf M 1960 Limitations and possibilities for improvement of photovoltaic energy converters. Part I: considerations for Earth's surface operation *Proc. IRE* vol 48, pp 1246–960
[4] Trupke T, Green M A and Wurfel P 2002 Improving solar cell efficiencies by up-conversion of sub-band-gap light *J. Appl. Phys.* **92** 4117–22
[5] Trupke T, Green M A and Wurfel P 2002 Improving solar cell efficiencies by down-conversion of high-energy photons *J. Appl. Phys.* **92** 1668–74
[6] Luque A and Martí A 2003 Theoretical limits of photovoltaic energy conversion *Handbook of Photovoltaic Science and Engineering* ed A Luque and S Hegedus (Chichester: Wiley) ch 4
[7] Luque A, Marti A and Cuadra L 2002 Thermodynamics of solar energy conversion in novel structures *Physica* E **14** 107–14
[8] Sze S M 1981 *Physics of Semiconductor Device* 2nd edn (New York: Wiley) ch 2
[9] Luque A, Martí A and Cuadra L 2001 Thermodynamic consistency of sub-bandgap absorbing solar cell proposals *IEEE Trans. Electron. Devices* **48** 2118–24
[10] Cuadra L, Martí A and Luque A 2000 Modeling of the absorption coefficient of the intermediate band solar cell *Proc. 16th European Photovoltaic Solar Energy Conf.* (London: James & James) pp 15–21
[11] Martí A, Cuadra L and Luque A 2002 Quasi drift-diffusion model for the quantum dot intermediate band solar cell *IEEE Trans. Electron. Devices* **49** 1632–9
[12] Shockley W and Queisser 1961 Detailed balance limit of efficiency of p–n junction solar cells *J. Appl. Phys.* **32** 510–19
[13] Martí A, Balenzategui J L and Reyna R M F 1977 Photon recycling and Shockley's diode equation *J. Appl. Phys.* **82** 4067–75
[14] Green M A 2001 Multiple band and impurity photovoltaic solar cells: general theory and comparison to tandem cells *Prog. Photovolt. Res. Appl.* **9** 137–44
[15] Wüerfel P 2001 Three and four bands solar cells at the International Workshop on Nanostructures in Photovoltaics, Dresden. (There is no printed version of this talk to our knowledge)
[16] Bremner S P, Honsberg C B and Corkish R 2000 Non-ideal recombination and transiport mechanisms in multiple band gap solar cells *Proc. 28th IEEE PVSC* (New York: IEEE) pp 1206–9
[17] Brown A S, Green M A and Corkish R P 2002 Limiting efficiency for a multi-band solar cell containing three and four bands *Physica* E **14** 121–5

[18] Luque A, Martí A and Cuadra L 2000 High efficiency solar cell with metallic intermediate band *Proc. 16th European Photovoltaic Solar Energy Conf.* (London: James & James) pp 59–61

[19] Martí A, Cuadra L and Luque A 1999 Quantum Dot Super Solar Cell, Actas de la Conferencia sobre Dispositivos Electrónicos, Madrid, ISBN: 84:00-07819-5, pp 363–6

[20] Martí A, Cuadra L and Luque A 2000 Quantum dot intermediate band solar cell *Proc. 28th Photovoltaics Specialist Conf.* (New York: IEEE) pp 940–3

[21] Datta S 1984 Quantum phenomena *Modular Series on Solid State Devices* vol VIII (Reading, MA: Addison-Wesley)

[22] Galindo A and Pascual P 1978 *Mecanica Cuantica* ch 4, Ed. Alhambra S.A.

[23] Weisbuch C and Vinter B 1991 *Quantum Semiconductor Structures: Fundamentals and Applications* (San Diego, CA: Academic) ch 1

[24] Cuadra L, Martí A and Luque A 2000 Type II broken band heterostructure quantum dot to obtain a material for the intermediate band solar cell *Physica* E **14** 162–5

[25] Bimberg D, Grudmann M and Ledentsov N N 1999 *Quantum Dot Heterostructures* ch 1 (Chichester: Wiley)

[26] Landau L D and Lifshitz E M 2000 *Quantum Mechanics* 3rd edn (Oxford: Butterworth–Heinemann) (original in English from 1958) ch XII

[27] Ballentine L E 1998 *Quantum Mechanics, A Modern Development* (Singapore: World Scientific) ch 12

[28] Pankove J I 1971 *Optical Processes in Semiconductors* (New York: Dover) ch 6

[29] Ridley B K 1999 *Quantum Processes in Semiconductors* 4th edn (Oxford: Clarendon) chs 1 and 2

[30] Luque A *et al* 2002 Progress towards the practical implementation of the intermediate band solar cell *29th IEEE PVSC (New Orleans, LA)* (New York: IEEE) pp 1190–3

[31] Mukai K and Sugawara M 1999 The phonon bottleneck effect in quantum dots *Self-Assembled InGaAs/GaAs Quantum Dots* ed M Sugawara (San Diego, CA: Academic) ch 5

[32] Harrison P 2000 *Quantum Wells, Wires And Dots* (New York: Wiley) pp 289–95

[33] Shik A 1997 *Quantum Wells, Physics And Electronics Of Two-Dimensional Systems* (Singapore: World Scientific) pp 35–6

[34] Bimberg D, Grundmann M and Ledenstov N N 1999 *Quantum Dot Heterostructures* (Chichester: Wiley) pp 154–5

[35] Loehr J P and Manasreh M O 1993 *Semiconductor Quantum Wells And Superlattices For Long-Wavelength Infrared Detectors* ed M O Manasreh (Boston, MA: Artech House) ch 2

[36] Tablero C, Wahnón P, Cuadra L, Martí A, Fernández J and Luque A 2002 Efficiencies of half-filled intermediate band solar cell designed by first principles calculations *Proc. 17th European Photovoltaic Solar Energy Conf.* (London: James & James) pp 296–9

[37] Wahnón P and Tablero C 2002 Ab-initio electronic structure calculations for metallic intermediate band formation in photovoltaic materials *Phys. Rev.* B **65** 1–10

[38] Tablero C and Wahnon P 2003 Ab-initio analysis of electronic density for metallic intermediate band formation in photovoltaic materials *Appl. Phys. Lett.* **82** 151–3

[39] Fernández J J, Tablero C and Wahnón C 2003 Development of the exact exchange scheme using a basis set framework *Int. J. Quantum Chem.* **91** 157–64

[40] Datta S 1989 *Quantum Phenomena* (Reading, MA: Addison Wesley)

[41] Cuadra L, Martí A and Luque A 2000 Tecnología de puntos cuánticos para la obtención de un semiconductor de banda intermedia *Proc. XV Simposium Nacional de la Unión Científica Internacional de Radio—URSI* pp 555–6

[42] Sugawara M 1999 Self-assembled InGaAs/GaAs quantum dots *Semiconductors and Semimetals* **60** (London: Academic)

[43] Grundmann M *et al* 1995 *Phys. Status Solidi* **188** 249

[44] Ledentsov N N *et al* 1996 Direct formation of vertically coupled quantum dots in Stranski–Krastanow growth *Phys. Rev.* B **54** 8743–50

[45] Cuadra L, Martí A, Luque A, Stanley C R and McKee A 2001 Strain considerations for the design of the quantum dot intermediate band solar cell in the $In_xGa_{1-x}As/Al_yGa_{1-y}As$ material system *Proc. 17th European PVSC* (London: James & James) pp 98–101

[46] Stanley C Private communication

[47] Merz J L, Barabási A L, Furdyna J K, WIlliams R S 1999 Nanostructure self-assembly as an emerging technology *Future Trends In Microelectronics, The Roas Ahead* ed S Luryi, J Xu and A Zaslavski (New York: Wiley)

[48] Leon R, Lobo C, Chin T P, Woodall J M, Fafard S, Ruvimov S, Liliental-Weber Z and Stevens Kalceff M A 1998 Self-forming InAs/GaP quantum dots by direct island growth *Appl. Phys. Lett.* **72** 1356–8

[49] Bimberg D *et al* 1996 InAs–GaAs quantum pyramid lasers: *in situ* growth, radiative lifetimes and polarization properties *Japan. J. Appl. Phys.* **35** 1311–19

[50] Ridley B K 1999 *Quantum Processes in Semiconductors* 4th edn (Oxford: Clarendon) ch 2

[51] Martí A, Cuadra L and Luque A 2002 Quasi drift-diffusion model for the quantum dot intermediate band solar cell *IEEE Trans. Electron. Devices* **49** 1632–9

[52] Cuadra L, Martí A and Luque A 2001 Modelling of the intermediate band width in the quantum dot intermediate band solar cell *Proc. Conferencia de Dispositivos Electrónicos* pp 193–6 Dep. Legal GR-133/2001, Impresión Plácido Cuadros. Gonzalo Callas 13, Granada

[53] Martí A, Cuadra L and Luque A 2002 Design constrains of the quantum dot intermediate band solar cell *Physica* E **14** 150–7

[54] Martí A, Cuadra L and Luque A 2000 Quantum dot intermediate band solar cell *Proc. 28th Photovoltaics Specialist Conf.* (New York: IEEE) pp 940–3

[55] Martí A, Cuadra L and Luque A 2001 Partial filling of a quantum dot intermediate band for solar cells *IEEE Trans. Electron. Devices* **48** 2394–9

[56] Harrison P 2000 *Quantum Wells, Wires And Dots* (Chichester: Wiley) ch 2

[57] Martí A, Cuadra L and Luque A 2001 Analysis of the space charge region of the quantum dot intermediate band solar cell photovoltaics for the 21st century *Proc. 199th Electrochemical Society Meeting* (Pennington: The Electrochemical Society) pp 46–60

Chapter 8

Multi-interface novel devices: model with a continuous substructure

Z T Kuznicki
Laboratoire de Physique et d'Application des Semi-conducteurs
PHASE, CNRS-STIC UPR 292, 23 rue du Loess, F-67037
Strasbourg cedex 2, France

8.1 Introduction

There is little hope of finding a breakthrough in today's photovoltaic (PV) conversion efficiency of inorganic materials if one stays with the well-understood macroscopic bulk approach. Such solar cell operation is based on the one-stage fundamental process of optical electron–hole generation. In Si-based solar cells, which dominate PV applications, two energy- and volume-dependent maladjustments degrade this process:

- the first concerns the incident sunlight (containing multi-energy photons) and electron–hole extraction (at a fixed energy)—this effect appears in energy space; and
- the second consists of the spatial distribution of the photon absorption inside the conversion volume but a fixed extraction site (the metal-semiconductor contact)—this effect appears in geometrical space.

Because of these two effects, there are losses in the carrier thermalization and collection. Potential ways to enhance the conversion efficiency consist of

- reducing the PV conversion volume (geometrical space) by requiring large absorbance; and/or
- adapting the conversion events, for example, by a multi-level multi-stage process (energy space).

In a conventional solar cell, there are two metastable light-generated carrier populations that are collected at the average carrier energy. The quasi-Fermi levels

165

refer to the metastable carrier populations if their concentrations have a Fermi–Dirac distribution.

To improve the conversion efficiency, new mechanisms making use of more than the two metastable populations are needed, as suggested in [1, 2]. These carrier populations must not reach equilibrium with each other or with the lattice before they are extracted to a region of the device where the average energy of the carrier population is close to equilibrium.

In general, practical realizations will represent alternatives to or supplement a single pn-junction device. Probably the best solution will be a superposition of conventional PV conversion with suitable additional mechanisms.

There are several effects which have demonstrated their usefulness as they introduce multiple intermediate energy levels into a semiconductor or improve the solar cell *collection efficiency* (CE). Their specificity lies in their nanometric scale where it is possible to conceive specific structures and tunnel-like or electric-field-enhanced conduction. In this way, new mechanisms when more than two metastable carrier populations are present in the same device seem to be possible. If the losses inherent in the previously mentioned energy- and volume-dependent effects could be avoided totally, a solar cell conversion limit of 86% could be reached. This value corresponds to the theoretical performance under monochromatic light.

The multi-interface device concept [3] is based on the idea that energy interactions and geometrical space could be better adjusted in a locally modified conventional PV material or from foreign insertions into it. In the latter case, the conventional material plays the role of a conducting matrix. Local transformations depend on nearby active interfaces. Such a concept leads to nanoscale substructures (active interfaces with their active transition zones) whose activity is strongly dependent on their position within the initial conventional device.

In a solar cell with several carrier subpopulations, the collection and contacting mechanisms are altered compared to a conventional single pn-junction. In the device in which the active zones (absorber, converter) are imbedded in a conducting volume (conductor), the carrier extraction can take a multistage course of action where each successive stage represents a transition from a subpopulation with a higher specific temperature to that with a lower specific temperature. This leads to a multilevel multistage process that continues up to a stable average energy for the carrier that is nearly equal to the equilibrium Fermi level. When there are only two device contacts at the lowest quasi-Fermi level separation, high-energy carriers can generate multiple low-energy carriers. Original approaches concerned Auger or impact ionization solar cells [2, 4–6].

A novel contacting mechanism that appears to be particularly well-adapted for this purpose has been recently proposed in the form of an energy-selective contact taking the form of a so-called Würfel membrane [7]. A single device can use one or any combination of these contacting schemes.

Low dimensionality active zones are able to fit the characteristic lengths of PV conversion quantum effects. The energy space can be better adapted to a new multistage conversion mechanism by additional energy levels, which can be introduced within the forbidden bandgap of a transformed semiconductor. Certain analogies to these exist in the floral or microbial worlds [8].

Because of the practical importance of conventional Si photovoltaics, we concentrate here on a single Si material modification, namely the insertion of a nanoscale Si-layered system. There are methods of material engineering which seem to be able to give the desired results. One of these methods consists of ion beam implantation with subsequent treatment.

First, we recall recent nanoscale improvements in optoelectronic Si behaviour. Next, we describe a model multi-interface cell that allows a clear distinction between the electronic and optical features. Then we will describe the self-consistent calculation of the CE in the presence of a *carrier collection limit* (CCL). A continuous nanoscale layered system buried within the emitter is at the origin of a CCL [9], because the CCL delimits an electronically inactive surface layer, totally blocking the minority holes in their movement towards the collecting pn-junction. No blockade exists for the majority electrons.

Finally we describe two new mechanisms appearing in such a device, namely enhanced absorption and photogeneration with more than one electron per photon.

8.2 Novelties in Si optoelectronics and photovoltaics

Si optoelectronics evolves rapidly and our perception of Si has changed over the last ten years because the properties of Si are sensitive to its structure at the nanometric scale. This opens new perspectives on Si and its interaction with visible light thanks to Si radiative states and the formation of radiative states in Si nanocrystals at their oxide interface (and oxygen-rich environment) [10].

Certain existing analogies with the multi-interface nanoscale Si transformations can be cited:

— superlattices—where two known materials give a third one with a totally new behaviour [11],
— gettering effects—where a damaged material can be improved,
— scale effects—where space confinement or specific new structures lead to a new behaviour as, for example, porous materials [12–18] and
— spatial localization of conversion activity; examples of existing devices are dye cells [19] (which can serve as an interesting model for original approaches on Si).

Still others can be added: (1) the Ross and Nozik multilevel approach [20], (2) the energy selective membrane of Würfel [7] and (3) certain post-implantation transformations (epitaxy, defects, etc.).

Modifications offer a new approach for improving the performance of a Si device:

- from the material point of view, a differentiation among device components according to their main functions of *absorber, converter, transporter/conductor*;
- from the optical point of view, change in the absorption, transparency, reflectivity, and optical effective lengths; and
- from the electronic point of view, type of electronic conductivity (uni-, bipolar transport, drift/diffusion behaviour, injection effects, etc).

8.2.1   Enhanced absorbance

In practice, an *absorbing substructure* can be extremely thin if its absorbance (product of absorption coefficient times substructure thickness, $\alpha \times d_\delta$) is large enough. Indeed, Si absorbance can be increased by *damaged or amorphous* Si insertions (d-Si or a-Si) in the crystalline (c-Si) bulk. But the exploitation of systems containing these insertions can be useful only when good CE can be kept. To retain this, the conservation of excess carrier lifetime and of effective free carrier path length are necessary. The device volume is then spliced on a Si-based absorber and on the surrounding Si conductor [3]. In the limiting case, the substructure takes the form of a thin absorber layer (with an injection mechanism) and the rest of the device assumes the photocarrier separation, electronic transport and carrier collection tasks.

One of the main results of our work on multi-interface devices, described later, is that a nanolayer system with relatively large absorbance has been inserted within a Si device emitter, at the same time conserving good electronic quality.

A contradiction concerning the characteristic path lengths, that have to be as short as possible from the electronic point of view and as long as possible from the optical point of view, can be overcome using light trapping for the optical performance.

8.2.2   Enhanced conversion

For improving the yield of Si solar cells, an enhancement of the absorbance is necessary but not sufficient. To be more efficient it has to be combined with a new multistage conversion mechanism. The previously mentioned multilayer emitter should be completed by converter components/dots/layers (close to or unified with the absorber layer) able to convert the energetic photons more efficiently than usual. Hot carriers could generate additional electron–hole pairs, for example, instead of being thermalized. Thus, an increased absorption can result in a spatially confined conversion in the Si device. Because of additional generation and good transport yield, this type of improvement appears mainly in the increased short-circuit current. Certain effects of this type have been observed experimentally and will be described at the end of this chapter.

8.3 Active substructure and active interfaces

The concept of an active substructure (nanoscale absorber and converter layer) with its own electric field at its limiting edges gives more freedom for the geometrical design and device processing than one based on the pn-junction space-charge. Certain proposed modifications use a pn-junction electric field to exploit the extrinsic levels introduced by ion beam implantation but both the built-in electric field and the generation intensity are too small to show any measurable improvement [21]. A theoretical approach based on a δ-doping type complex interface [22] indicated certain ways to improve the optical and electronic performances of the Si single-crystal cell simultaneously. We can enumerate some advantageous features:

- There are strong enough built-in electric fields close to the numerous recombination centres present in the active substructure. There is no design limitation from substructure damage density. Because these fields screen the substructure from the surrounding bulk, the equilibrium and excess minority carriers generated outside the substructure cannot reach the recombination centres. The result is that their effective lifetime is little affected by the presence of numerous damages.
- This screening effect can be reinforced or even largely surpassed by a bandgap offset from the a-Si/c-Si hetero-interface. In the n-type emitters investigated later, a particularly interesting asymmetric offset is possible. When the substructure has a greater bandgap than the surrounding material, the valence band takes the whole bandgap offset (perfectly blocking the minority hole transport). Simultaneously the conduction band is flat throughout the substructure (providing good majority electron conduction). Thus for photocarriers generated below the continuous n-type substructure (visible, near and medium IR light), the electronic transport might be totally preserved.
- The excess minority carriers generated within the active substructure are drawn out by an edge's electric field before their extremely fast recombination, so the effective minority carrier lifetime of the damaged substructure is extended.
- In an extreme case, electronic transport across the damaged substructure seems to be possible for short wavelength generation in front of the substructure when the substructure is thin enough for majority carrier conduction (a substitution effect, when the lifetime of minority carriers is longer than the transit time of majority carriers through the substructure).
- There is a possibility of increased light trapping because of additional interfaces.

The practical realization of an active substructure within a Si material can be obtained by the superposition of two differently structured layers, amorphized and crystalline, forming a three-layer sandwich. Amorphized layers can be buried within crystalline Si by ion beam implantation and subsequent thermal treatment.

8.4   Active substructure by ion implantation

Many aspects of implantation processing and its material modifications are known [23]. The first application of implantation processing to solar cell fabrication appeared at the beginning of the 1980s [24, 25]. In that case, the texture-etched front surface was ion-implanted to form the emitter. The best PV performance obtained for a conventional cell was about 18% [25].

In contrast to this, in the case of a multi-interface novel device (MIND) [3], the implantation is primarily used to modify the Si material. New mechanisms of subgap optical absorption and carrier generation can appear. To be useful the processing requires knowledge and control of implantation-induced damages. The factors include:

- interface morphology and transition zones,
- structural relaxation and built-in strain control in material,
- defects introduced by the implantation process and their annihilation or control and
- useful and useless activity of defects and their amplification or quenching.

For dose rates applied in a so-called self-annealing implantation, both kinetic and thermal effects are important. They change drastically the pre-existing state. The crystalline/amorphized phase transition can occur when the free energy of the damaged region reaches that of the amorphous state [26, 27]. In the region extending from the surface down to $0.8R_p$ (where R_p is the implantation range) a vacancy-rich zone is found, while between R_p and $2R_p$ an interstitial-rich zone is formed.

An amorphization takes place when the number of displaced atoms per unit volume reaches the atomic concentration, i.e. all atoms are displaced. On the atomic level, the as-implanted state is the result of a competition between the instantaneous amorphizing effect of the beam and its self-annealing behaviour, which can be of thermal or non-thermal origin. If the crystallinity of the surface layer is to be preserved, low dose rates are recommended.

If more defect reactions are possible either simultaneously or consecutive to the implantation, there are ways to enhance or suppress some of them by choosing a proper heat cycle. This can be approximated in defect engineering by the usually used Boltzmann factor

$$F_i \approx A_i \exp(-\Delta E_i / kT) \qquad (8.1)$$

where F_i is the ith defect reaction depending on the thermal treatment sequence (temperature, thermal budget, etc), A_i (proportional intensity) and ΔE_i (characteristic energy) of ith defect can be determined among reactions [23].

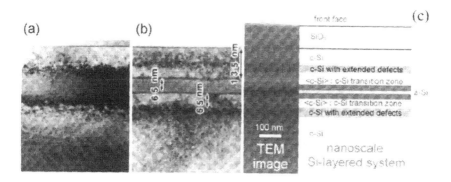

Figure 8.1. Transformation of a-Si/c-Si hetero-interfaces realized by controlled recrystallization of the amorphized phase. X-TEM images of a buried substructure formed by P$^+$-ion implantation: (*a*) as-implanted, (*b*) after thermal treatment [27] and (*c*) nanoscale Si-layered system of a test device (LP-06-C9) formed by ion-implantation and annealing within a Si single-crystal [9]. Each recrystallized layer around the a-Si is divided into two parts: one with and one without local strain (the strained stratum is denoted ⟨c-Si⟩) [3].

Populations of states enter into the kinetics, which can be in equilibrium or have a time dependence.

In an as-implanted sample, the a-Si/c-Si interfaces are rough and composed of a mismatched superposition of single-crystal and amorphized cluster-like grains containing hundreds of atoms with relatively wide transition zones between the two different Si phases (amorphous and crystalline), see figures 8.1(*a*) and 8.2(*a*).

During the thermal treatment, a planar nanostructure is formed by solid-state epitaxy (figures 8.1(*b*) and (*c*)). Two effects are observed: *recrystallization* of the a-Si phase (thickness <100 nm) and *dilution of the crystalline inclusions* inside the amorphized substructure [27].

The layer thickness is reduced and there is a total homogenization of the amorphized material with interface flattening. Optimized transition zones between the two materials are only a few atomic layers thick. The edges of a planar nanostructure realized with 180 keV P-ion implantation followed by a 500 °C thermal treatment are shown in figure 8.2(*b*).

To be 'active' the interfaces have to play a role in the modification of the optoelectronic behaviour. This can be obtained by the introduction or modulation of the local mechanical strain modifying transition zone material (lattice constant, defect states), the introduction of the band model discontinuity and/or the presence of a built-in electric field. All these effects are present in our MIND model structures and are described in more detail in the following paragraphs.

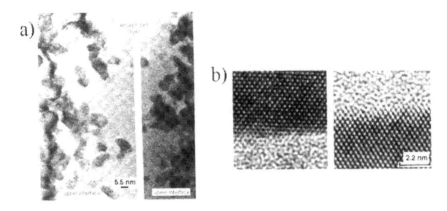

Figure 8.2. (*a*) Two as-implanted transition zones, upper (left image) and lower (right image) visualized on a scale adapted to the size of crystalline inclusions within a-Si [9]. (*b*) Annealed transition zones of two a-Si/c-Si hetero-interfaces, upper (left image) and lower (right image). The crystalline order/disorder transition is on the atomic scale.

Several potentialities lie in interface transition zones which form an integral part of the active substructures. These particularly interesting and not sufficiently explored mesoscopic transformations have to produce a large enough optoelectronic effect and be combined with useful defects, which should play the role of active device components.

Nanolayered Si systems obtained by implantation have been investigated theoretically and experimentally by *high-energy* (5 MeV) Si-ion implantation processing [28]. High-energy implantation allows better insight then *medium-energy* (and *low-energy*) into the transformed structures because of its near amorphization/recrystalization kinetic equilibrium during processing. In this case a *fine-layered nanostructure* is obtained, which does not need any further thermal treatment to show a similar interface quality as the medium-energy (180 keV) amorphization (figure 8.3), which requires a post-implantation thermal treatment.

An *ab initio* simulation can be made for a well-defined amorphization or recrystallization mechanism, such as high-energy processing. Simulating rapid freezing processes of liquid Si by molecular dynamics and Monte Carlo codes can generate realistic atomic configurations of the c-Si/a-Si hetero-interface [29]. Motooka *et al* have recently shown the relatively good agreement of such a simulation with experimental data [28].

Because the structural quality and the abruptness of the a-Si/c-Si transition zones can be excellent [28, 29], the problem of interface recombination states should be resolved.

We concentrate here only on transformations of the UV and visible light performance following a medium-energy insertion of a suitable nanoscale Si-layered system within single-crystal Si.

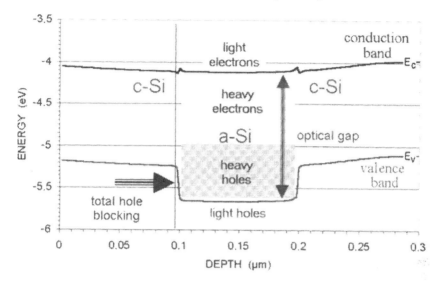

Figure 8.3. Energy diagram of the MIND emitter obtained from a SimWindows simulation taking into account experimental data (optical bandgaps, doping levels, etc.). Because of the offset of the valence band, minority hole penetration from the upper c-Si layer into a-Si layer is excluded (continuous barrier height >0.4 eV!).

8.4.1 Hetero-interface energy band offset

The two a-Si/c-Si hetero-interface transition zones contain built-in electric and strain fields. The acting forces show coulombian behaviour; in the second they have atomic bonding (covalent) behaviour. This suggests that the penetration depths of the strain on the single-crystal sides, where the covalent forces are strong in comparison with those of simple coulombic behaviour, are small. Experimental investigations by x-ray spectroscopy and electron energy loss spectra (EELS) show an 8–10 nm thick strained layer on the c-Si side [30].

Several heterojunctions are now applied, especially in II–VI and GaAs photovoltaics but in rather specific ways. The simplest Anderson model of an abrupt semiconductor heterojunction [31] includes the possibility of *energy discontinuities* in the conduction band and the valence band at the junction interface because of differences in *electron affinity* and *bandgap* between the two semiconductors. The *doping density* fixing the Fermi levels within bandgaps also plays an important role in the electronic behaviour of the junction transition zone.

8.4.1.1 Energy band offset

A mobility bandgap divides the generated carrier populations (electrons and holes) of the amorphous layer into two subpopulations: fast (light) carriers of the conduction and valence bands (optical or mobility bandgap) and slow (heavy) carriers of bandgap states with a hopping conduction. In the short-circuit mode, the bandgap states (e.g. the valence band tail) of the amorphized layer seem to be totally occupied because of a steady-state light generation and the relatively slow hopping transport. Because of the competition of transit time constants, only the light holes have a possibility of leaving the amorphous layer and being collected. Thus, only the optical bandgap of the amorphous material has to be taken into account, as indicated in figure 8.3.

8.4.1.2   *Carrier transmission across an interface*

This depends on an energy-dependent transmission coefficient, the number of occupied states on the emitting side and the number of vacant states on the receiving side [32]. The net carrier flow depends on the difference in the occupancy on either side multiplied by the transmission coefficient. Since the minority carrier density is non-degenerate, the corresponding functions reduce to exponentials of the same arguments that are proportional to minority carrier densities.

A corresponding simulation of the energy diagram of the n-type Si crystalline/amorphous interface shows that an abrupt change in optical bandgap produces a discontinuity in the energy of only the *valence band edge*, presenting a selective barrier to minority-carrier transport. The conduction bands remain practically flat. So, in this particular case, the band model can be adapted for a useful selective conduction of majority–minority carriers. The asymmetric conduction of electrons and holes requires a suitably doped a-Si and c-Si superposition (a light degeneration doping density, see figure 8.10(d)).

The CE of holes generated at the surface zone is reduced by a Boltzmann factor corresponding to the barrier, i.e. $\exp(-E/kT)$ where E is the barrier height, which is greater than 0.4 eV. A simulation by Stangl *et al* [33] leads to similar results.

8.4.2   Built-in electric field

One of the previously mentioned concepts entrusted the role of the *built-in electric field* to carrier collection of the pn-junction [34]. A built-in electric field can be obtained in a semiconductor structure in different ways. To obtain the effect used for PV conversion, Bube [35] enumerated single-junctions, heterojunctions, buried or surface junctions and Schottky barriers. In Si photovoltaics, the single pn-junction is now the most widely used selective membrane for photocarrier separation. Another single-junction (LH, low–high) is also largely used as the so-called back surface field (BSF) and high–low emitter (HLE) protecting minority carriers against surface recombination states.

In the pn-junction, the maximal value of the electric field $E_{\max}$ is limited by the both donor and acceptor doping densities N_D and N_A (because of the electric performance) and the transition zone thickness:

$$E_{max} = \frac{1}{d_T}\Phi_B = \frac{1}{d_T}\frac{kT}{q}\ln\frac{N_A N_D}{n_i^2} \tag{8.2}$$

where Φ_B is the pn potential barrier height, q is the electron charge and n_i is the intrinsic concentration. The transition zone thickness d_T is defined by the doping profiles. From the excess carrier lifetime viewpoint this expression contains a contradiction. The optimal field has to be as high and extended as possible, i.e. largest doping concentrations, but the doping density is limited by the equilibrium minority carrier lifetime and electric field penetration depth. So even in the ideal abrupt pn interface, the electric field cannot be strong enough to draw out photogenerated minority carriers from a damaged or amorphized substructure fast and far enough from recombination centres.

Because of the limited spatial distribution of the pn electric field, the inserted substructure should be as close as possible to the collecting junction interface, although it cannot reduce E_{max}. The thickness of a typical pn space-charge is a fraction of a micrometre. Outside this region the field benefit rapidly becomes negligible.

An eventual superposition of fields from the collecting pn-junction and from the inserted substructure edges is dominated, in general, by the hetero-interface field. Because the continuous substructure always has one edge polarized in the opposite direction to that of the pn-junction, its resultant built-in field can be negligible or even inversed. Some aspects of this problem are treated later, particularly in the discussion of the CCL.

8.4.2.1 *LH-type electric field*

As seen earlier, a new MIND concept treats the interface as an elementary device component. Used here, the notion of an LH-type interface signifies the same n- or p-type doping on both interface sides differentiated from each other only by doping density. The design based on a single collection junction is made possible by new possibilities offered by LH-type hetero-interfaces and their transition zones.

Because a real hetero-interface is a rather complicated structure, a simpler example of an LH-type single-junction can aid in understanding the substructure concept and its electrical behaviour [36]. In this approach, both interfaces of a thin flat substructure are close to each other on the nanometric scale. The doping profile changes abruptly and the impurity density inside the substructure has the largest possible value. This leads to the well-known δ-doping profile [21, 37]. Such an LH single-junction is in close proximity to the damaged region, allowing the photocarriers to be drawn out and injected into the material of a good electronic performance. Simultaneously, free carriers from the substructure outside are screened against recombination centres of the damaged material.

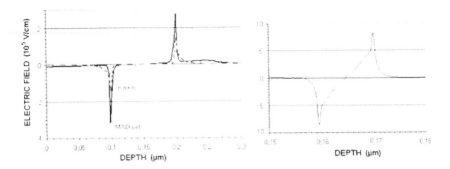

Figure 8.4. Electric field through c-Si/a-Si hetero-interfaces (right) and an n–n$^+$–n homostructure (left) obtained by SimWindows simulation for thick and thin substructures: a maximal value of $\sim 8 \times 10^5$ V cm^{-1} has been obtained for a 10 nm a-Si layer.

The corresponding electric field simulation is shown in figure 8.4 where the results are presented for a relatively thick and a relatively thin amorphization. The latter one simulates the electric field of the test device investigated in this work.

8.4.3   Built-in strain field

A structural relaxation process during the thermal treatment leads to a non-uniform strain distribution in the neighbourhood of a buried amorphized layer. This gives a built-in strain field at the mesoscopic level, which is caused by the density difference between the c-Si and a-Si phases. A local parallel strain appears on both sides of the c-Si/a-Si hetero-interface: compression in the a-Si material and extension in the c-Si material. Structural relaxation occurs perpendicular to the hetero-interface, leading to a surface displacement of tens of Å [3,23]. Except for the substructures with their transition zones, the whole device bulk is free of post-implantation strain.

Bragg diffraction combined with x-ray topography shows a very ordered crystalline transformation near the a-Si/c-Si hetero-interface [38]. As shown in [39] for an as-implanted structure and in [26] for an annealed structure, the stress varies from a smaller value in the as-implanted thicker transition zone to a larger value in the annealed thinner transition zone. On this basis, the local bidimensional stress distribution can be drawn schematically as in figure 8.5. This gives a new PV material in the form of a c-Si compressed layer. This new Si material is called ⟨c-Si⟩ here.

The local strain, with an average value 250–300 MPa, has been measured in our model cells by μ-Raman [40] after their annealing (figure 8.6).

Recently it has been reported [41] that a form of strained Si enhances the electronic behaviour (drive current) of microprocessors, while adding only 2% to the cost of a processed wafer. By changing the Si lattice structure to allow faster

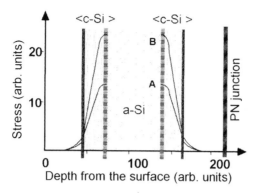

Figure 8.5. The distribution of mechanical strains in an as-implanted (A) and an annealed (B) sample. The distributions are normalized to the substructure thickness [3].

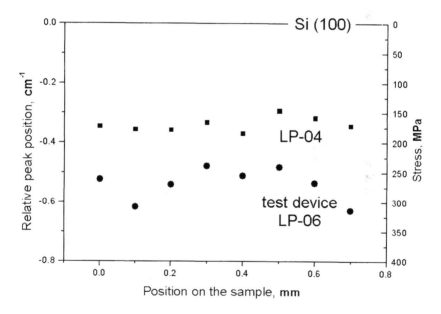

Figure 8.6. Distribution of mechanical strain in experimental samples LP-04 and LP-06 measured by μ-Raman [40].

electron flow, a 1% change in Si atom spacing gives a 10–20% increase in drive current. This can be done with no deterioration in terms of short channel effect or junction leakage. Strained silicon ($\langle$c-Si$\rangle$) takes advantage of increased carrier mobility through its Si lattice.

8.4.4   Defects

Differences in structural relaxation modify the post-implantation gettering conditions and activation energies of persistent defects. These modifications have been observed by optical absorption and corresponding photocurrent on amorphized samples which had been treated at temperatures largely above those permitted for free amorphized layers. After processing, one part of the device bulk can be totally free of post-implantation defects and another part uniformly filled with them.

The best method for improving the carrier collection even in the presence of d-Si or a-Si substructures consists in the creation of strong and extended built-in electric fields. These localized fields are close to the active substructures and offset any electronic deteriorations caused by defect activity [42]. In this way the reduced active space allows for crystal damage because of charge carrier injection from the absorber layer into the surrounding material upon excitation by incident photons. This injection process can be considered the heart of a new effect similar to that known for injection solar cells.

8.5   Model of multi-interface solar cells

What is important for Si photovoltaics is that nanoscale modifications, producing an increased absorption and allowing better exploitation solar spectrum, are possible. MIND model cells [3, 42] represent an example of this approach where the material is profoundly modified by the presence of interfaces, which are considered as *elementary device components*. In this case the term 'interface' means homo- and hetero-structures on a nanometric scale with their neighbouring transition zones. The interface is expected to play an active role in the light-to-electricity conversion, which cannot be achieved by a super(juxta)position of single materials in a *multilayer structure*. In such an approach the *material* plays the secondary role of an inactive *matrix* in which nanoscale transformations are operative.

The constant characteristic feature of the devices investigated is their buried amorphized substructure inserted within n-type c-Si. The substructure is delimited at its front and back edges by abrupt a-Si/c-Si hetero-interfaces giving a nanoscale Si-layered system. The multi-interface system consists of more than a simple *superposition of two different Si phases*/layers as exists, for example, in the Heterojunction with Intrinsic Thin-layer (HIT) design [43, 44]. Different Si phases are used instead for their ability to form active interfaces. One of the most important features concerns perfect defect screening. The effect is directly visible on the device conduction as a function of the thermal treatment that completely transforms the electronic transport across a c-Si/a-Si interface. In this way defects can be assimilated with the Si solar cell operation. Other specific and original features are linked with strain and doping profiles.

In any case, the first stage of local modification consists of a post-implantation structural unification, i.e. flattening of interfaces and bidimensional structural homogenization of the transition zones [45]. The whole device processing requires bandgap, defect, strain and nanoscale engineering of active substructures and active interfaces. Figure 8.7 illustrates possible multi-interface designs [3, 42] and schematizes different *active components* as *a function of the investigated property: structural, electronic, and optical.*

All test structures have been produced on single-crystal Si containing the following layers successively down from the surface (see figure 8.1(*c*)): (1) crystalline relaxed, (2) crystalline strained, (3) amorphized, (4) crystalline strained and (5) crystalline relaxed. A more detailed investigation shows that the upper crystalline relaxed layer consists of two parts: one non-recrystallized (top) and another one recrystallized (lower). In the bottom part of the non-recrystallized layer a zone of extended post-implanted defects can be distinguished.

The so-called upper emitter represents a dead zone (figure 8.8) being electronically blocked from the bottom device by the CCL [45, 46], whose position is indicated in the figure. The continuous substructure generates an LH-type built-in electric field at its front and back edges that is combined with a valence band offset. This leads to specific effects including

- the appearance of the CCL, which blocks the surface PV conversion;
- the creation of a minority carrier reservoir, which can improve the surface layer performance when the surface is electronically well-passivated; and
- an increased open-circuit voltage (except for the LH-type barrier voltage drop)[1].

The lower emitter, placed under the CCL, is composed of amorphized Si, crystalline strained and crystalline relaxed layers. A zone of extended post-implanted defects can be also distinguished within the upper part of the crystalline relaxed layer.

If desired, even two or three substructures could be inserted in the same device. Single substructures could be devoted to a given wavelength range of the solar spectrum: UV, visible, IR. The multi-interface design respects the three rules set out in [1]: (1) the existence of multiple metastable carrier populations, (2) good transport properties (except CCL barrier) and (3) spectral (energy) selectivity.

In the test devices investigated in this work, certain of these effects (normally considered separately) are observed simultaneously thanks to UV- and visible-sensitive substructures inserted within the emitter. One of the perceived and earlier reported effects arising in such a system is a better spectral response in the IR. This is possible because long wavelength photons are only weakly absorbed in the surface dead zone so the main absorption appears under the

---

[1] The potential drop across the inserted substructure is 0.25 V (or more) [46]. The measured V_{oc} of the best MIND models is about 0.58 V, so the total barrier is 0.83 V.

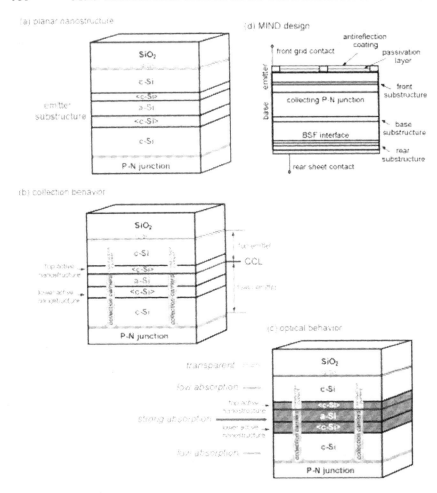

Figure 8.7. Nanoscale Si layered system buried in the emitter (not to scale): (*a*) structure of the system obtained by ion implantation and thermal annealing, (*b*) collection properties of generated photocarriers and (*c*) absorption of the incident light in different strata of the emitter. Only photons penetrating behind the CCL are able to generate collectable carriers. (*d*) Example of a multi-interface solar cell design with several inserted substructures [45].

inserted substructure. Corresponding results have been presented in our earlier papers as [31].

Although up to now the model cells have not broken any efficiency records, when they are optimized they could offer other important lessons for definitive cell structure design. The approach has certain commercial and environmental advantages: it is compatible with a long-lasting, sustainable development. All the materials used are abundantly available, low temperatures ($\leq 500\,^\circ$C) can be applied for all process steps required for device production and the energy demand for its fabrication would be quite low. However, there is still a long way to go.

8.5.1 Collection efficiency and internal quantum efficiency

In the MIND model cell, the electronic decoupling of the surface zone from the rest of the device by a specific CCL plane is near perfect. This allows a separate investigation of the optoelectronic behaviour of the bottom part of the wafer only.

The CE represents the ratio of the number of electrons collected in the whole device to the number of photons absorbed in this same device including its dead layers. It can be expressed in relation to the attenuation length z [47]:

$$CE = \int_0^w g(z, x) f(x) \, dx \tag{8.3}$$

where the integration is made over the total sample thickness w, $f(x)$ is the *collection probability* (a generated carrier is either collected or it recombines) that expresses all losses by recombination and $g(z, x)$ is the *generation rate*.

The internal quantum efficiency (IQE) represents the ratio of the number of electrons collected in a given volume to the number of photons absorbed in the same volume. If the collection probability is 1.0, the IQE represents the generation properties of the material.

8.5.2 Generation rate

The rate of generation by incident light for a homogeneous material is given by

$$g(z, x) = -\frac{d\Phi(z, x)}{dx} = \alpha(z)\Phi_0 e^{-\alpha(z)x} \tag{8.4}$$

where Φ_0 is the flux transmitted by the front face of the device.

It can be different from the absorption rate (for example, because of absorption by free carriers). But in classic PV conversion, the difference is so small that it is negligible.

If there are new generation mechanisms, their effect appears only in the generation rate according to

$$g(z, x) = g_0(z, x) + \Delta g(z, x) \tag{8.5}$$

where $g_0(z, x)$ is the classical generation rate and $\Delta g(z, x)$ comes from the new effects, such as optical confinement or secondary generation.

8.5.3 Carrier collection limit

A part of the light flux penetrating into the wafer is absorbed without participating in PV conversion (in particular the UV and blue spectra) because of a CCL

Table 8.1. Parameters used in a simulation of the CCL effect in figure 8.8.

Parameter	Simulation 1	Simulation 2
upper c-Si thickness	55 nm	70 nm
upper (c-Si) thickness	45 nm	30 nm
a-Si thickness	100 nm	100 nm
CCL depth	150 nm	115 nm
	(half of a-Si is inactivated)	(15% of a-Si is inactivated)

potential barrier. The barrier transparency to carriers is defined by the Boltzmann factor: $\exp(-\Delta E / kT) = 5 \times 10^{-8}$. In the case of cells with a continuous substructure, it is the minority carriers (holes) that flow towards the collecting pn-junction. The extremes of conduction bands from a-Si and c-Si are at the same height, so the value of ΔE is given by the difference in the bandgaps, respectively, between amorphous and crystalline: $\Delta E \geq 1.54 - 1.12 = 0.42$ eV.

Figure 8.8(*a*) shows the effect of the CCL (and corresponding enhanced absorbance) in a good conventional cell performance. The effect is greatly amplified for the shortest wavelengths, which are totally absorbed before attaining the CCL.

8.5.4   Surface reservoir

In the planar geometry used, the energy diagram discontinuity and built-in electric field of the lower amorphous/crystalline hetero-interface is directed in the same way as the collecting pn-junction reverse polarization. This enhances minority carrier transport towards the pn-junction and preserves carriers from substructure recombination states. The extended defects of the lower emitter are saturated at the usual light intensities. A PC-1D simulation (with a minority carrier lifetime ≥ 100 μs) of a multi-interface planar structure with negligible surface recombination (flat energy levels) with and without illumination is shown in figure 8.8(*b*).

The upper amorphous/crystalline interface is directed oppositely to the collecting pn-junction and an accumulation of minority holes appears because of a CCL blocking their natural movement. The minority holes stored in the dead zone reservoir recombine in several ways: (1) at the front face, (2) within the front c-Si layer and (3) at the upper c-Si/a-Si hetero-interface. Two different majority populations can be distinguished in the recombination: one of the same photogeneration (fast recombination) and another having crossed the substructure from the collecting junction. In the test devices, the latter component is negligible. Probably the minority carrier lifetime within the upper c-Si layer is too short.

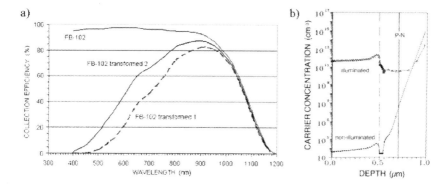

Figure 8.8. (*a*) Example of CE with transformed curves and without the CCL. The simulation is based on a conventional cell (FB-102) with parameters from table 8.1. (*b*) Carrier reservoir within the dead zone (PC-1D simulation). The PN-junction depth: 0.7 μm; parameters of the buried amorphization: depth 0.5 μm, thickness 40 nm, max doping level 10^{20} cm^{-3} [48].

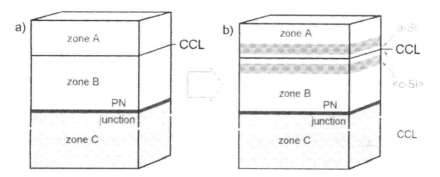

Figure 8.9. (*a*) Three collection zones in the presence of a CCL and (*b*) the same with the two ⟨c-Si⟩ layers having an increased absorbance.

8.5.5 Collection zones

In the presence of a CCL, three zones with different collection properties can be distinguished (figure 8.9(*a*)):

- *zone* A—dead zone, consisting of the surface strata lying above the CCL, from which it is practically impossible to collect generated photocarriers;
- *zone* B—intermediate zone (lying between the CCL and the pn collection junction), from where it is possible to collect photogenerated carriers but where the absorbed light flux is amputated strongly by *zone* A; and
- *zone* C—lower zone (lying between the pn-junction and the rear cell face), where the collection is nearly perfect (almost the same as for a classical

cell); for IR wavelengths *zone* A being very thin does not perturb much the absorption of *zone* C.

8.5.6   Impurity band doping profile

A double P^+-ion implantation, using energies of 180 and 15 keV, produced n-doping of the emitter, with a highly doped amorphized layer and a surface LH-type doping profile. From the electronic viewpoint, cells with an inserted δ-doping layer or a buried amorphization are analogous (see figure 8.10). Both substructures are limited by edges blocking minority hole flow.

Figure 8.10(*b*) shows the P profile measured by secondary ion mass spectroscopy (SIMS) and calculated by the stopping and range of ions in the matter SRIM code. At 180 keV the implantation dose was $0.1–1 \times 10^{15}$ cm^{-2}, the ion current $1.0 \ \mu A$ cm^{-2} and the thermal treatment for 15–30 min at 500 °C. The doping profile does not change during the applied annealing conditions.

The spreading resistance profile (SRP) shows (figure 8.10) a relatively good doping ionization for devices obtained exclusively by ion implantation (figure 8.10(*c*)). In the case of the P prediffused test device (figure 8.10(*d*)) the ionization rate is even greater because the doping process takes place at a relatively high temperature of 850 °C. Simultaneously the doping concentration due to implantation is one order of magnitude smaller than its diffusion counterpart. The balance between substitutional (mainly diffused) and interstitial (mainly implanted) P atoms is largely in favour of the substitution.

The active doping concentration corresponds to a *low degeneration* leading to the formation of an *impurity band* close to the conduction band (the Fermi level lies near the bottom of the conduction band).

8.5.7   Uni- and bipolar electronic transport in a multi-interface emitter

The current density contains different components (diffusion, drift, unipolar, bipolar). It is generally accepted that minority carriers, generated by light in the conventional emitter and diffusing to the pn-junction, are responsible for the current in a single-junction cell. The majority current in a conventional emitter is carried by drift since the product of *majority carrier density* multiplied by *mobility* is usually orders of magnitude higher in the whole emitter space than that of minority carriers. Briefly, conduction in a conventional emitter is controlled by a *unipolar minority carrier transport*, which consists of the diffusion component.

Detailed balance calculations for pn single-junctions avoid the spatial dependence of transport by assuming flat quasi-Fermi levels. This assumption, which is equivalent to infinite mobility, removes spatial dependencies. Once generated, carriers can reach any region of the device independently of their generation site.

The composition of the current within the multi-interface emitter with a continuous amorphization depends on the depth. Several sub-zones of specific

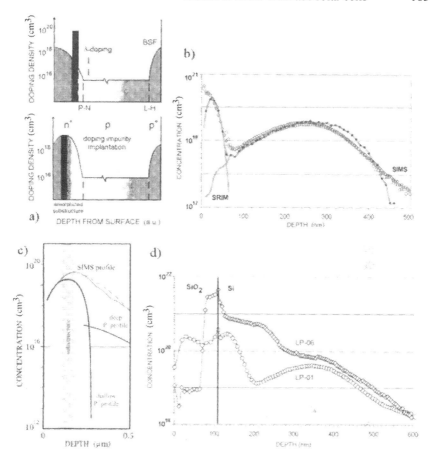

Figure 8.10. (*a*) Comparison of doping profiles for two concepts having similar electronic behaviour: δ-doping layer and buried amorphization [22, 37, 42]. (*b*) P double-profile (implantation energies: 15 and 180 keV) were measured by SIMS and calculated by the Monte Carlo simulation code SRIM. (*c*) SIMS and SRP used for the evaluation of the doping activation rate of the implanted sample. (*d*) Comparison of doubly implantated and prediffused (with amorphization) doping profiles.

conductivity, each dominated by their main conduction mechanism, appear (see table 8.2).

Main carrier components (and dominating generation and conduction effects) depend on the emitter depth. In zone I, corresponding to the device's superficial layer, there is a large fundamental absorption of shortest wavelengths, which generates an important non-equilibrium minority carrier concentration largely surpassing the thermal concentration present deep in the emitter. In a conventional cell, the minority current goes back by diffusion through the zone

Table 8.2. Different electronic transport zones composing a conventional and a MIND emitter.

Zone	Description	Cell type
I	shortest wavelength photo generation	conventional & MIND
II	front transport zone	conventional
II	front carrier reservoir	MIND
III	blocking substructure barrier	MIND
IV	damage substructure	MIND
V	lower emitter transport zone	conventional & MIND

II. Even with its own generation, zone II plays the role of a transporter where the first photocarrier separation appears. This separation is provoked by a built-in electric field induced by the gradient of the minority carrier density. The effect is measured by a displacement of the quasi-Fermi levels. This idealized image neglects surface recombination centres, because in modern well-passivated devices the surface recombination is strongly attenuated.

The habitual *minority diffusion current* flowing in zone II of a conventional cell does not exist in the multi-interface test device. Zone IV conserves a good carrier collection as in conventional cells. In certain conditions the majority drift current of the test device substitutes for the component of blocked minority carriers. To be useful across the substructure, the mechanism requires a relatively short transit time. Indeed, the conditions seem to be favourable because of the flat conduction band but in reality the recombination in zone II is quicker than the electron flow through the substructure.

Photocarriers generated deep enough are separated in the pn-junction (as in a conventional solar cell). They recombine only after travelling through an external circuit, causing a photogenerated current. The lower emitter (lying under the CCL) evacuates a bipolar current composed of (1) minority carriers diffusing from the substructure towards the collecting junction and (2) majority carriers drifting from the collecting junction towards the substructure. It corresponds to the transformation of usual unipolar to a new bipolar electronic transport.

On the basis of spectral response data, one can determine the value of the short-circuit current as a function of the photon penetration depth. However, the attenuation length is undetermined without knowing the corresponding absorbances that cannot be obtained by a direct measurement.

8.5.8 Absorbance in presence of a dead zone

The active layer lies between the CCL and the rear face of the device. Optoelectronic activity of strata forming the cell can be 'sampled' by the attenuation length according to the wavelength of the incident light.   For wavelengths absorbed almost completely in the thickness of the wafer (UV and visible)

$$\Phi_A \approx \Phi_0 = \Phi_{A-z} + \frac{1}{e}\Phi_0 \qquad (8.6)$$

where Φ_A, Φ_0, Φ_{A-z} are respectively the photon flux absorbed in all the wafer, photon flux transmitted by the surface and the photon flux absorbed within the attenuation length, so that

$$\Phi_{A-z} = \left(1 - \frac{1}{e}\right)\Phi_0. \qquad (8.7)$$

It is useful to distinguish the part of the light flux absorbed in the dead zone Φ_{zD} and outside of it. The part of the flux absorbed outside the attenuation length is given by

$$\Phi_{A-zD} = \left(1 - \frac{1}{e}\right)\Phi_0 - \Phi_{zD}. \qquad (8.8)$$

The flux transmitted by the cell's front face is divided, according to the absorption depth, into three parts: (1) Φ_{zD}, absorbed in the dead zone; (2) $(1 - \frac{1}{e})\Phi_0$, absorbed within the attenuation length (for $z \geq z_D$); and (3) $\frac{1}{e}\Phi_0$, absorbed beyond z. Only the last two parts can generate 'collectable' carriers.

8.5.9   Self-consistent calculation

Due to the lack of knowledge, certain optoelectronic features can be deduced with a self-consistent calculation that takes into account the available experimental data for (1) the spectral response and optical reflection, (2) c-Si and equivalent a-Si absorption coefficients (measured independently by photothermal deflection spectroscopy (PDS) and the constant photocurrent method (CPM)), (3) the thickness of the dead zone (obtained by electron beam induced current (EBIC)), (4) the depth and thickness of the amorphized layer (by transmission electron microscopy (X-TEM)) and (5) the depth of the pn-junction (by SIMS and SRP) as well as the relation between the dead zone absorbance and the zone B collection efficiency. Other experimental data for verification of the results obtained can be used.

An iterative calculation for different wavelengths is based on a just balance between photons absorbed *in front of the CCL* and carriers collected exclusively *below the CCL*. This is a *photogeneration/collection balance*. The method does not require any initial information on the position of the CCL, which appears as a result of the calculation. The necessary short-circuit currents are given directly by spectral response. In place of knowing the absorbance of each stratum of the dead zone, one can be satisfied with its overall absorbance value.

The *photogeneration/collection balance* leads to an increased dead zone aborbance. By analogy a similar effect is expected below the CCL. So the

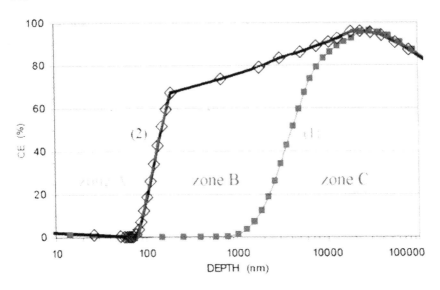

Figure 8.11. CE with an increased surface absorbance. There is good agreement of the simulated CCL (self-consistent calculation) at a depth of about 200 nm with the corresponding dead zone (EBIC measurement) at 210 nm. Zone A represents a surface dead zone, zones B and C have good collection probabilities. Spectral response of sample CB-A-0503 was measured at the FhG ISE of Freiburg.

nanoscale Si-layered system is at the origin of a significant enhancement of the emitter absorption [9].

A comparison of two dead zones obtained on the basis of two optical models is shown in figure 8.11. One homogeneous layer with its c-Si absorption coefficient leads to the particularly deep active zone (curve 1). It signifies that photocarriers can be collected only from the strata buried deeper than 1 μm from the surface, i.e. far behind the collecting pn-junction at 0.5 μm depth. This is an evident paradox because the best collection appears normally at the pn-junction space-charge region. The implantation damage is confined in the implantation range of 0.5 μm.

Curve 2 is obtained on the basis of a self-consistent calculation using a two-layer optical model where the about 100 nm thick surface layer conserves the c-Si absorption coefficient and the about 50 nm thick ⟨c-Si⟩ layer lying below it has an absorption coefficient of amorphized Si material which has been measured by PDS by CPM [49].

The CE fitting leads unavoidably to a definite CCL position and an *increased surface absorbance*. A single-crystal transition zone under tensile stress may be responsible for this increase. An increase in absorption in tensile-stressed Si has been reported previously [50].

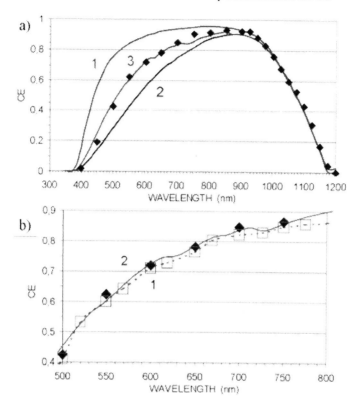

Figure 8.12. (*a*) Fitting of the experimental CE of a test device with a continuous buried amorphization. (*b*) Comparison of experimental CEs of the test device obtained with more measured points (light filters) (symbols and discontinuous curve 1 to guide the eye) and the self-consistent calculation (continuous curve 2).

8.6 An experimental test device

To answer certain fundamental questions, test devices have been fabricated. One example is our LP-06-C9 sample, which was fabricated from a FZ Si, $\langle 100 \rangle$ oriented, B-doped an experimental test device single crystal. The wafer was polished on only the front face. A buried amorphous layer was obtained by 180 keV P^+ implantation at room temperature (implantation current of 1 μA, implantation dose $\leq 10^{15}$ cm^{-2}) in the wafer with a prediffusion P profile (about 10^{20} cm^{-3} in the surface zone). The buried amorphization was annealed at a temperature between 450 and 550 °C during a time to be optimized. The method of manufacture (implantation) transforms a thin surface layer $d \leq 0.5$ μm of the wafer. So the effects investigated are limited in depth and easy to localize. TEM of the test device is shown in figure 8.1(*c*).

Figure 8.12 shows the main result from the spectral response and its *fitting* taking into account the activity of CCL at 165 nm depth (as determined by TEM images[2] (figure 8.1(c)) and EBIC measurements). The measured CE has been fitted with the self-consistent calculation on the basis of a three-layer structure (similar to that in figure 8.7(c)). As previously mentioned, the two-layer optical model of the surface dead zone uses the two absorption functions ($\alpha = f(\lambda)$) of classical c-Si [51] and amorphized material [49].

Three calculated CEs for the test device are compared with experimental values: (1) with only single-crystal absorbance at the surface layer (see figure 8.12(a), curve 1); (2) with the surface absorbance of a five-layer system (two c-Si, a-Si and two transition zones $\langle$c-Si$\rangle$) obtained by self-consistent calculation (see figure 8.12(a), curve 2); and (3) fitting with the increased surface absorbance of curve 2 and an additional low-energy generation [23] (see figure 8.12(a), curve 3).

Curves 1 and 2 in figure 8.12(a) represent a conventional one-stage PV conversion. The reference value for curve 2 has been established for monochromatic UV light of 400 nm. Curve 3 in figure 8.12(a) represents a multistage process [2] where the conventional PV conversion is reinforced by new mechanisms.

The spectral response of figure 8.12(a) was remeasured six months later with more filters. The CE of the same sample (LP-06-C9) was remeasured one year later in FhG ISE in Freiburg, Germany to verify its performance (figure 8.12(b)). Details of the simulated curve 2 in figure 8.12(b) reproduce quite well the characteristic 'bumps' of the experimental curve 1 in figure 8.12(b) in the range lying between 520 and 750 nm.

8.6.1 Enhanced internal quantum efficiency

The deduction of new effects, as can be observed in many samples, is relatively simple and based on a *comparison of the optical and electronic properties of the same semiconductor volume.* For certain wavelengths the number of collected electrons is larger than that of corresponding photons (only those absorbed in the lower active zone). This result can be explained by a multistage PV conversion at the amorphous/crystalline transition zones. They are the only regions junction at 0.5 μm depth modified in comparison with a conventional solar cell. The rate of the *multistage generation is proportional to wavelength*, being greater for more energetic photons and disappearing totally for $\lambda \geq 900$ nm.

Figure 8.13(a) illustrates the difference between the experimental and corresponding simulated CE from figure 8.12(b). The number of collected electrons per photon absorbed in the active zone lying under the CCL is shown in figure 8.13(b) [9].

---

[2] TEM images show a thin (10 nm) a-Si layer at this depth buried within the single-crystal Si. The CCL corresponds to the position of the upper a-Si/c-Si interface.

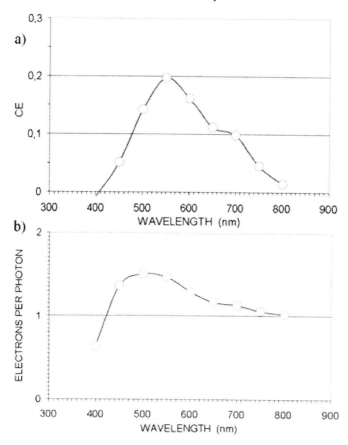

Figure 8.13. (*a*) Difference between an experimental and corresponding simulated CE for a continuous buried amorphization for a surface absorbance in the UV (400 nm filter). (*b*) Number of collected electrons per photon absorbed at the active zone as a function of wavelength on the basis of calculations illustrated in figure 8.12(*b*).

This experimental approach suggests that it would be informative to make a solar cell demonstrator without a CCL where it would be possible to collect photocarriers generated in the surface zone.

8.6.2   Sample without any carrier collection limit (CCL)

A simple experimental check with a *sample without any CCL*, in the form of a two-layer amorphous/crystalline structure obtained by total surface amorphization by P$^+$ ions implanted at 150 keV and thermal treatment, has been made (figure 8.14). The result of the corresponding simulation for the test device free of a CCL is presented for comparison. Values of all parameters are the same

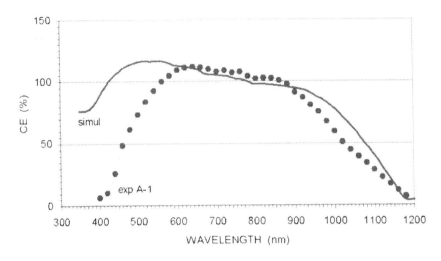

Figure 8.14. Experimental CE of a sample without a CCL (symbols). The CE simulated for the test device with the CCL suppressed (continuous curve); parameters obtained from the fit of the test device with continuous amorphization.

as those for the curve 3 of figure 8.12(*b*). The test device has not been optimized for performance—its modification was made with the least possible single-crystal perturbation, i.e. the thinnest possible buried layer (figure 8.1(*c*)).

The poor performance in the IR of the (A-1) sample with the *totally amorphized-surface* can be explained by the short diffusion length of the minority carriers and the small optical confinement (full Al metallization of the rear face). The large drop of CE of A-1 in the UV is caused by the large recombination rate at the amorphized surface. A thickened superficial zone has been totally deactivated.

The lesser performance of the test device (LP-06-C9) in the 600–900 nm range (figure 8.14) most probably results from the small amorphized layer thickness (figure 8.1(*c*)). The experimental error of the reference measurements is smaller than the difference between the two curves. Too small a thickness could be at the origin of a local stress relaxation and, as a consequence, a proportionally smaller activity than that calculated for the interface transition zone.

8.7 Concluding remarks and perspectives

A self-consistent calculation to fit the MIND model cell spectral response allows observing three effects: (1) the appearance of the CCL, (2) increased surface absorbance and (3) additional local generation. Certain perhaps less obvious particularities of the continuous nanoscale Si-layered system resulting from the buried amorphization can be noted:

- Strain and electric fields lead to local transformations of the original single-crystal Si that produce new optoelectronic attributes.
- The short-circuit current in the UV and blue range is, first and foremost, influenced by a valence band offset. Recombination caused by post-implantation defects is of second order.
- For certain wavelengths, the number of collected electrons can be greater than the photons absorbed in the active zone, i.e. one photon can create more than one electron–hole pair.

The multi-interface approach modifies a semiconductor, such as single-crystal Si, by making use of geometrical confinement. This allows the introduction of multiple energy levels and a 'new' band structure leading to a multilevel multistage PV conversion. A good 'tuning' of device energy levels to the whole solar spectrum should be possible. In its simplest form it can be obtained by a superposition of only two mechanisms with different energy thresholds: larger primary (conventional conversion stage) and smaller secondary (new conversion stage). The secondary mechanism combined with confinement characteristics can be at the origin of multiple subpopulations distinguished from each other by their average energy. It seems reasonable to suppose that high efficiencies can be reached without a large number of materials as in tandem cells.

Another advantage of the multi-interface design is that it can more closely approach efficiency limits under less than ideal conditions. This could render the devices more tolerant to non-idealities. For example, the key design criterion of single-junction solar cells that is the recombination minimization be attenuated. A new physical mechanism may increase the conversion performance above that of a single-junction by introducing additional energy levels while degrading the recombination properties. It may lead to an increase in the corresponding theoretical efficiency limit.

It could be possible to make an efficient Si solar cell based on the condition that all photocarriers, including those generated in the surface zone, be collected. Further development of multi-interface cell operation should permit the collection of photocarriers regardless of their generation site.

Acknowledgments

The author wishes to express his thanks to his PhD student Marc Ley for his aid with the measurements and simulations, U Zammit of the Universita di Roma for his PDS measurements, J Pelant of the Charles University of Prague for CPM measurements, G Ehret and G Würtz of the IPCMS of Strasbourg for their TEMs, D Ballutaud of LBSP for EBICs, A Pape for criticism and the personnel of the PHASE Laboratory for their experimental aid.

References

[1] Honsberg C B, Bremner S P and Corkish R 2002 *Physica* E **14** 126–41

[2] Kuznicki Z T, Capot F and de Unamuno S 1998 *Proc. 2nd World Conf. on Photovolatic Energy Conversion (WCPEC) (Vienna, 6–10 July)* (Ispra: European Commission) pp 80–3

[3] Kuznicki Z T 1996 *Proc. E-MRS First Polish–Ukrainian Symposium 'New Photovoltaic Materials for Solar Cells' (Cracow-Przegorzaly, 21–22 October)* pp 58–78
Kuznicki Z T 1997 *Proc. 26th IEEE Photovoltaic Spec. Conf. (Anaheim, CA, 29 September–3 October)* (New York: IEEE) pp 291–4

[4] Robbins D J and Landsberg P T 1980 *J. Chem. Phys.* **13** 2425–39

[5] Werner J H, Kolodinski S and Queisser H J 1994 *Phys. Rev. Lett.* **72** 3851–4

[6] Werner J H, Brendel R and Queisser H J 1994 *Proc. IEEE First World Conf. Photovoltaic Energy Conversion, IEEE Catalog number 94CH3365-4* (New York: IEEE) pp 1742–5

[7] Würfel P 1997 *Solar Energy Mater. Solar Cells* **46** 43–56

[8] Markvart T 2000 *Prog. Quantum Electron.* **24** 107–86

[9] Kuznicki Z T 2002 *Appl. Phys. Lett.* **81** 4853–5

[10] Liang-Sheng Liao, Xi-Mao Bao, Ning-Sheng Li, Xiang-Qin Zheng and Nai-Ben Min 1996 *J. Non-Cryst. Solids* **198–200** 199–204

[11] Pavesi L, dal Negro L, Mazzoleni L, Franzo G and Priolo F 2000 *Nature* **408** 440–4

[12] Lu Z H, Lockwood D J and Baribeau J M 1996 *Solid-State Electron.* **40** 197–201

[13] Canham L T 1990 *Appl. Phys. Lett.* **57** 1046–50

[14] Cullis A G and Canham L T 1991 *Nature* **353** 335–8

[15] Hirschman K D, Tsybeskov L, Duttagupta S P and Fauchet P M 1996 *Nature* **384** 338–41

[16] Wilson W L, Szajowski P F and Brus L E 1993 *Science* **262** 1242–4

[17] Dal-Negro L, Pavesi L, Pucker G, Franzo G and Priolo F 2001 *Opt. Mater.* **17** 41–4

[18] Nassiopoulos A G, Grigoropoulos S and Papadimitriou D 1996 *Appl. Phys. Lett.* **69** 2267–9

[19] Cahen D, Grätzel, Guillemoles J E and Hodes G 2001 *Electrochemistry of Nanomaterials* (Weinheim: Wiley) pp 201–27

[20] Ross R T and Nozik A J 1982 *J. Appl. Phys.* **53** 3812–18

[21] Zundel M K, Csaszar W and Endros A L 1995 *Appl. Phys. Lett.* **67** 3945

[22] Kuznicki Z T 1993 *J. Appl. Phys.* **74** 2058–63

[23] Ziegler J F 1992 *Handbook of Ion Implantation Technology* (Amsterdam: North-Holland)

[24] Douglas E C and D'Aiello R V 1980 *IEEE Trans. Electron. Devices* **ED-27** 792–802

[25] Spitzer M B, Keavney C J, Tobin S P and Milstein J B 1984 *17th IEEE PVSC (Golden, CO, 1–4 May)* (New York: IEEE) pp 398–402

[26] Kuznicki Z T, Thibault J, Chautain-Matys F and de Unamuno S 1996 *Proc. E-MRS First Polish–Ukrainian Symposium 'New Photovoltaic Materials for Solar Cells' (Cracow-Przegorzaly, 21–22 October)* pp 99–107

[27] Kuznicki Z T, Thibault J, Chautain-Matys F, Wu L, Sidibé S, de Unamuno S, Bonarski J T, Swiatek Z and Ciach R 1998 E-MRS Spring Meeting, Strasbourg, France, 16–19 June 1998, published in 1999 *Nucl. Instrum. Meth. Phys. Res.* B **147** 126–41

[28] Motooka T, Nisihira K, Munetoh S, Moriguchi K and Shintani A 2000 *Phys. Rev.* B **61** 8537–40
[29] Weber B, Stock D M and Gärtner K 2000 *Mater. Sci. Eng.* B **71** 212–18
[30] Kuznicki Z T, Ley M, J. Thibault J and Bouchet D 2001 E-MRS Spring Meeting, Strasbourg, France, 5–8 June
[31] Anderson R L 1962 *Solid-State Electron.* **5** 341–51
[32] Green M A 1997 *J. Appl. Phys.* **81** 268–71
[33] Stangl R, Froitzheim A, Elstner L and Fuhs W 2001 *Proc. 17th EPVSECE (22–26 October 2001, Munich)* (Munich and Florence: WIP and ETA) pp 1387–90
[34] Wolf M 1960 *Proc. IRE* **48** 1246–63
[35] Bube R H 1992 *Photoelectric Properties of Semiconductors* (Cambrige: Cambrige University Press)
[36] Kuznicki Z T 1991 *J. Appl. Phys.* **69** 6526–41
[37] Kuznicki Z T, Lipinski M and Muller J-C 1993 *Proc. 23th IEEE Photovoltaic Spec. Conf. (Louisville, May 10–14)* pp 327–31
[38] Swiatek Z 2002 *PhD Thesis* Institut of Metallurgy and Materials of Polish Academy of Sciences Cracow
[39] Milita S and Servidori M 1996 *J. Appl. Phys.* **79** 8278–84
[40] Reif J 2002 Brandenburgische Technische Universität Cottbus private communication
[41] 2002 European Semiconductor www.eurosemi.eu.com
[42] Kuznicki Z T 1994 French patent no 94 08885 of July 12, and 1999 USA patent no 5,935,345 of August 10
[43] Tanaka M, Toguchi M, Matsuyama T, Sawada T, Tsuda S, Nakano S, Hanafusa H and Kuwano Y 1992 *Japan. J. Appl. Phys.* **31** 3518–22
[44] Taguchi M, Kawamoto K, Tsuge S, Baba T, Sakata H, Moriziane M, Uchihashi K, Nakamura N, Kiyama S and Oota O 2000 *Prog. Photovolt. Res. Appl.* **8** 503–13
[45] Kuznicki Z T and Ley M 2001 *International Workshop on Nanostructures in Photovoltaics* Max Planck Institute for the Physics of Complex Systems, Dresden, July 28–August 4
[46] Ley M and Kuznicki Z T 2001 *E-MRS Spring Meeting (Strasbourg, 5–8 June)* published in 2002 *Solar Energy Mater. Solar Cells* **72** 613–19
[47] Sinkkonen J, Ruokolainen J, Uotila P and Hovinen A 1995 *Appl. Phys. Lett.* **66** 206–8
[48] Kuznicki Z T, Wu L and Muller J-C 1994 *Proc. 12th European Photovolatic Solar Energy Conference and Exhibition (Amsterdam, 11–15 April)* (Bedford: H S Stephens & Associates) pp 552–5
[49] Kuznicki Z T and Ley M 2003 *Appl. Phys. Lett.* **82** 4241–3
[50] Leroy B 1987 *Phil. Mag.* B **55** 159–99 and references therein
[51] Green M A and Keevers M J 1995 *Prog. Photovolt. Res. Appl.* **3** 189–92

Chapter 9

Quantum dot solar cells

A J Nozik
Center for Basic Sciences, National Renewable Energy Laboratory, Golden, CO 80401, USA and Department of Chemistry, University of Colorado, Boulder, CO 80309, USA

9.1 Introduction

The two most important factors that determine the cost of delivered photovoltaic (PV) power are conversion efficiency and the initial cost per unit area of the PV system. To achieve very low cost PV power (a few US cents/kWh) and, hence, introduce PV power on a massive scale, it is necessary to develop cells that either have ultra-high conversion efficiency (50–60%) with moderate cost (US$100–150 m^{-2}) or moderate efficiency (about 10%) compensated by ultra-low cost (<US$2 m^{-2}). We have been investigating the possibility of achieving ultra-high conversion efficiency in single bandgap cells by utilizing high energy electrons (termed hot electrons and created by absorption of solar photons larger than the bandgap) before these high energy electrons convert their excess kinetic energy (equal to the difference between the photogenerated electron energy and the conduction band energy) to heat through phonon emission. The formation of discrete quantized levels in semiconductor quantum dots affects the relaxation dynamics of high energy electrons and holes and could enhance the conversion efficiency by allowing electrical free energy to be extracted from the energetic electrons and/or holes before they relax to their lowest electronic state and produce heat.

As is well known, the maximum thermodynamic efficiency for the conversion of unconcentrated solar irradiance into electrical free energy in the radiative limit, assuming detailed balance, a single threshold absorber and thermal equilibrium between electrons and phonons, was calculated by Shockley and Queisser in 1961 [1] to be about 31%; this analysis is also valid for the conversion

196

to chemical free energy [2,3]. This efficiency is attainable in semiconductors with bandgaps ranging from about 1.25 to 1.45 eV.

However, the solar spectrum contains photons with energies ranging from about 0.5 to 3.5 eV. Photons with energies below the semiconductor bandgap are not absorbed, while those with energies above the bandgap create charge carriers (electrons and holes) with a total excess kinetic energy equal to the difference between the photon energy and the bandgap. This excess kinetic energy creates an effective temperature for an ensemble of photogenerated carriers that can be much higher than the lattice temperature; such carriers are called 'hot electrons and hot holes' and their initial temperature upon photon absorption can be as high as 3000 K with the lattice temperature at 300 K. In bulk semiconductors the division of this kinetic energy between electrons and holes is determined by their effective masses, with the carrier having the lower effective mass receiving more of the excess energy [4]. Thus,

$$\Delta E_e = (h\nu - E_G)[1 + m_e^*/m_h^*]^{-1} \qquad (9.1)$$

$$\Delta E_h = (h\nu - E_G) - \Delta E_e \qquad (9.2)$$

where ΔE_e is the energy difference between the conduction band and the initial energy of the photogenerated electron and ΔE_h is the energy difference between the valence band and the photogenerated hole (see figure 9.1). However, in quantum dots (QDs), the distribution of excess energy is determined by the quantized energy level structure in the QDs and the associated selection rules for the optical transitions [5].

In the Shockley–Queisser analysis, a major factor limiting the conversion efficiency to 31% is that the absorbed photon energy above the semiconductor bandgap is lost as heat through electron–phonon scattering and subsequent phonon emission, as the carriers relax to their respective band edges (bottom of conduction band for electrons and top of valence band for holes) (see figure 9.1) and equilibrate with the phonons. The main approach to reduce this loss and increase efficiency above the 31% limit has been to use a stack of cascaded multiple pn-junctions in the absorber with bandgaps better matched to the solar spectrum; in this way higher energy photons are absorbed in the higher bandgap semiconductors and lower energy photons in the lower bandgap semiconductors, thus reducing the overall heat loss due to carrier relaxation via phonon emission. In the limit of an infinite stack of bandgaps perfectly matched to the solar spectrum, the ultimate conversion efficiency at one-sun intensity increases to about 66%. For practical purposes, the stacks have been limited to two or three pn-junctions; actual efficiencies of about 32% have been reported in PV cells with two cascaded pn-junctions. Other approaches to exceed the Shockley–Queisser limit include hot carrier solar cells [4,6,7], solar cells producing multiple electron–hole pairs per photon through impact ionization [8,9], multiband and impurity solar cells [10,11] and thermophotovoltaic/thermophotonic cells [10]. Here, we will only discuss hot carrier and impact ionization solar cells and the

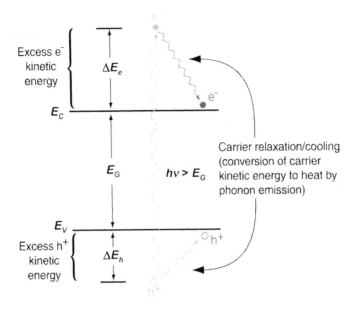

Excess e⁻ kinetic energy

ΔE_e

E_c

E_G

$h\nu > E_G$

Carrier relaxation/cooling (conversion of carrier kinetic energy to heat by phonon emission)

E_V

Excess h⁺ kinetic energy

ΔE_h

Figure 9.1. Hot carrier relaxation/cooling dynamics in semiconductors (from [4]. Reprinted with permission).

effects of size quantization in semiconductor QDs on the carrier dynamics that control the probability of these processes.

There are two fundamental ways to utilize the hot carriers for enhancing the efficiency of photon conversion. One way produces an enhanced photovoltage and the other way produces an enhanced photocurrent. The former requires that the carriers be extracted from the photoconverter before they cool [6,7], while the latter requires the energetic hot carriers to produce a second (or more) electron–hole pair through impact ionization [8, 9]—a process that is the inverse of an Auger process whereby two electron–hole pairs recombine to produce a single highly energetic electron–hole pair. In order to achieve the former, the rates of photogenerated carrier separation, transport and interfacial transfer across the semiconductor interface must all be fast compared to the rate of carrier cooling [7,12–14]. The latter requires that the rate of impact ionization (i.e. inverse Auger effect) is greater than the rate of carrier cooling and forward Auger processes.

Hot electrons and hot holes generally cool at different rates because they generally have different effective masses; for most inorganic semiconductors electrons have effective masses that are significantly lighter than holes and consequently cool more slowly. Another important factor is that hot carrier cooling rates are dependent upon the density of the photogenerated hot carriers (i.e. the absorbed light intensity) [15–17]. Here, most of the dynamical effects

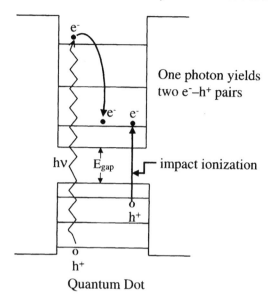

One photon yields
two e⁻–h⁺ pairs

impact ionization

Quantum Dot

Figure 9.2. Enhanced photovoltaic efficiency in quantum dot solar cells by impact ionization (inverse Auger effect) (from Nozik A J 2002 *Physica* E **14** 115–20. Reprinted with permission).

we will discuss are dominated by electrons rather than holes; therefore, we will restrict our subsequent discussion primarily to the relaxation dynamics of photogenerated electrons.

Finally, in recent years it has been proposed [7, 12, 13, 18–21] and experimentally verified in some cases [4, 22, 23] that the relaxation dynamics of photogenerated carriers may be markedly affected by quantization effects in the semiconductor (i.e. in semiconductor quantum wells, quantum wires, quantum dots, superlattices and nanostructures). That is, when the carriers in the semiconductor are confined by potential barriers to regions of space that are smaller than or comparable to their de Broglie wavelength or to the Bohr radius of excitons in the semiconductor bulk, the relaxation dynamics can be dramatically altered; specifically the hot carrier cooling rates may be dramatically reduced, and the rate of impact ionization could become competitive with the rate of carrier cooling [4] (see figure 9.2).

9.2   Relaxation dynamics of hot electrons

Excitation with a single wavelength produces initial carrier populations of electrons and holes that are nearly monoenergetic; they may not be perfectly mono-energetic because of possible multiplicities of hole states that are available for the optical transition. This possibility is strongest for semiconductors exhibiting quantum confinement [24]. In either case, the initial carrier distributions are not Boltzmann-like and the first step toward establishing equilibrium is for the hot carriers to interact separately amongst themselves and with the initial population of cold carriers through their respective carrier–carrier collisions and inter-valley scattering to form separate Boltzmann distributions of electrons and holes. These two Boltzmann distributions can then be separately assigned an electron and hole temperature that reflects the distributions of kinetic energy in the respective charge carrier populations. If photon absorption produces electrons and holes with initial excess kinetic energies at least kT above the conduction and valence bands, respectively, then both initial carrier temperatures are always above the lattice temperature and the carriers are called hot carriers. This first stage of relaxation or equilibration occurs very rapidly (<100 fs) [15, 16], and this process is often referred to as carrier thermalization (i.e. formation of a thermal distribution described by Boltzmann statistics).

After the separate electron and hole populations come to equilibrium amongst themselves in less than 100 fs, they are still not yet in equilibrium with the lattice. The next step of equilibration is for the hot electrons and hot holes to equilibrate with the semiconductor lattice. The lattice temperature is the ambient temperature and is lower than the initial hot electron and hot hole temperatures. Equilibration of the hot carriers with the lattice is achieved through carrier–phonon interactions (phonon emission) whereby the excess kinetic energy of the carriers is transferred from the carriers to the phonons; the phonons involved in this process are the longitudinal optical (LO) phonons. This may occur by each carrier undergoing separate interactions with the phonons or in an Auger process where the excess energy of one carrier type is transferred to the other type, which then undergoes the phonon interaction. The phonon emission results in cooling of the carriers and heating of the lattice until the carrier and lattice temperatures become equal. This process is termed carrier cooling but some researchers also refer to it as thermalization; however, this latter terminology can cause confusion with the first stage of equilibration that just establishes the Boltzmann distribution amongst the carriers. Here, we will restrict the term thermalization to the first stage of carrier relaxation and we will refer to the second stage as carrier cooling (or carrier relaxation) through carrier–phonon interactions.

The final stage of equilibration results in complete relaxation of the system: the electrons and holes can recombine, either radiatively or non-radiatively, to produce the final electron and hole populations that existed in equilibrium in the dark before photoexcitation. Another important possible pathway following photoexcitation of semiconductors is for the photogenerated electrons and holes to undergo spatial separation. Separated photogenerated carriers can subsequently produce a photovoltage and a photocurrent (PV effect) [25–27]; alternatively, the separated carriers can drive electrochemical oxidation and reduction reactions (generally labelled redox reactions) at the semiconductor surface (photoelectrochemical energy conversion) [28]. These two processes form

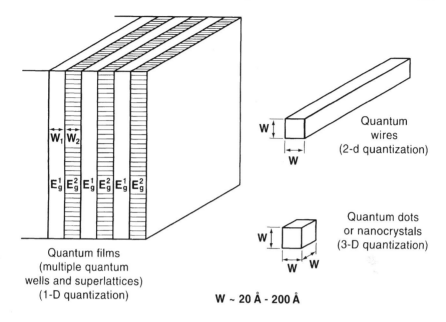

Quantum wires
(2-d quantization)

Quantum dots
or nanocrystals
(3-D quantization)

Quantum films
(multiple quantum
wells and superlattices)
(1-D quantization)

W ~ 20 Å - 200 Å

Figure 9.3. Three types of quantization configurations. The type depends upon the dimensionality of carrier confinement (from [87]. Copyright © 1995 John Wiley & Sons Inc. This material is used by permission of John Wiley & Sons Inc.).

the basis for devices/cells that convert radiant energy (e.g. solar energy) into electrical [25–27] or chemical free energy (PV cells and photoelectrochemical cells, respectively) [28].

9.2.1 Quantum wells and superlattices

Semiconductors show dramatic quantization effects when charge carriers are confined by potential barriers to small regions of space where the dimensions of the confinement are less than their de Broglie wavelength; the length scale at which these effects begin to occur range from about 10 to 50 nm for typical semiconductors (groups IV, III–V, II–VI). In general, charge carriers in semiconductors can be confined by potential barriers in one spatial dimension, two spatial dimensions, or in three spatial dimensions. These regimes are termed quantum films, quantum wires and quantum dots, respectively (see figure 9.3). Quantum films are also more commonly referred to simply as quantum wells (QWs).

One-dimensional quantum wells, hereafter called quantum films or just quantum wells, are usually formed through epitaxial growth of alternating layers of semiconductor materials with different bandgaps. A single quantum well (QW) is formed from one semiconductor sandwiched between two layers of a second semiconductor having a larger bandgap; the centre layer with the smaller bandgap semiconductor forms the QW while the two layers sandwiching the centre layer create the potential barriers. Two potential wells are actually formed in the QW structure: one well is for conduction band electrons, the other for valence band holes. The well depth for electrons is the difference (i.e. the offset) between the conduction band edges of the well and barrier semiconductors, while the well depth for holes is the corresponding valence band offset. If the offset for either the conduction or valence bands is zero, then only one carrier will be confined in a well.

Multiple QW structures consist of a series of QWs (i.e. a series of alternating layers of wells and barriers). If the barrier thickness between adjacent wells prevents significant electronic coupling between the wells, then each well is electronically isolated; this type of structure is termed a *multiple quantum well* (MQW). However, if the barrier thickness is sufficiently thin to allow electronic coupling between wells (i.e. there is significant overlap of the electronic wavefunctions between wells), then the electronic charge distribution can become delocalized along the direction normal to the well layers. This coupling also leads to a broadening of the quantized electronic states of the wells; the new broadened and delocalized quantized states are termed *minibands* (see figure 9.4). A multiple QW structure that exhibits strong electronic coupling between the wells is termed a *superlattice*. The critical thickness at which miniband formation just begins to occur is about 40 Å [29, 30]; the electronic coupling increases rapidly with decreasing thickness and miniband formation is very strong below 20 Å [29]. Superlattice structures yield efficient charge transport normal to the layers because the charge carriers can move through the minibands; the narrower the barrier, the wider the miniband and the higher the carrier mobility. Normal transport in MQW structures (thick barriers) require thermionic emission of carriers over the barriers, or if electric fields are applied, field-assisted tunnelling through the barriers [31].

9.2.1.1   *Measurements of hot electron cooling dynamics in QWs and superlattices*

Hot electron cooling times can be determined from several types of time-resolved photoluminescence (PL) experiments. One technique involves hot luminescence nonlinear correlation [32–34], which is a symmetrized pump-probe type of experiment. Figure 2 of [32] compares the hot electron relaxation times as a function of the electron energy level in the well for bulk GaAs and a 20-period MQW of GaAs/$Al_{0.38}Ga_{0.62}As$ containing 250 Å GaAs wells and 250 Å $Al_{0.38}Ga_{0.62}As$ barriers. For bulk GaAs the hot electron relaxation time varies from about 5 ps near the top of the well to 35 ps near the bottom of the well. For the MQW the corresponding hot electron relaxation times are 40 ps and 350 ps.

Another method uses time-correlated single-photon counting to measure PL lifetimes of hot electrons. Figure 9.5 shows 3D plots of PL intensity as a function

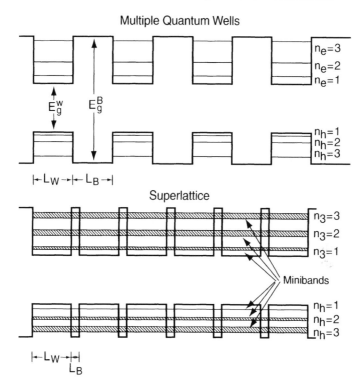

Figure 9.4. Difference in electronic states between multiple quantum well structures (barriers >40 Å) and superlattices (barriers <40 Å); miniband formation occurs in the superlattice structure, which permits carrier delocalization (from [87]. Copyright © 1995 John Wiley & Sons Inc. This material is used by permission of John Wiley & Sons Inc.).

of energy and time for bulk GaAs and a 250 Å/250 Å GaAs/Al$_{0.38}$Ga$_{0.62}$As MQW [17]. It is clear from these plots that the MQW sample exhibits much longer-lived hot luminescence (i.e. luminescence above the lowest $n = 1$ electron to heavy-hole transition at 1.565 eV) than bulk GaAs. Depending upon the emitted photon energy, the hot PL for the MQW is seen to exist beyond times ranging from hundreds to several thousand ps. However, the hot PL intensity above the bandgap (1.514 eV) for bulk GaAs is negligible over most of the plot: it is only seen at the very earliest times and at relatively low photon energies.

Calculations were performed [17] on the PL intensity *versus* time and energy data to determine the time dependence of the quasi-Fermi level, electron temperature, electronic specific heat and, ultimately, the dependence of the characteristic hot-electron cooling time on electron temperature.

The cooling, or energy-loss, rate for hot electrons is determined by LO phonon emission through electron–LO-phonon interactions. The time constant

(*a*)

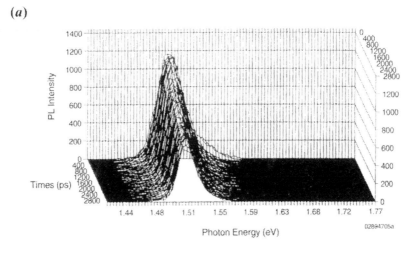

(*b*)

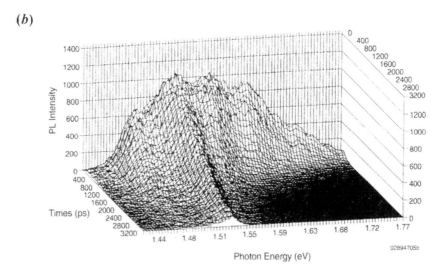

Figure 9.5. Three-dimensional plots of PL intensity *versus* time and photon energy for (*a*) bulk GaAs and (*b*) 250 Å GaAs/250 Å Al$_{0.38}$Ga$_{0.62}$As MQW (from [17]. © 1993 by the American Physical Society).

characterizing this process can be described by the following expression [35–37]:

$$P_{\mathrm{e}} = -\frac{\mathrm{d}E}{\mathrm{d}t} = \frac{\hbar\omega_{\mathrm{LO}}}{\tau_{\mathrm{avg}}}\exp(-\hbar\omega_{\mathrm{LO}}/kT_{\mathrm{e}}) \qquad (9.3)$$

where P_{e} is the power loss of electrons (i.e. the energy-loss rate), $\hbar\omega_{\mathrm{LO}}$ is the LO phonon energy (36 meV in GaAs), T_e is the electron temperature, and τ_{avg} is the time constant characterizing the energy-loss rate.

The electron energy-loss rate is related to the electron temperature decay rate through the electronic specific heat. Since at high light intensity the electron distribution becomes degenerate, the classical specific heat is no longer valid. Hence, the temperature- and density-dependent specific heat for both the QW and bulk samples need to be calculated as a function of time in each experiment so that τ_{avg} can be determined.

The results of such calculations (presented in figure 9.2 of [17]) show a plot of τ_{avg} *versus* electron temperature for bulk and MQW GaAs at high and low carrier densities. These results show that at a high carrier density $[n \sim (2-4) \times 10^{18}$ cm$^{-3}]$, the τ_{avg} values for the MQW are much higher ($\tau_{avg} = 350$–550 ps for T_e between 440 and 400 K) compared to bulk GaAs ($\tau_{avg} = 10$–15 ps over the same T_e interval). However, at a low carrier density $[n \sim (3-5) \times 10^{17}$ cm$^{-3}]$ the differences between the τ_{avg} values for bulk and MQW GaAs are much smaller.

A third technique to measure cooling dynamics is PL up-conversion [17]. Time-resolved luminescence spectra were recorded at room temperature for a 4000 Å bulk GaAs sample at the incident pump powers of 25, 12.5 and 5 mW. The electron temperatures were determined by fitting the high-energy tails of the spectra; only the region which is linear on a semilogarithmic plot was chosen for the fit. The carrier densities for the sample were 1×10^{19}, 5×10^{18} and 2×10^{18} cm^{-3}, corresponding to the incident excitation powers of 25, 12.5 and 5 mV, respectively. Similarly, spectra for the MQW sample were recorded at the same pump powers as the bulk. Figure 9.6 shows τ_{avg} for bulk and MQW GaAs at the three light intensities, again showing the much slower cooling in MQWs (by up to two orders of magnitude).

The difference in hot electron relaxation rates between bulk and quantized GaAs structures is also reflected in time-integrated PL spectra. Typical results are shown in figure 9.7 for single-photon counting data taken with 13 ps pulses of 600 nm light at 800 kHz focused to about 100 μm with an average power of 25 mW [38]. The time-averaged electron temperatures obtained from fitting the tails of these PL spectra to the Boltzmann function show that the electron temperature varies from 860 K for the 250 Å/250 Å MQW to 650 K for the 250 Å/17 Å superlattice, while bulk GaAs has an electron temperature of 94 K, which is close to the lattice temperature (77 K). The variation in the electron temperatures between the quantized structures can be attributed to differences in electron delocalization between MQWs and SLs and the associated non-radiative quenching of hot electron emission.

As shown earlier, the hot carrier cooling rates depend upon photogenerated carrier density: the higher the electron density, the slower the cooling rate. This effect is also found for bulk GaAs but it is much weaker compared to quantized GaAs. The most generally accepted mechanism for the decreased cooling rates in GaAs QWs is an enhanced 'hot phonon bottleneck' [39–41].

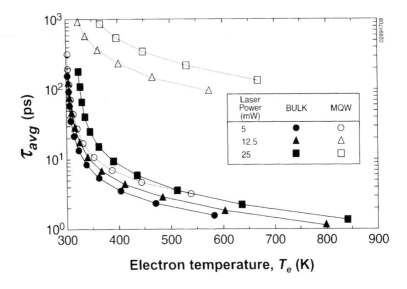

Figure 9.6. Time constant for hot-electron cooling (τ_{avg}) *versus* electron temperature for bulk GaAs and GaAs MQWs at three excitation intensities (from [17]. © 1993 by the American Physical Society).

In this mechanism a large population of hot carriers produces a non-equilibrium distribution of phonons (in particular, optical phonons which are the type involved in the electron–phonon interactions at high carrier energies) because the optical phonons cannot equilibrate fast enough with the crystal bath; these hot phonons can be re-absorbed by the electron plasma to keep it hot. In QWs the phonons are confined in the well and they exhibit slab modes [40], which enhance the 'hot phonon bottleneck' effect.

9.2.2   Relaxation dynamics of hot electrons in quantum dots

As previously discussed, slowed hot electron cooling in QWs and superlattices that is produced by a *hot* phonon bottleneck requires very high light intensities in order to create the required photogenerated carrier density of greater than about 1×10^{18} cm^{-3}. This required intensity, possible with laser excitation, is many orders of magnitude greater than that provided by solar radiation at the earth's surface (maximum solar photon flux is about 10^{18} cm^{-2} s^{-1}; assuming a carrier lifetime of 1 ns and an absorption coefficient of 1×10^5 cm^{-1}, this translates into a photoinduced electron density of about 10^{14} cm^{-3} at steady state). Hence, it is not possible to obtain slowed hot carrier cooling in semiconductor QWs and superlattices with solar irradiation via a *hot* phonon bottleneck effect; solar

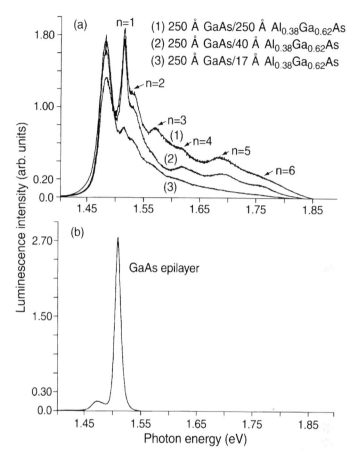

Figure 9.7. (*a*) Time-integrated PL spectra for MQWs and SLs showing hot luminescence tails and high energy peaks arising from hot electron radiative recombination from upper quantum levels. (*b*) Equivalent spectrum for bulk GaAs showing no hot luminescence (from [38]. © 1990 with permission from Elsevier).

concentration ratios greater than 10^4 would be required, resulting in severe practical problems.

However, the situation with 3D confinement in quantum dots (QDs) is potentially more favourable. In the QD case, slowed hot electron cooling is theoretically possible even at arbitrarily low light intensity; this effect is simply called a 'phonon bottleneck', without the qualification of requiring hot phonons (i.e. a non-equilibrium distribution of phonons). Furthermore, there is a possibility that the slowed cooling could make the rate of impact ionization (inverse Auger effect) an important process in QDs [42]. PL blinking in QDs (intermittent PL as a function of time) has been explained [43,44] by an Auger process whereby if two electron–hole pairs are photogenerated in a QD, one pair recombines and transfers its recombination energy to one of the remaining charge carriers, ionizing it over the potential barrier at the surface into the surface region. This creates a charged QD that quenches radiative emission after subsequent photon absorption; after some time the ionized electron can return to the QD core and the PL is turned on again. Since this Auger process can occur in QDs, the inverse Auger process, whereby one high energy electron–hole pair (created from a photon with $h\nu > E_g$) can generate two electron–hole pairs, can also occur in QDs [42]. The following discussion will present a discussion of the hot carrier cooling dynamics.

9.2.2.1 *Phonon bottleneck and slowed hot electron cooling in quantum dots*

The first prediction of slowed cooling at low light intensities in quantized structures was made by Boudreaux *et al* [7]. They anticipated that cooling of carriers would require multiphonon processes when the quantized levels are separated in energy by more than phonon energies. They analysed the expected slowed cooling time for hot holes at the surface of highly doped n-type TiO_2 semiconductors, where quantized energy levels arise because of the narrow space charge layer (i.e. depletion layer) produced by the high doping level. The carrier confinement in this case is produced by the band bending at the surface; for a doping level of 1×10^{19} cm^{-3}, the potential well can be approximated as a triangular well extending 200 Å from the semiconductor bulk to the surface and with a depth of 1 eV at the surface barrier. The multiphonon relaxation time was estimated from

$$\tau_c \sim \omega^{-1} \exp(\Delta E / kT) \qquad (9.4)$$

where τ_c is the hot carrier cooling time, ω is the phonon frequency, and ΔE is the energy separation between quantized levels. For strongly quantized electron levels, with $\Delta E > 0.2$ eV, τ_c could be > 100 ps according to equation (9.4).

However, carriers in the space charge layer at the surface of a heavily doped semiconductor are only confined in one dimension, as in a quantum film. This quantization regime leads to discrete energy states which have dispersion in k-space [45]. This means the hot carriers can cool by undergoing inter-state transitions that require only one emitted phonon followed by a cascade of single-phonon intra-state transitions; the bottom of each quantum state is reached by intra-state relaxation before an inter-state transition occurs. Thus, the simultaneous and slow multiphonon relaxation pathway can be bypassed by single-phonon events and the cooling rate increases correspondingly.

More complete theoretical models for slowed cooling in QDs have been proposed by Bockelmann and co-workers [20, 46] and Benisty and co-workers [19, 21]. The proposed Benisty mechanism [19, 21] for slowed hot carrier cooling and phonon bottleneck in QDs requires that cooling only occurs via LO phonon emission. However, there are several other mechanisms by which hot electrons can cool in QDs. Most prominent among these is the Auger mechanism [47].

Here, the excess energy of the electron is transferred via an Auger process to the hole, which then cools rapidly because of its larger effective mass and smaller energy level spacing. Thus, an Auger mechanism for hot electron cooling can break the phonon bottleneck [47]. Other possible mechanisms for breaking the phonon bottleneck include electron–hole scattering [48], deep level trapping [49] and acoustical-optical phonon interactions [50, 51].

9.2.2.2 *Experimental determination of relaxation/cooling dynamics and a phonon bottleneck in quantum dots*

Over the past several years, many investigations have been published that explore hot electron cooling/relaxation dynamics in QDs and the issue of a phonon bottleneck in QDs. The results are controversial and it is quite remarkable that there are so many reports that both support [22, 23, 52–69] and contradict [49, 70–83] the prediction of slowed hot electron cooling in QDs and the existence of a phonon bottleneck. One element of confusion that is specific to the focus of this chapter is that while some of these publications report relatively long hot electron relaxation times (tens of ps) compared to what is observed in bulk semiconductors, the results are reported as being not indicative of a phonon bottleneck because the relaxation times are not excessively long and PL is observed [84–86] (theory predicts a very long relaxation lifetime for excited carriers for the extreme, limiting condition of a phonon bottleneck; thus, the carrier lifetime would be determined by non-radiative processes and PL would be absent). However, since the interest here is on the rate of relaxation/cooling compared to the rate of electron separation and transfer, and impact ionization, we consider that slowed relaxation/cooling of carriers has occurred in QDs if the relaxation/cooling times are greater than 10 ps (about an order of magnitude greater than that for bulk semiconductors). This is because electron separation and transport and impact ionization can be very fast (sub-ps). For solar fuel production, previous work that measured the time of electron transfer from bulk III–V semiconductors to redox molecules (metallo-cenium cations) adsorbed on the surface found that electron transfer times can also be sub-ps to several ps [87–90]; hence, photoinduced hot carrier separation, transport and transfer can be competitive with electron cooling and relaxation if the latter is greater than about 10 ps. Impact ionization rates can also be in the sub-ps regime [91, 92].

In a series of papers, Sugawara *et al* [57, 58, 60] have reported slow hot electron cooling in self-assembled InGaAs QDs produced by Stranski–Krastinow (SK) growth on lattice-mismatched GaAs substrates. Using time-resolved PL measurements, the excitation-power dependence of PL, and the current dependence of the electroluminescence spectra, these researchers report cooling times ranging from 10 ps to 1 ns. The relaxation time increased with electron energy up to the fifth electronic state. Also, Mukai and Sugawara [93] have recently published an extensive review of phonon bottleneck effects in QDs, which concludes that the phonon bottleneck effect is indeed, present in QDs.

Gfroerer *et al* [69] report slowed cooling of up to 1 ns in strain-induced GaAs QDs formed by depositing tungsten stressor islands on a GaAs QW with AlGaAs barriers. A magnetic field was applied in these experiments to sharpen and further separate the PL peaks from the excited state transitions and, thereby, determine the dependence of the relaxation time on level separation. The authors observed hot PL from excited states in the QD, which could only be attributed to slow relaxation of excited (i.e. hot) electrons. Since the radiative recombination time is about 2 ns, the hot electron relaxation time was found to be of the same order of magnitude (about 1 ns). With higher excitation intensity sufficient to produce more than one electron–hole pair per dot, the relaxation rate increased.

A lifetime of 500 ps for excited electronic states in self-assembled InAs/GaAs QDs under conditions of high injection was reported by Yu *et al* [64]. PL from a single GaAs/AlGaAs QD [67] showed intense high-energy PL transitions, which were attributed to slowed electron relaxation in this QD system. Kamath *et al* [68] also reported slow electron cooling in InAs/GaAs QDs.

QDs produced by applying a magnetic field along the growth direction of a doped InAs/AlSb QW showed a reduction in the electron relaxation rate from 10^{12} to 10^{10} s^{-1} [59,94].

In addition to slow electron cooling, slow hole cooling was reported by Adler *et al* [65,66] in SK InAs/GaAs QDs. The hole relaxation time was determined to be 400 ps based on PL risetimes, while the electron relaxation time was estimated to be less than 50 ps. These QDs only contained one electron state but several hole states; this explained the faster electron cooling time since a quantized transition from a higher quantized electron state to the ground electron state was not present. Heitz *et al* [61] also report relaxation times for holes of about 40 ps for stacked layers of SK InAs QDs deposited on GaAs; the InAs QDs are overgrown with GaAs and the QDs in each layer self-assemble into an ordered column. Carrier cooling in this system is about two orders of magnitude slower than in higher dimensional structures.

All of these studies on slowed carrier cooling were conducted on self-assembled SK type of QDs. Studies of carrier cooling and relaxation have also been performed on II–VI CdSe colloidal QDs by Klimov *et al* [52, 76], Guyot-Sionnest *et al* [55], Ellingson *et al* [23] and Blackburn *et al* [22]. The Klimov group first studied electron relaxation dynamics from the first-excited 1P to the ground 1S state in CdSe QDs using interband pump-probe spectroscopy [76]. The CdSe QDs were pumped with 100 fs pulses at 3.1 eV to create high energy electron and holes in their respective band states, and then probed with fs white light continuum pulses. The dynamics of the interband bleaching and induced absorption caused by state filling was monitored to determine the electron relaxation time from the 1P to the 1S state. The results showed very fast 1P to 1S relaxation, on the order of 300 fs, and was attributed to an Auger process for electron relaxation which bypassed the phonon bottleneck. However, this experiment cannot separate the electron and hole dynamics from each other. Guyot-Sionnest *et al* [55] followed up these experiments using fs infrared pump-probe spectroscopy. A visible pump beam creates electrons and holes in the respective band states and a subsequent IR beam is split into an IR pump and an IR probe beam; the IR beams can be tuned to monitor only the intraband transitions of the electrons in the electron states and, thus, can separate electron dynamics from hole dynamics. The experiments were conducted with CdSe QDs that were coated with different capping molecules (TOPO, thiocresol and pyridine), which exhibit different hole-trapping kinetics. The rate of hole trapping increased in the order: TOPO, thiocresol and pyridine. The results generally show a fast relaxation component (1–2 ps) and a slow relaxation component ($\approx$200 ps). The relaxation times follow the hole-trapping ability of the different capping molecules and are longest for the QD systems with the fastest hole trapping caps; the slow component dominates the data for the pyridine cap, which is attributed to its faster hole-trapping kinetics.

These results [55] support the Auger mechanism for electron relaxation, whereby the excess electron energy is rapidly transferred to the hole which then relaxes rapidly through its dense spectrum of states. When the hole is rapidly removed and trapped at the QD surface, the Auger mechanism for hot electron relaxation is inhibited and the relaxation time increases. Thus, in these experiments, the slow 200 ps component is attributed to the phonon bottleneck, most prominent in pyridine-capped CdSe QDs, while the fast 1–2 ps component reflects the Auger relaxation process. The relative weight of these two processes in a given QD system depends upon the hole-trapping dynamics of the molecules surrounding the QD.

Klimov *et al* further studied carrier relaxation dynamics in CdSe QDs and published a series of papers on the results [52, 53]; a review of this work was also recently published [54]. These studies also strongly support the presence of the Auger mechanism for carrier relaxation in QDs. The experiments were done using ultrafast pump-probe spectroscopy with either two or three beams. In the former, the QDs were pumped with visible light across its bandgap (hole states to electron states) to produce excited state (i.e. hot) electrons; the electron relaxation was monitored by probing the bleaching dynamics of the resonant HOMO to LUMO transition with visible light, or by probing the transient IR absorption of the 1S to 1P intraband transition, which reflects the dynamics of electron occupancy in the LUMO state of the QD. The three-beam experiment was similar to that of Guyot-Sionnest *et al* [55] except that the probe in the experiments of Klimov *et al* is a white light continuum. The first pump beam is at 3 eV and creates electrons and holes across the QD bandgap. The second beam is in the IR and is delayed with respect to the optical pump; this beam re-pumps electrons that have relaxed to the LUMO back up in energy. Finally, the third beam is a broadband white light continuum probe that monitors photoinduced interband absorption changes over the range of 1.2 to 3 eV. The experiments were done with two different caps on the QDs: a ZnS cap and a pyridine cap. The results showed that with the ZnS-capped CdSe the relaxation time from the 1P to 1S state was about 250 fs, while for the pyridine-capped CdSe, the relaxation time increased to 3 ps. The increase in the latter experiment was attributed to a phonon bottleneck produced by rapid hole trapping by the pyridine, as also proposed by Guyot-Sionnest *et al* [55]. However, the time scale of the phonon bottleneck induced by hole trapping by pyridine caps on CdSe that were reported by Klimov *et al* was not as great as that reported by Guyot-Sionnest *et al* [55].

Recently, results from the National Renewable Energy Laboratory (NREL) [22, 23] were reported for the electron cooling dynamics in InP QDs where the QD surface was modified to affect hole trapping and also where only electrons were injected into the QD from an external redox molecule (sodium biphenyl) so that holes necessary for the Auger cooling mechanism were not present in the QD [22]. For InP, HF etching was found to passivate electronic surface states but not hole surface states [95, 96]; thus holes can become localized at the surface in both etched and unetched TOPO-capped QDs, and the dynamics associated with these two samples will not deviate significantly. The relaxation was found to be bi-exponential and suggests the presence of two subsets of quantum dots within the sample [22, 23]. Since etching has been shown to passivate hole traps inefficiently, it is proposed that two subsets of quantum dots are probed in the experiment: one subset in which the hole and electron are efficiently confined to the interior of the nanocrystal (hole trap absent; exciton confined to the QD core), and one subset in which the hole is localized at the surface of the QD on a phosphorous dangling bond (hole trap present; charge-separated QD) [22, 23].

With the electron and hole confined to the QD core, strong electron–hole interaction leads to efficient, fast relaxation via the Auger mechanism and in QDs, where the hole is localized at the surface, the increased spatial separation inhibits the Auger process and results in slower relaxation. The data imply that hole trapping at the intrinsic surface state occurs in less than 75 fs [22].

To further investigate the mechanisms involved in the intraband relaxation, experiments were conducted at NREL in which only electrons are present in the QDs and holes are absent. Sodium biphenyl is a very strong reducing agent which has been shown to inject electrons successfully into the conduction band of CdSe QDs [97, 98], effectively bleaching the 1S transition and allowing an IR-induced transition to the $1P_e$ level. Sodium biphenyl was, therefore, used to inject electrons into the 1S electron level in InP QDs [22]. This $1S_e$ electron may be excited to the $1P_e$ level with an IR pump and its relaxation dynamics back to the ground 1S state monitored. Time-resolved, IR-induced transitions in n-type (electron injected) InP QDs show that he relaxation of the excited electrons from the 1P to the 1S level can be fit to a single exponential, with an average time constant of 3.0 ps, corresponding to a relaxation rate of 0.092 eV ps^{-1}; in neutral 50 Å TOP/TOPO-capped InP QDs, the relaxation shows a large 400 fs component indicative of fast electron cooling. These experiments confirm that in the absence of a core-confined hole, electronic relaxation is slowed by about an order of magnitude. However, it should be noted that the relaxation rate in the absence of a hole is close to the relaxation rate with the hole localized at the surface. This is surprising and raises the question of why electron cooling in the absence of

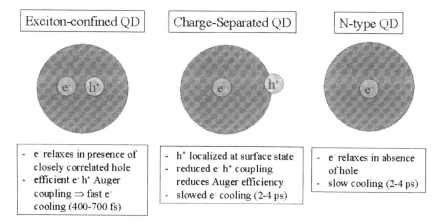

Figure 9.8. Different electron–hole configurations in a quantum dot and the resulting relaxation/cooling dynamics.

a hole is slower. Possible explanations have been proposed [99] including that (1) positive counter ions of the oxidized sodium biphenyl are adsorbed on the QD surface and behave like a trapped hole in producing a significant Coulomb interaction with the electron to permit Auger cooling and (2) an enhanced Huang–Rhees parameter occurs in charged QDs and enhances multiphonon relaxation. A summary of all of these experiments investigating the effects of electron cooling on electron–hole separation is shown in figure 9.8.

In contradiction to the results previously discussed, many other investigations exist in the literature in which a phonon bottleneck was apparently not observed. These results were reported for both self-organized SK QDs [49, 70–83] and II–VI colloidal QDs [76, 78, 80]. However, in several cases [61, 84, 86] hot electron relaxation was found to be slowed but not sufficiently for the authors to conclude that this was evidence of a phonon bottleneck. For the issue of hot electron transfer, this conclusion may not be relevant since in this case one is not interested in the question of whether the electron relaxation is slowed so drastically that non-radiative recombination dominates and quenches photoluminescence but rather whether the cooling is slowed sufficiently so that impact ionization can occur or excited state electron transport and transfer can occur across the semiconductor interface before cooling. For this purpose the cooling time need only be increased above about 10 ps, since carrier separation, transport and transfer, as well as impact ionization, can occur within this time scale [87–92].

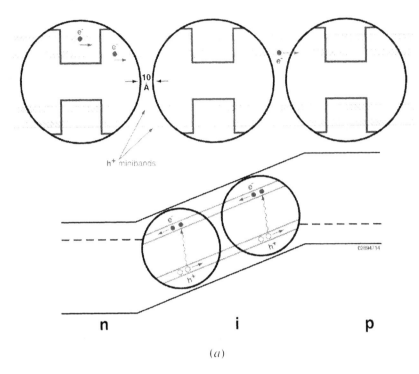

(a)

Figure 9.9. Configurations for quantum dot solar cells. (*a*) A QD array used as a photoelectrode for a photoelectrochemical or as the i-region of a p–i–n photovoltaic cell. (*b*) QDs used to sensitize a nanocrystalline film of a wide bandgap oxide semiconductor (namely TiO$_2$) to visible light. This configuration is analogous to the dye-sensitized solar cell where the dye is replaced by QDs. (*c*) QDs dispersed in a blend of electron- and hole-conducting polymers. In configurations (*a*)–(*c*), the occurrence of impact ionization could produce higher photocurrents and higher conversion efficiency. In (*a*), enhanced efficiency could be achieved either through impact ionization or hot carrier transport through the minibands of the QD array resulting in a higher photopotential (from Nozik A J 2002 *Physica* E **14** 115–20. Reprinted with permission).

9.3   Quantum dot solar cell configuration

The two fundamental pathways for enhancing the conversion efficiency (increased photovoltage [6, 7] or increased photocurrent [8, 9] can be accessed, in principle, in three different QD solar cell configurations; these configurations are shown in figure 9.9 and they are described here. However, it is emphasized that these potential high efficiency configurations are conceptual and there is no experimental evidence yet that demonstrates actual enhanced conversion efficiencies in any of these systems.

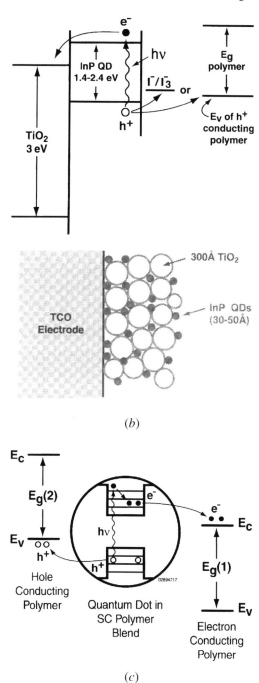

Figure 9.9. (Continued)

9.3.1   Photoelectrodes composed of quantum dot arrays

In this configuration, the QDs are formed into an ordered 3D array with inter-QD spacing sufficiently small such that strong electronic coupling occurs and minibands are formed to allow long-range electron transport (see figure 9.9(a)). The system is a 3D analogue to a 1D superlattice and the miniband structures formed therein [4] (see figure 9.4). The delocalized quantized 3D miniband states could be expected to slow the carrier cooling and permit the transport and collection of hot carriers to produce a higher photopotential in a PV cell or in a photoelectrochemical cell where the 3D QD array is the photoelectrode [100]. Also, impact ionization might be expected to occur in the QD arrays, enhancing the photocurrent (see figure 9.2). However, hot electron transport/collection and impact ionization cannot occur simultaneously: they are mutually exclusive and only one of these processes can be present in a given system.

Significant progress has been made in forming 3D arrays of both colloidal [101–103] and epitaxial [104] II–VI and III–V QDs. The former have been formed via evaporation and crystallization of colloidal QD solutions containing a uniform QD size distribution; crystallization of QD solids from broader size distributions lead to close-packed QD solids but with a high degree of disorder. Concerning the latter, arrays of epitaxial QDs have been formed by successive epitaxial deposition of epitaxial QD layers; after the first layer of epitaxial QDs is formed, successive layers tend to form with the QDs in each layer aligned on top of each other [104, 105]. Theoretical and experimental studies of the properties of QD arrays are currently under way. Major issues are the nature of the electronic states as a function of inter-dot distance, array order *versus* disorder, QD orientation and shape, surface states, surface structure/passivation and surface chemistry. Transport properties of QD arrays are also of critical importance and they are under investigation.

9.3.2   Quantum dot-sensitized nanocrystalline TiO$_2$ solar cells

This configuration is a variation of a recent promising new type of PV cell that is based on dye-sensitization of nanocrystalline TiO$_2$ layers [106–108]. In this latter PV cell, dye molecules are chemisorbed onto the surface of 10–30 nm-size TiO$_2$ particles that have been sintered into a highly porous nanocrystalline 10–20 μm TiO$_2$ film. Upon photoexcitation of the dye molecules, electrons are injected very efficiently from the excited state of the dye into the conduction band of the TiO$_2$, affecting charge separation and producing a PV effect. The cell circuit is completed using a non-aqueous redox electrolyte that contains I$^-$/I$_3^-$ and a Pt counter electrode to allow reduction of the adsorbed photooxidized dye back to its initial non-oxidized state (via I$_3^-$ produced at the Pt cathode by reduction of I$^-$).

For the QD-sensitized cell, QDs are substituted for the dye molecules; they can be adsorbed from a colloidal QD solution [109] or produced *in situ*

[110–113] (see figure 9.9(*b*)). Successful PV effects in such cells have been reported for several semiconductor QDs including InP, CdSe, CdS and PbS [109–113]. Possible advantages of QDs over dye molecules are the tunability of optical properties with size and better heterojunction formation with solid hole conductors. Also, as discussed here, a unique potential capability of the QD-sensitized solar cell is the production of quantum yields greater than one by impact ionization (inverse Auger effect) [42]. Efficient inverse Auger effects in QD-sensitized solar cells could produce much higher conversion efficiencies than are possible with dye-sensitized solar cells.

9.3.3 Quantum dots dispersed in organic semiconductor polymer matrices

Recently, photovoltaic effects have been reported in structures consisting of QDs forming junctions with organic semiconductor polymers. In one configuration, a disordered array of CdSe QDs is formed in a hole-conducting polymer—MEH-PPV (poly(2-methoxy,5-(2′-ethyl)-hexyloxy-p-phenylenevinylene) [114]. Upon photoexcitation of the QDs, the photogenerated holes are injected into the MEH-PPV polymer phase, and are collected via an electrical contact to the polymer phase. The electrons remain in the CdSe QDs and are collected through diffusion and percolation in the nanocrystalline phase to an electrical contact to the QD network. Initial results show relatively low conversion efficiencies [114, 115] but improvements have been reported with rod-like CdSe QD shapes [116] embedded in poly(3-hexylthiophene) (the rod-like shape enhances electron transport through the nanocrystalline QD phase). In another configuration [117], a polycrystalline TiO$_2$ layer is used as the electron conducting phase and MEH-PPV is used to conduct the holes; the electron and holes are injected into their respective transport mediums upon photoexcitation of the QDs.

A variation of these configurations is to disperse the QDs into a blend of electron and hole-conducting polymers (see figure 9.9(*c*)). This scheme is the inverse of light-emitting diode structures based on QDs [118–122]. In the PV cell, each type of carrier-transporting polymer would have a selective electrical contact to remove the respective charge carriers. A critical factor for success is to prevent electron–hole recombination at the interfaces of the two polymer blends; prevention of electron–hole recombination is also critical for the other QD configurations mentioned earlier.

All of the possible QD-organic polymer PV cell configurations would benefit greatly if the QDs can be coaxed into producing multiple electron–hole pairs by the inverse Auger/impact ionization process [42]. This is also true for all the QD solar cell systems described earlier. The various cell configurations simply represent different modes of collecting and transporting the photogenerated carriers produced in the QDs.

9.4   Conclusion

The relaxation dynamics of photoexcited electrons in semiconductor quantum dots can be greatly modified compared to the bulk form of the semiconductor. Specifically, the cooling dynamics of highly energetic (hot) electrons created by absorption of supra-bandgap photons can be slowed by at least one order of magnitude (4–7 ps *versus* 400–700 fs). This slowed cooling is caused by a so-called 'phonon bottleneck' when the energy spacing between quantized levels in the quantum dot is greater than the LO-phonon energy, thus inhibiting hot electron relaxation (cooling) by electron–phonon interactions. In order to produce the slowed hot electron cooling via the phonon bottleneck, it is necessary to block an Auger process that could bypass the phonon bottleneck and allow fast electron cooling. The Auger cooling process involves the transfer of the excess electron energy to a hole, which then cools rapidly because of its higher effective mass and closely-spaced energy levels. Blocking the Auger cooling is achieved by rapidly removing the photogenerated hole before it undergoes Auger scattering with the photogenerated electron or by injecting electrons into the LUMO level (conduction band) of the QD from an external electron donating chemical species and then exciting these electrons with an IR pulse. Slowed electron cooling in QDs offers the potential to use QDs in solar cells to enhance their conversion efficiency. In bulk semiconductors, the hot electrons (and holes) created by absorption of supra-bandgap photons cool so rapidly to the band edges that the excess kinetic energy of the photogenerated carriers is converted to heat and limits the theoretical Shockley–Quiesser thermodynamic conversion efficiency to about 32% (at one sun). Slowed cooling in QDs could lead to their use in solar cell configurations wherein impact ionization (the formation of two or more electron–hole pairs per absorbed photon) or hot electron separation, transport and transfer can become significant, thus producing enhanced photocurrents or photovoltages and corresponding enhanced conversion efficiencies with thermodynamics limits of 66% (one sun). Three configurations for QD solar cells have been described here that would produce either enhanced photocurrent or photovoltage.

Acknowledgments

This work was supported by the US Department of Energy, Office of Science, Office of Basic Energy Sciences, Division of Chemical Sciences, Geosciences and Biosciences and the Photovoltaics Program of the Office of Energy Efficiency and Renewable Energy. Vital contributions have been made by Olga Mićić, Randy Ellingson, Jeff Blackburn, Garry Rumbles, Phil Ahrenkiel, and Alexander Efros.

References

[1]   Shockley W and Queisser H J 1961 *J. Appl. Phys.* **32** 510

[2] Ross R T 1966 *J. Chem. Phys.* **45** 1

[3] Ross R T 1967 *J. Chem. Phys.* **46** 4590

[4] Nozik A J 2001 *Annu. Rev. Phys. Chem.* **52** 193

[5] Ellingson R J, Blackburn J L, Yu P, Rumbles G, Micic O I and Nozik A J 2002 *J. Phys. Chem.* **106** 7758

[6] Ross R T and Nozik A J 1982 *J. Appl. Phys.* **53** 3813

[7] Boudreaux D S, Williams F and Nozik A J 1980 *J. Appl. Phys.* **51** 2158

[8] Landsberg P T, Nussbaumer H and Willeke G 1993 *J. Appl. Phys.* **74** 1451

[9] Kolodinski S, Werner J H, Wittchen T and Queisser H J 1993 *Appl. Phys. Lett.* **63** 2405

[10] Green M A 2001 *Third Generation Photovoltaics* (Sydney: Bridge Printery)

[11] Luque A and Marti A 1997 *Phys. Rev. Lett.* **78** 5014

[12] Nozik A J, Boudreaux D S, Chance R R, and Williams F 1980 *Charge Transfer at Illuminated Semiconductor-Electrolyte Interfaces (Advances in Chemistry 184)* ed M Wrighton (New York: ACS) p 162

[13] Williams F E and Nozik A J 1984 *Nature* **311** 21

[14] Nozik A J 1980 *Phil. Trans. R. Soc.* A **295** 453

[15] Pelouch W S, Ellingson R J, Powers P E, Tang C L, Szmyd D M and Nozik A J 1992 *Phys. Rev.* B **45** 1450

[16] Pelouch W S, Ellingson R J, Powers P E, Tang C L, Szmyd D M and Nozik A J 1992 *Semicond. Sci. Technol.* **7** B337

[17] Rosenwaks Y, Hanna M C, Levi D H, Szmyd D M, Ahrenkiel R K and Nozik A J 1993 *Phys. Rev.* B **48** 14 675

[18] Williams F and Nozik A J 1978 *Nature* **271** 137

[19] Benisty H, Sotomayor-Torres C M and Weisbuch C 1991 *Phys. Rev.* B **44** 10 945

[20] Bockelmann U and Bastard G 1990 *Phys. Rev.* B **42** 8947

[21] Benisty H 1995 *Phys. Rev.* B **51** 13281

[22] Blackburn J L, Ellingson R J, Micic O I and Nozik A J 2003 *J. Phys. Chem.* **107** 102

[23] Ellingson R J, Blackburn J L, Nedeljkovic J M, Rumbles G, Jones M, Fu H and Nozik A J 2003 *Phys. Rev.* B **67** 75 308

[24] Stanton C J, Bailey D W, Hess K and Chang Y C 1988 *Phys. Rev.* B **37** 6575

[25] Pankove J I 1975 *Optical Processes in Semiconductors* (New York: Dover)

[26] Sze S 1981 *Physics of Semiconductor Devices* (New York: Wiley)

[27] Green M A 1992 *Solar Cells* (Kensington: The University of New South Wales)

[28] Nozik A J 1978 *Annu. Rev. Phys. Chem.* **29** 189

[29] Dingle R, ed. 1987 *Applications of Multiquantum Wells, Selective Doping and Superlattices (Semiconductors and Semimetals 24)* (New York: Academic)

[30] Peterson M W, Turner J A, Parsons C A, Nozik A J, Arent D J, Van Hoof C, Borghs G, Houdre R and Morkoc H 1988 *Appl. Phys. Lett.* **53** 2666

[31] Parsons C A, Thacker B R, Szmyd D M, Peterson M W, McMahon W E and Nozik A J 1990 *J. Chem. Phys.* **93** 7706

[32] Edelstein D C, Tang C L and Nozik A J 1987 *Appl. Phys. Lett* **51** 48

[33] Xu Z Y and Tang C L 1984 *Appl. Phys. Lett.* **44** 692

[34] Rosker M J, Wise F W and Tang C L 1986 *Appl. Phys. Lett* **49** 1726

[35] Ryan J F, Taylor R A, Tuberfield A J, Maciel A, Worlock J M, Gossard A C and Wiegmann W 1984 *Phys. Rev. Lett.* **53** 1841

[36] Christen J and Bimberg D 1990 *Phys. Rev.* B **42** 7213

[37] Cai W, Marchetti M C and Lax M 1986 *Phys. Rev.* B **34** 8573

[38] Nozik A J, Parsons C A, Dunlavy D J, Keyes B M and Ahrenkiel R K 1990 *Solid State Commun.* **75** 297

[39] Lugli P and Goodnick S M 1987 *Phys. Rev. Lett.* **59** 716

[40] Campos V B, Das Sarma S and Stroscio M A 1992 *Phys. Rev.* B **46** 3849

[41] Joshi R P and Ferry D K 1989 *Phys. Rev.* B **39** 1180

[42] Nozik A J 1997 unpublished manuscript

[43] Nirmal M, Dabbousi B O, Bawendi M G, Macklin J J, Trautman J K, Harris T D and Brus L E 1996 *Nature* **383** 802

[44] Efros A L and Rosen M 1997 *Phys. Rev. Lett.* **78** 1110

[45] Jaros M 1989 Quantum wells, superlattices, quantum wires and dots *Physics and Applications of Semiconductor Microstructures* (New York: Oxford University Press) p 83

[46] Bockelmann U and Egeler T 1992 *Phys. Rev.* B **46** 15574

[47] Efros A L, Kharchenko V A and Rosen M 1995 *Solid State Commun.* **93** 281

[48] Vurgaftman I and Singh J 1994 *Appl. Phys. Lett.* **64** 232

[49] Sercel P C 1995 *Phys. Rev.* B **51** 14532

[50] Inoshita T and Sakaki H 1992 *Phys. Rev.* B **46** 7260

[51] Inoshita T and Sakaki H 1997 *Phys. Rev.* B **56** R4355

[52] Klimov V I, Mikhailovsky A A, McBranch D W, Leatherdale C A and Bawendi M G 2000 *Phys. Rev.* B **61** R13349

[53] Klimov V I, McBranch D W, Leatherdale C A and Bawendi M G 1999 *Phys. Rev.* B **60** 13740

[54] Klimov V I 2000 *J. Phys. Chem.* B **104** 6112

[55] Guyot-Sionnest P, Shim M, Matranga C and Hines M 1999 *Phys. Rev.* B **60** R2181

[56] Wang P D, Sotomayor-Torres C M, McLelland H, Thoms S, Holland M and Stanley C R 1994 *Surf. Sci.* **305** 585

[57] Mukai K and Sugawara M 1998 *Japan. J. Appl. Phys.* **37** 5451

[58] Mukai K, Ohtsuka N, Shoji H and Sugawara M 1996 *Appl. Phys. Lett.* **68** 3013

[59] Murdin B N, *et al* 1999 *Phys. Rev.* B **59** R7817

[60] Sugawara M, Mukai K and Shoji H 1997 *Appl. Phys. Lett.* **71** 2791

[61] Heitz R, Veit M, Ledentsov N N, Hoffmann A, Bimberg D, Ustinov V M, Kop'ev P S and Alferov Z I 1997 *Phys. Rev.* B **56** 10435

[62] Heitz R, Kalburge A, Xie Q, Grundmann M, Chen P, Hoffmann A, Madhukar A and Bimberg D 1998 *Phys. Rev.* B **57** 9050

[63] Mukai K, Ohtsuka N, Shoji H and Sugawara M 1996 *Phys. Rev.* B **54** R5243

[64] Yu H, Lycett S, Roberts C and Murray R 1996 *Appl. Phys. Lett.* **69** 4087

[65] Adler F, Geiger M, Bauknecht A, Scholz F, Schweizer H, Pilkuhn M H, Ohnesorge B and Forchel A 1996 *Appl. Phys.* **80** 4019

[66] Adler F, Geiger M, Bauknecht A, Haase D, Ernst P, Dörnen A, Scholz F and Schweizer H 1998 *J. Appl. Phys.* **83** 1631

[67] Brunner K, Bockelmann U, Abstreiter G, Walther M, Böhm G, Tränkle G and Weimann G 1992 *Phys. Rev. Lett.* **69** 3216

[68] Kamath K, Jiang H, Klotzkin D, Phillips J, Sosnowski T, Norris T, Singh J and Bhattacharya P 1998 *Inst. Phys. Conf. Ser.* **156** 525

[69] Gfroerer T H, Sturge M D, Kash K, Yater J A, Plaut A S, Lin P S D, Florez L T, Harbison J P, Das S R and Lebrun L 1996 *Phys. Rev.* B **53** 16474

[70] Li X-Q, Nakayama H and Arakawa Y 1998 Phonon decay and its impact on carrier

relaxation in semiconductor quantum dots *Proc. 24th Int. Conf. Phys. Semicond.* ed D Gershoni (Singapore: World Scientific) p 845

[71] Bellessa J, Voliotis V, Grousson R, Roditchev D, Gourdon C, Wang X L, Ogura M and Matsuhata H 1998 Relaxation and radiative lifetime of excitons in a quantum dot and a quantum wire *Proc. 24th Int. Conf. Phys. Semicond.* ed D Gershoni (Singapore: World Scientific) p 763

[72] Lowisch M, Rabe M, Kreller F and Henneberger F 1999 *Appl. Phys. Lett.* **74** 2489

[73] Gontijo I, Buller G S, Massa J S, Walker A C, Zaitsev S V, Gordeev N Y, Ustinov V M and Kop'ev P S 1999 *Japan. J. Appl. Phys.* **38** 674

[74] Li X-Q, Nakayama H and Arakawa Y 1999 *Japan. J. Appl. Phys.* **38** 473

[75] Kral K and Khas Z 1998 *Phys. Status Solidi* B **208** R5

[76] Klimov V I and McBranch D W 1998 *Phys. Rev. Lett.* **80** 4028

[77] Bimberg D, Ledentsov N N, Grundmann M, Heitz R, Boehrer J, Ustinov V M, Kop'ev P S and Alferov Z I 1997 *J. Lumin.* **72–74** 34

[78] Woggon U, Giessen H, Gindele F, Wind O, Fluegel B and Peyghambarian N 1996 *Phys. Rev.* B **54** 17681

[79] Grundmann M *et al* 1996 *Superlatt. Microstruct.* **19** 81

[80] Williams V S, Olbright G R, Fluegel B D, Koch S W and Peyghambarian N 1988 *J. Mod. Opt.* **35** 1979

[81] Ohnesorge B, Albrecht M, Oshinowo J, Forchel A and Arakawa Y 1996 *Phys. Rev.* B **54** 11532

[82] Wang G, Fafard S, Leonard D, Bowers J E, Merz J L and Petroff P M 1994 *Appl. Phys. Lett.* **64** 2815

[83] Sandmann J H H, Grosse S, von Plessen G, Feldmann J, Hayes G, Phillips R, Lipsanen H, Sopanen M and Ahopelto J 1997 *Phys. Status Solidi* B **204** 251

[84] Heitz R *et al* 1998 *Physica* E **2** 578

[85] Li X-Q and Arakawa Y 1998 *Phys. Rev.* B **57** 12285

[86] Sosnowski T S, Norris T B, Jiang H, Singh J, Kamath K and Bhattacharya P 1998 *Phys. Rev.* B **57** R9423

[87] Miller R D J, McLendon G, Nozik A J, Schmickler W and Willig F 1995 *Surface Electron Transfer Processes* (New York: VCH)

[88] Meier A, Selmarten D C, Siemoneit K, Smith B B and Nozik A J 1999 *J. Phys. Chem.* B **103** 2122

[89] Meier A, Kocha S S, Hanna M C, Nozik A J, Siemoneit K, Reineke-Koch R and Memming R 1997 *J. Phys. Chem.* B **101** 7038

[90] Diol S J, Poles E, Rosenwaks Y and Miller R J D 1998 *J. Phys. Chem.* B **102** 6193

[91] Bude J and Hess K 1992 *J. Appl. Phys.* **72** 3554

[92] Harrison D, Abram R A and Brand S 1999 *J. Appl. Phys.* **85** 8186

[93] Mukai K and Sugawara M 1999 *The Phonon Bottleneck Effect in Quantum Dots (Semiconductors and Semimetals 60)* ed R K Willardson and E R Weber (San Diego, CA: Academic) p 209

[94] Murdin B N *et al* 1998 Suppression of LO phonon scattering in quasi quantum dots *Proc. 24th Int. Conf. Phys. Semicond.* ed D Gershoni (Singapore: World Scientific) p 1867

[95] Langof L, Ehrenfreund E, Lifshitz E, Micic O I and Nozik A J 2002 *J. Phys. Chem.* B **106** 1606

[96] Micic O I, Nozik A J, Lifshitz E, Rajh T, Poluektov O G and Thurnauer M C 2002 *J. Phys. Chem.* **106** 4390

[97]   Shim M and Guyot-Sionnest P 2000 *Nature* **407** 981
[98]   Shim M, Wang C and Guyot-Sionnest P J 2001 *J. Phys. Chem.* **105** 2369
[99]   Efros A L and Nozik A J 2003 *J. Phys. Chem.* B to be published
[100] Nozik A J 1996 unpublished manuscript
[101] Murray C B, Kagan C R and Bawendi M G 2000 *Annu. Rev. Mater. Sci.* **30** 545
[102] Micic O I, Ahrenkiel S P and Nozik A J 2001 *Appl. Phys. Lett.* **78** 4022
[103] Micic O I, Jones K M, Cahill A and Nozik A J 1998 *J. Phys. Chem.* B **102** 9791
[104] Sugawara M 1999 *Self-Assembled InGaAs/GaAs Quantum Dots (Semiconductors and Semimetals 60)* ed M Sugawara (San Diego, CA: Academic) p 1
[105] Nakata Y, Sugiyama Y and Sugawara M 1999 *Semiconductors and Semimetals* vol 60, ed M Sugawara (San Diego, CA: Academic) p 117
[106] Hagfeldt A and Grätzel M 2000 *Acc. Chem. Res.* **33** 269
[107] Moser J, Bonnote P and Grätzel M 1998 *Coord. Chem. Rev.* **171** 245
[108] Grätzel M 2000 *Prog. Photovolt.* **8** 171
[109] Zaban A, Micic O I, Gregg B A and Nozik A J 1998 *Langmuir* **14** 3153
[110] Vogel R and Weller H 1994 *J. Phys. Chem.* **98** 3183
[111] Weller H 1991 *Ber. Bunsen-Ges. Phys. Chem.* **95** 1361
[112] Liu D and Kamat P V 1993 *J. Phys. Chem.* **97** 10 769
[113] Hoyer P and Könenkamp R 1995 *Appl. Phys. Lett.* **66** 349
[114] Greenham N C, Poeng X and Alivisatos A P 1996 *Phys. Rev.* B **54** 17 628
[115] Greenham N C, Peng X and Alivisatos A P 1997 A CdSe Nanocrystal/MEH-PPV Polymer Composite Photovoltaic *Future Generation Photovoltaic Technologies: First NREL Conf.* ed R Mcconnell (New York: AIP) p 295
[116] Huynh W U, Peng X and Alivisatos P 1999 *Adv. Mater.* **11** 923
[117] Arango A C, Carter S A and Brock P J 1999 *Appl. Phys. Lett.* **74** 1698
[118] Dabbousi B O, Bawendi M G, Onitsuka O and Rubner M F 1995 *Appl. Phys. Lett.* **66** 1316
[119] Colvin V, Schlamp M and Alivisatos A P 1994 *Nature* **370** 354
[120] Schlamp M C, Peng X and Alivisatos A P 1997 *J. Appl. Phys.* **82** 5837
[121] Mattoussi H, Radzilowski L H, Dabbousi B O, Fogg D E, Schrock R R, Thomas E L, Rubner M F and Bawendi M G 1999 *J. Appl. Phys.* **86** 4390
[122] Mattoussi H, Radzilowski L H, Dabbousi B O, Thomas E L, Bawendi M G and Rubner M F 1998 *J. Appl. Phys.* **83** 7965

Chapter 10

Progress in thermophotovoltaic converters

Bernd Bitnar[1], Wilhelm Durisch[1], Fritz von Roth[1], Günther Palfinger[1], Hans Sigg[1], Detlev Grützmacher[1], Jens Gobrecht[1], Eva-Maria Meyer[2], Ulrich Vogt[2], Andreas Meyer[3] and Adolf Heeb[4]

[1] *Paul Scherrer Institut CH-5232 Villigen PSI, Switzerland*
[2] *EMPA CH-8600 Dübendorf, Switzerland*
[3] *Solaronix SA CH-1170 Aubonne, Switzerland*
[4] *Hovalwerk AG FL-9490 Vaduz, Liechtenstein*

10.1   Introduction

Thermophotovoltaics (TPV) is a technique to convert radiation from a synthetic emitter into electricity by the use of photocells. In TPV there is a larger flexibility in achieving a high efficiency compared with solar photovoltaics (PV), where the solar spectrum as well as the radiation power is naturally fixed when no concentrating system is used. The emitter spectrum, however, can be widely varied by the choice of emitter material, its surface structure and its temperature. The radiation density on the cell surface depends on the distance to the emitter, as well as on the emission characteristics. In solar PV, the efficiency is increased by improving the photocell, especially by reducing the thermalization losses, which occur due to the absorption of the broad-band solar spectrum. Third-generation PV photocell designs often use a band structure with several transitions to adapt the cell to the Sun's broad-band spectrum. In TPV the system's efficiency is optimized by spectrally matching the emitter to the photocells. By using conventional single bandgap photocells, a high efficiency is achieved for radiation from a selective emitter with a narrow-band radiation spectrum located closely above the bandgap of the cells.

In the past, therefore, TPV was discussed with respect to converting solar radiation into electricity. A solar TPV system uses a light concentrator to focus

223

the sunlight onto the selective emitter. The emitter converts the broad-band solar irradiation into a narrow-band radiation spectrum matched to the bandgap of the photocells. In an idealized case, the system's efficiency can resemble that of the photocell for monochromatic irradiation. The temperature of the emitter is limited by the stability of the used material. Therefore, the emitter has a much lower temperature than the sun. In this case, the emission of infrared radiation with photon energies smaller than the bandgap of the photocells is an important loss mechanism. So far, experimental solar TPV systems have exhibited only poor efficiencies.

TPV attains more interest due to its capability to convert heat into electricity. In a combustion-driven TPV furnace, gas, oil or biomass can be fired to heat the emitter. An attractive application is combined heat and power generation, where the heat of the exhaust gas as well as the heat accrued from cell cooling is used for room heating or warm water production. TPV is, therefore, highly suitable for integration into residential heating systems.

TPV was investigated for military electricity production with silent transportable generators. Another application field, which will be discussed for the future, is its use for electricity production in deep space probes. In this case the emitter is heated by a radioactive source. A good overview of TPV developments so far is given in [1].

10.2   TPV based on III/V low-bandgap photocells

The convertible radiation power of an emitter strongly depends on its temperature and the energy of the cell's bandgap. The temperature is limited by the thermal stability of the materials used and, for commercial applications, by the emission of NO_x in the combustion gas. For the latter reason, the maximum emitter temperature is about 1800 K. To convert the radiation of such an emitter efficiently into electricity, GaSb photocells with a bandgap of 0.72 eV were used in several experimental TPV systems. A well-developed TPV system is the Midnight Sun stove from JX Crystals (USA), described in [2]. This system uses a SiC broad-band emitter, an infrared filter and GaSb photocells. The TPV generator produces 100 W of electrical power. The thermal power of the burner is about 7 kW. The system is part of a small stove for room heating. Several of these TPV stoves have been manufactured and field tested.

An improved TPV generator using GaSb photocells has been designed and this is described in [3]. To increase the system's efficiency, a selective emitter made from antireflection-coated metal with an emission spectrum well matched to the GaSb bandgap has been developed. The burner is equipped with a recuperator to reduce drastically the heat loss in the exhaust gas. However, in [3], only the experimental tests of single components of the system are described.

For TPV applications with lower emitter temperatures around 1300 K, photocells with even lower bandgaps are required. In [4] the development of a monolithic interconnected module (MIM) from GaInAs/InAsP tandem cells is described. The bandgaps of these materials are 0.74 and 0.55 eV, which are suitable for relatively cool broad-band emitters. A possible application is their use with a radionuclide heat source as an electricity generator for deep space vehicles.

Two disadvantages of low-bandgap III/V semiconductors are that they contain toxic elements like Sb and As and their production process is very expensive. They may be used for special applications where a high system efficiency is required but widespread use in residential furnaces seems to be doubtful. Ge with a bandgap of 0.66 eV could be an alternative to GaSb. Relevant progress has been made in the development of Ge photocells specified for TPV [5]. But so far, their efficiency remains lower compared to GaSb and no TPV system working with Ge cells has so far been reported.

10.3   TPV in residential heating systems

Natural gas and oil fired furnaces are widely used for room heating and sanitary water production in the household. An integrated TPV system can convert part of the thermal energy into electricity. As a first application, this electricity can be used to supply the electrical components such as the circulation water pump, air fan and electronics of the heating system for an electrically autonomous operation. Such a TPV heater can operate in rural sites, where no electricity grid is available, or during natural disasters, when the electricity grid fails. Typically, a conventional residential gas heating system with a maximum thermal power of 20 kW for a one-family house needs 100–150 W of electrical power.

If the TPV system has a higher electrical power, excess electricity can be supplied into the house grid or, by using an inverter, into the external mains. TPV systems can help to use energy more efficiently, because in a co-generating TPV furnace, electricity is generated with 100% efficiency from the additional consumed gas compared with 40% efficient electricity production in conventional power stations including power transmission losses. Another advantage is the combination of a TPV furnace with a PV system, because both techniques have ambivalent production periods. During the day in summer, PV generates its maximum power; in the winter, however, the TPV heating system has its production peak. Both systems can use the same grid inverter.

To be used in heating furnaces, the TPV system has to be cost effective, safe in case of malfunction and ecologically inoffensive. The system efficiency is less important, if the TPV is only used for an electrically self-powered operation of the heater. Si solar cells in combination with a suitable selective emitter easily fulfil these requirements.

A TPV boiler for supplying room heat and hot water using Si photocells has been developed [6]. This system uses an alumina fibre matrix emitter coated with Yb_2O_3 providing a selective emission peak centred at 1.27 eV photon energy.

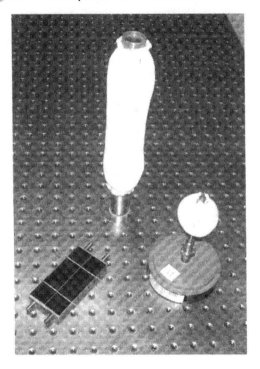

Figure 10.1. TPV components: two selective Yb_2O_3 emitters and Si photocells (RWE Schott Solar, Germany) glued onto a water-cooled metal block.

This fits well into the spectral response of the high efficiency solar cells used to deliver energy. A system with a thermal power of 25 kW achieved 0.8% efficiency resulting in a maximum electrical power of about 200 W. Due to the use of low-consumption electrical components 142 W surplus electrical power remained in steady-state operation to be supplied into the house grid.

Another approach using a Yb_2O_3 mantle emitter and Si solar cells is presented in [7] and [8]. The technology was applied for a small prototype system with 2 kW thermal power, which achieved a record system efficiency of 2.4% using high efficiency Si solar cells from the University of New South Wales (Australia) (UNSW). A larger demonstration system with 12–20 kW thermal power achieved a maximum efficiency of 1.0%. Relatively inexpensive solar cells from RWE Schott Solar (Germany) were used. To judge the market potential of this TPV technology, a cost estimate is given in [9]. The result of the estimation gives an electricity cost of about 25 €cent/kWh for the existing technology with a cost reduction potential for future technical improvements of about a factor of four. Therefore, TPV could offer economic electricity.

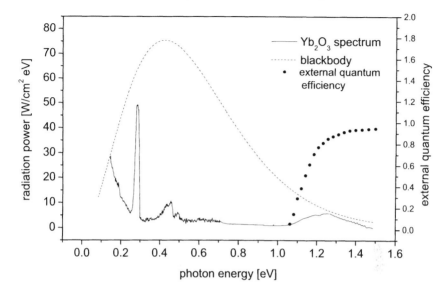

Figure 10.2. Radiation spectrum of the Yb_2O_3 mantle emitter (left-hand axis) and external quantum efficiency of the SH2 photocell (right-hand axis). A blackbody spectrum with $T = 1735$ K is shown for comparison.

10.4 Progress in TPV with silicon photocells

10.4.1 Design of the system and a description of the components

The most important components of our TPV systems, the selective emitter and the Si photocell, are shown in figure 10.1. The emitters consist of a Yb_2O_3 mantle structure, similar to incandescent mantles. For their production, a raw-cotton mantle is impregnated with Yb-nitrate, followed by hydrolysis and a baking-out step. A detailed description of the emitter fabrication process is given in [7]. Figure 10.1 shows two emitters with different sizes, which were fixed around perforated burner tubes. During burning the gas flows through the holes of the burner tube and penetrates through the emitter. The flame is fixed at the emitter structure resulting in an emitter temperature close to the temperature of the out-flowing flame gas. The thin structure is mechanically fragile, so that care has to be taken when mounting the emitter in a TPV system.

The photocells are monocrystalline Si solar cells SH2 from RWE Schott Solar (Germany). The cells are cut into 25 mm wide stripes, which are series connected and glued onto metal blocks with drillings for cooling water (see figure 10.1). A 100 μm thick ethylvinylacetate (EVA) foil covers the front side of the cells to protect them against mechanical damage or condensed water [8]. The cell features a very low reflectance and a high spectral response for low-energy radiation around 1.2 eV.

Figure 10.3. Photocell generators for the small prototype (left) and the demonstration system (right).

Figure 10.2 shows the radiation power spectrum of our mantle emitter structure (left-hand axis) together with the external quantum efficiency of the SH2 photocell (right-hand axis). It is clearly visible that the selective peak with its maximum at 1.27 eV can be almost completely converted by the photocell. The emission below the selective peak is low, which is demonstrated by the comparison with the blackbody spectrum for the emitter temperature of 1735 K. Two emission lines at 0.28 and 0.45 eV originate from CO_2 and OH bonds of water in the hot combustion gas. The selectivity of this emitter, defined as convertible radiation power for Si photocells ($E > 1.1$ eV) relative to the total emitted radiation, is 20%. The result of a detailed characterization of the Yb_2O_3 mantle emitter is given in [10].

The selectivity of the emitter is high enough, so that no additional filter is necessary to operate our TPV system. To prevent the cell surface from contact with the hot exhaust gas, a simple quartz tube is placed between the emitter and the cells.

10.4.2   Small prototype and demonstration TPV system

Two TPV systems were built using these components. Figure 10.3 shows a photograph of the photocell generators of the small prototype and large demonstration system. Both cylindrical generators have similar designs. The

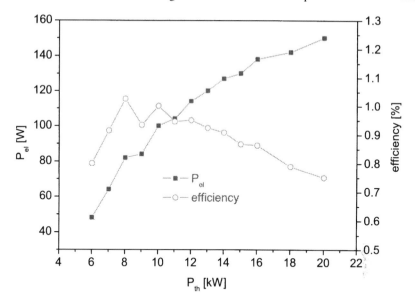

Figure 10.4. Electrical power P_{el} (left-hand axis) and system efficiency (right-hand axis) as a function of the thermal input power P_{th} of the TPV demonstration system.

single cooling blocks are mounted into a metal housing. Flexible tubes connect the cooling channels in the blocks. In the large generator, the photocells are arranged in three annuli of series connected cells with a total cell area of $2100 \, cm^2$.

The small prototype system operates with a methane burner with 0.7–2.5 kW thermal power and the small emitter, which is shown in figure 10.1. A quartz tube is placed around the emitter and protects the small photocell generator shown in figure 10.3. Two gold reflectors close the system in the axial direction. The upper reflector has a central hole with a 40 mm diameter through which the exhaust gas escapes. The small prototype was built with several modifications, which are described in [11]. With the SH2 photocell generator and a 2 kW butane burner, a system efficiency of 1.5% was achieved. With a generator consisting of the higher efficient UNSW cells, the system's efficiency was increased to 2.4%, producing 48 W of electrical power.

The demonstration system is kept as simple as possible and operates without axial reflectors. A 6–20 kW methane burner heats the large cylindrical emitter shown in figure 10.1. The photocell generator is protected by a quartz tube. Every cell annulus is connected to a grid inverter and the generated electrical power is supplied into the grid. The electrical output power is measured with a three-channel power analyzer.

Figure 10.4 shows the resulting system efficiency and electrical DC power output P_{el} as a function of the thermal power P_{th}. P_{el} increases with P_{th}. The

Figure 10.5. The photograph shows the Hoval TPV prototype heating system in preparation for a long-term test.

system's efficiency reaches a maximum around 1.0% between 8 and 12 kW thermal power. Scattering in the curves result from poor power tracking of the grid inverters. Even though the emitter temperature increases with increasing P_{th} producing a higher radiation power, the system's efficiency drops. The reason for this descent is the decreasing fill factor of the photocells. The SH2 photocells are designed for non-concentrated solar illumination. In the TPV demonstration system at 20 kW thermal power the electrical power density of the cells is 75 mW cm^{-2}. This is about a factor of five higher than under AM 1.5g solar irradiation. Under this high illumination level, the fill factor is limited by the series resistance of the contacts.

To run a typical commercial heating system independently from the electricity grid, an electrical power of about 120 W is needed. Our TPV demonstration system generates enough electrical power at a thermal power of at least 13 kW. At $P_{th} = 20$ kW an electrical power of 150 W was achieved, corresponding to a surplus of 30 W.

10.4.3 Prototype heating furnace

To study the suitability of our TPV technology in residential heating systems, as well as to investigate the long-term stability of the mantle emitter, a prototype

TPV furnace was built. Figure 10.5 shows the Hoval prototype gas heating system. A cell generator similar to the one in the demonstration system was built into the centre of the furnace. A fan blows the gas/air mixture from the top of the system into the burner tube with emitter, similar to the emitter shown in figure 10.1.

The electrical power of the TPV generator was measured with grid inverters and a power analyser. The gas flow, the NO_x and CO emissions in the exhaust and the photocell temperature were monitored with a data acquisition system.

In a first test, a Y_2O_3 mantle was mounted around the burner tube. The furnace was started and operated for 9 hr/day to investigate the long-term stability of the burner tube/mantle configuration. A stable operation over 54 hr burning time was achieved. Afterwards, the electricity output dropped and, after 95 hr, part of the burner tube had melted and the Y_2O_3 mantle was destroyed. The emitter radiation towards the burner tube seems to heat the tube too strongly and limits its long-term stability. In a subsequent experiment with a Y_2O_3 mantle and a burner tube which was protected with a heat-resistive kanthal web on its outer surface, stable operation over 108 hr was achieved.

In a second test, an electrical power of 64 W was achieved with a Yb_2O_3 emitter. The thermal power of the burner was 14 kW with an air number (ratio of combustion air volume to air volume necessary for a stochiometric combustion) of 1.09. The emission of NO_x was 85 ppm and the CO emission 10 ppm. The photocells were cooled with tap water resulting in a stable cell temperature of 32–39 °C. Unfortunately, the emitter in this experiment was mechanically damaged. In the future, it is expected to achieve at least a similar electricity output with the prototype furnace as with the demonstration system in the laboratory exhibited so far.

A cost estimate for this TPV system resulted in additional costs to those of a conventional heating system of 590 € [9]. Assuming a system lifetime of 20 years, an electricity price including the costs for gas of 0.25 € kWh^{-1} were calculated. In this calculation the most expensive components are the photocells followed by the quartz tube and the cell cooling. Therefore, it is planned to replace the quartz tube by Duran® glass and to simplify the cell cooling. A further improvement, in which concentrator photocells are used to reduce the system size, could achieve an electricity price of 0.10 € kWh^{-1}. Details of this cost estimation are given in [9].

10.4.4 Foam ceramic emitters

In order to obtain a higher mechanical stability for the emitter structures, porous foam ceramics were investigated. Early results of this investigation are presented in the following.

The foam emitters were manufactured using the polymer-sponge method [12]. In this method a polymer sponge is first impregnated with a ceramic slurry. The slurry is then dried and the sponge is vaporized by heating up to 600 °C in air.

The dried ceramic, which retains the foam structure of the sponge, is subsequently sintered to solidify the network.

A polyurethane sponge material was used as it can easily be formed into emitter tubes of a desired shape and size. The water-based slurry contained ceramic powder, water and a variety of organic substances including a binder, rheology-modifying additives and antifoaming and flocculating agents. While Al_2O_3, Zr_2O_3 and SiC are commonly used materials for foam ceramics, for the fabrication of TPV emitters, pure Yb_2O_3 was used as the main ceramic phase mixed with Al_2O_3, Y_2O_3, CeO_2 or Zr_2O_3 sintering additives.

Figure 10.6 shows the porous structure of a pure Yb_2O_3 foam sample with a porosity of 20 ppi (pores per inch) before and after sintering at 1700 °C. The dry ceramic (top photograph) forms a dense network without visible cracks in the struts. The porosity of the sintered foam (bottom picture) determines the gas flow resistance and the emission behaviour. To operate a foam emitter, which is placed around a burner tube similar to a mantle emitter, the gas flow resistance must not exceed a specific value, i.e. a certain minimum pore size is required. However, the pore size must not be so high as to cause the structure to become optically transparent and the pores should not form optical cavities, which can radiate as blackbody sources and, thus, destroy the selective emission behaviour of the material. Experimentally, 20 ppi foams were found to allow an easy ignition of the flame and to stabilize the flame in the emitter structure.

To improve the mechanical stability of the foams, they can be impregnated a second time with slurry after a first sintering step (re-infiltration), and sintered again to high temperature [12]. In a similar way, Al_2O_3 foams can be coated with Yb_2O_3 by using an Yb_2O_3 slurry for the second infiltration cycle. Such re-infiltrated Yb_2O_3 samples and Yb_2O_3-coated Al_2O_3 ceramics are shown in figure 10.7 together with pure Yb_2O_3 foams.

Emitter tubes to be tested with the small prototype system as well as samples with different compositions, which can be heated on a flat gas burner to compare their emission characteristics, have been produced. The results showed that up to 10 mol% additives could be added to the Yb_2O_3 without significantly influencing the selective emission. Therefore, the composition to be chosen should be the one that imparts the best mechanical and thermal stability. Subsequently, ceramic foam tubes were mounted into the small prototype system and ignited several times to study their thermal shock stability. Unfortunately, the pure Yb_2O_3 foam emitter cracked during these tests. The Yb_2O_3-coated Al_2O_3 foam, however, exhibited higher thermal shock stability and, consequently, this material was chosen as the subject for subsequent experiments.

Figure 10.8 shows the experimental set-up used to measure the radiation spectrum of the Yb_2O_3-coated Al_2O_3 foam emitter. The emitter tube was placed around the burner tube of a 2 kW butane burner. The radiation was coupled into a Bruker 113 FTIR spectrometer using a ZnS lens. By using an HgCdTe and an InGaAs detector, the photon energy range 0.1–1.4 eV was accessible. The spectrometer was calibrated absolutely by using an electrically heated graphite

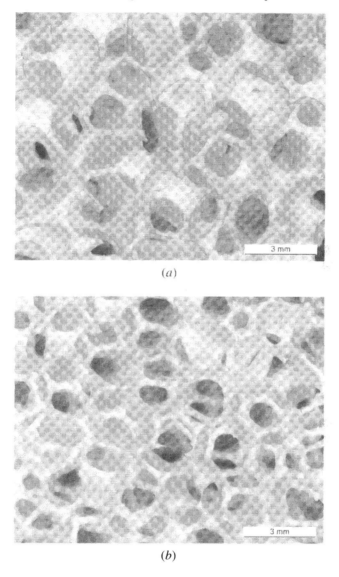

(*a*)

(*b*)

Figure 10.6. Porous structure of the Yb_2O_3 foam ceramic. A photograph of the green stage before sintering is shown top (*a*) and the sintered ceramic below (*b*). The shrinkage due to sintering is about 22%.

glowbar as a calibration source. Details of the spectroscopic method are described elsewhere [10].

The radiation spectra of the Yb_2O_3-coated Al_2O_3 foam emitter are compared to those of the Yb_2O_3 mantle emitter in figure 10.9. For technical reasons,

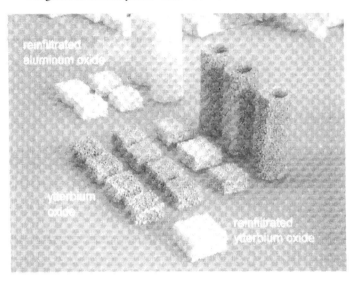

Figure 10.7. Foam ceramic samples and several ceramic emitter tubes.

the measurements of the higher and lower photon energies were carried out at different positions on the emitter surface and, consequently, the results are shown in separate graphs. The temperature at the two positions slightly differed, causing an offset between the two spectra.

On the left-hand spectrum, the convertible radiation peak appears at high photon energies. The peak emission of the foam is larger and shifted to higher photon energies compared to the mantle emitter, indicating a higher temperature for the foam emitter. The height of the CO_2 peak at 0.28 eV also points to the higher temperature of the foam emitter. The emission below 1.1 eV for the foam emitter is higher than that of the mantle, especially in the spectrum on the right at low photon energies. The foam also shows a strong emission below 0.3 eV, which is absent in the mantle emitter spectrum. Whether the strong infrared emission originates from the Al_2O_3 core or from blackbody emission in the pores remains to be investigated in further experiments.

Even though the foam emitter tube was 100 mm long, only the central region 30 mm in length shone brightly during operation, while the rest of the tube remained cooler. Because the radiation spectra were measured in the hot zone, a higher temperature in this region compared to the more uniformly heated mantle emitter is likely.

To study the suitability of the foam ceramic in a TPV system, the Yb_2O_3-coated Al_2O_3 foam emitter was mounted in a small prototype system (figure 10.10). The top opening of the emitter tube was covered with a ceramic plate. The cell temperature at $P_{th} = 2.0$ kW was 35 °C by cooling with 14 °C tap water, which is heated to 22 °C during operation. The photocells were heated

Figure 10.8. Spectroscopic set-up for the FTIR measurement of the foam ceramic emitter spectrum.

more strongly with the foam ceramic emitter than with the mantle due to the more intense infrared emission from the foam. In further experiments, attempts will be made to reduce the infrared emission by using pure Yb_2O_3 foam emitters or by changing the porosity of the foams.

The prototype system yielded 14 W electrical power at $P_{th} = 2.0$ kW, which corresponds to a system efficiency of 0.7%. Even though this efficiency is lower than that achieved with the mantle emitter, the suitability of the foam emitter was demonstrated in principle. We assume that the lower efficiency results from the non-uniform emitter temperature. To improve temperature uniformity, shorter emitter tubes will be manufactured and tested with adapted longer burner tubes.

10.5   Design of a novel thin-film TPV system

For large-scale application of TPV in residential heating systems, the cost of TPV is a crucial parameter. We suggest a novel design of a TPV generator operating

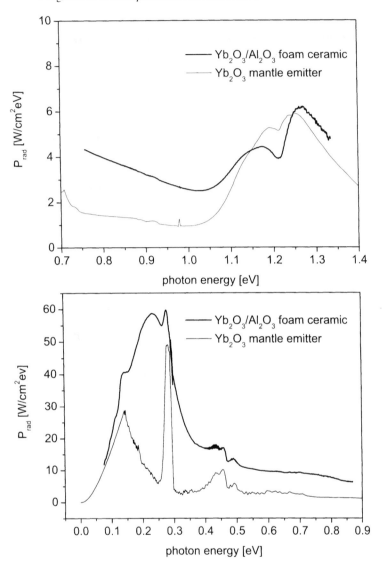

Figure 10.9. Radiation spectra of the Al$_2$O$_3$ foam ceramic tube coated with Yb$_2$O$_3$ in comparison with the emission spectrum of an Yb$_2$O$_3$ mantle. The selective emission peak is shown top and the emission at low photon energies in the bottom graph.

with thin-film CuInSe$_2$ (CIS) photocells, which may drastically reduce the costs of TPV in the future.

The bandgap of CIS is 1.0 eV but this can be increased up to 1.7 eV, when Ga is added. For commercial solar cells, Cu(In,Ga)Se$_2$ (CIGS) with a bandgap

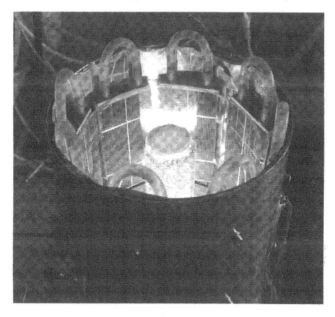

Figure 10.10. Ceramic foam emitter burning in the small TPV prototype system. The reflection of the emitter light in the surrounding quartz tube is clearly visible.

between 1.2 and 1.4 eV is used. For TPV applications with a selective Yb_2O_3 emitter, a lower Ga concentration can be chosen for an optimized match of the bandgap to the emitter radiation peak.

To investigate whether CIS/CIGS photocells are suitable for TPV, we used cell data from a 15.4% efficient CIS photocell published in [13] and inserted these into our simulation model of a TPV system, which is described in [7]. The simulation was for a system corresponding to the small prototype described earlier. The spectrum of the Yb_2O_3 mantle emitter was used. Gold reflectors were placed in the axial direction and the emitter radiation passes through a quartz tube. The external quantum efficiency as well as the CIS cell's fill factor of 0.73 from [13] was used. We computed a system efficiency of 2.0% at $P_{th} = 2.0$ kW. This is about 25% higher than by using the Si SH2 cell.

To study theoretically the influence of the bandgap on the TPV system performance, the electrical power as a function of the photon energy was calculated by varying the bandgap of the cell. The electrical power was calculated from the published external quantum efficiency, open-circuit voltage V_{OC} and fill factor multiplied by the radiation power of the Yb_2O_3 emitter at 1750 K. To vary the bandgap, we shifted the external quantum efficiency and corrected V_{OC} to about 900 mV eV^{-1} and the fill factor by 0.06 eV^{-1}. These correction factors were estimated from the CIGS cell compared to the CIS cell data published in [13]. Figure 10.11 gives the results of this calculation. It can be seen that

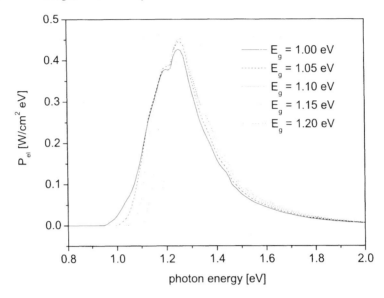

Figure 10.11. Calculated spectral electrical power of a TPV CIS/CIGS photocell for various bandgaps E_G.

the onset of electricity production is shifted to higher photon energies, when the bandgap increases. However, the electrical power at higher photon energies is enhanced with higher bandgaps, because V_{OC} and the fill factor are higher. A maximum integrated electrical power of 128 mW cm^{-2} is achieved with a bandgap of 1.05 eV.

Our novel TPV system design uses the advantages of CIS or CIGS thin-film photocells. A drawing of this patented system is shown in figure 10.12 [14]. The cylindrical system consists of an Yb_2O_3 selective emitter and a glass tube placed around it. On the outer surface of the tube an infrared reflective TCO filter, the thin-film photocell module and an electrically isolating layer are monolithically deposited. The outside of the tube with the layer stack is enclosed with cooling water.

In this design the photocells are fabricated as a monolithic module similar to the production of thin-film solar cell modules. The established technology of laser grooving can be used to separate and series connect single-cell stripes. The TCO filter can be used simultaneously as a conductive front layer and antireflection coating. Because the whole layer stack can be deposited and placed directly in the cooling water, mechanical assembly of cell modules and a cooling system are no longer necessary. This means the costs of the module assembly (see [9]) can be left out and a significant cost reduction potential is expected.

Unfortunately, CIGS solar cells have to be modified to be suitable for TPV. The requirements onCIGS TPV photocells are as follows.

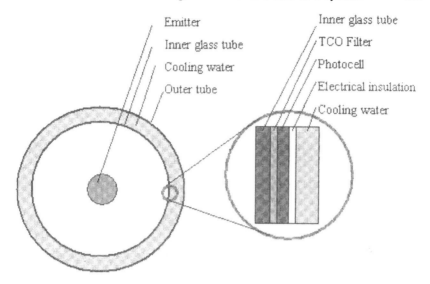

Figure 10.12. Top view drawing of the novel thin-film TPV system. In the inset a magnification of the cell module is shown.

- The bandgap has to be lowered by reducing the Ga content compared to conventional solar cells. The optimum bandgap for TPV is 1.05 eV.
- The cells should be optimized for an electrical power density of 100–200 mW cm^{-2}.
- In the design shown in figure 10.12, a superstrate cell configuration is needed. Conventional CIGS solar cells are manufactured on a substrate.
- The cells have to be deposited on the outer surface of a tube.

Various modifications of the novel thin-film design are possible. The TCO filter can be deposited on the inner surface of the glass tube if it is not used as conducting layer for the front contact of the cells. Instead of direct deposition onto the tube surface, the cells can be manufactured on a flexible substrate and bent around the tube. Due to the direct contact of the cells with the cooling water, it is expected that the cooling will be more effective than with crystalline cells glued onto metal blocks. A reduction in the system's diameter should be possible. The novel design can also be applied to several other thin-film photocells, like microcrystalline silicon or thin-film low bandgap materials.

Even though the novel thin-film TPV system was only theoretically designed, some components are already available, which might be used to build a first prototype in the future. Figure 10.13 shows, on the right-hand side, a water filter which was used in a precursor of the demonstration system [15]. A flexible thin-film cell could be placed between the glass tubes. The small quartz tube in figure 10.13 has an SnO$_2$ filter deposited on its inner surface, which could already

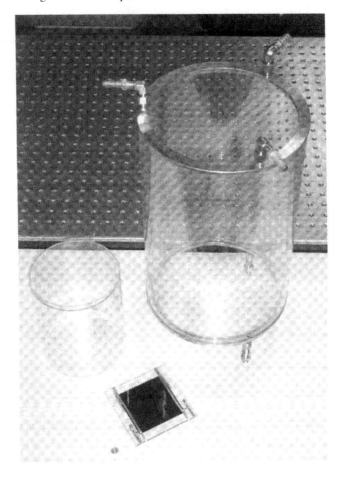

Figure 10.13. Photograph of some components, which can be used in a modified way in a future thin-film TPV system: water filter, SnO_2 filter tube and CIGS solar cell.

successfully be tested in the small TPV prototype system [7]. A small CIGS solar cell module is also shown in figure 10.13.

10.5.1   TPV with nanostructured SiGe photocells

An optimal match between the radiation spectrum of a selective emitter and the quantum efficiency of a photocell can be achieved, when the bandgap of the photocell is variable. A promising way to create a variable bandgap is the use of a quantum well structure. Due to quantum confinement in the wells, the bandgap of the well material can be adjusted by varying the thickness of the well.

In [16] this method was used to fabricate low-bandgap TPV photocells. A structure was developed, in which InGaAs quantum wells with 62% Ga content alternate with 45% Ga content wells. The strain due to the different lattice constants in the two well materials is compensated for and the effective lattice constant is matched to the InP substrate, so that a stack of 30 wells can be grown epitaxially. The well thickness determines the bandgap in the wells. A quantum efficiency starting at a photon energy of 0.69 eV was obtained with these cells. Due to the strain-balanced quantum well structure, the absorption edge could be shifted by about 0.06 eV compared to InGaAs lattice matched to InP. Thus, these cells are sensitized for radiation from a selective thulia emitter.

The bandgap of SiGe can be varied from the Ge bandgap of 0.66 eV to the Si bandgap of 1.12 eV depending on the Ge content. In thin strained layers even lower bandgaps can be achieved. SiGe is, therefore, a possible material for the production of TPV photocells, with bandgaps matched to a selective emitter made from Yb_2O_3 or Er_2O_3. In [17], incorporating a SiGe layer in the base of a Si solar cell to shift the absorption edge of the cell into the infrared is suggested. For this attractive application the SiGe layer has to be grown epitaxially without dislocations onto a Si substrate. But due to a lattice mismatch of 4.2% between Si and Ge, SiGe layers can only be grown up to a critical thickness onto Si substrates, depending on the Ge content [18]. A possible way is the growth of thin strained SiGe quantum well stacks. Figure 10.14 shows, as an example, a TEM micrograph of UHV-CVD grown $Si_{0.6}Ge_{0.4}$ quantum wells. Twenty SiGe layers 3 nm thick separated by 4 nm thick Si spacers were grown. The layers as well as the whole structure have a thickness close to the critical layer thickness. Therefore, the wells show a wavelike shape caused by the high strain in the structure but, fortunately, no dislocations were observed. The bandgap in SiGe quantum well structures strongly depends on strain effects and quantum confinement. In [19] the absorption coefficient in various SiGe quantum well structures was measured to study whether such structures are suitable to shift the absorption edge of Si photocells far enough into the infrared. Even though an absorption coefficient of 10 cm^{-1} at 0.85 eV was found for a quantum well structure with 45% Ge content, it is still about one order of magnitude too low for photocell applications.

Another way to incorporate SiGe nanostructures into Si photocells is the formation of Ge islands in the Stranski–Krastanov growth mode [20]. An AFM image of Ge quantum dots on a Si surface is shown in figure 10.15. The sample was grown by UHV-CVD at 520 °C. The dot density is 5×10^8 cm^{-2}. Higher densities above 10^{11} cm^{-2} were achieved by using Sb as a surface reactant in an MBE system [20]. A Si photocell with a stack of 75 Ge quantum dot layers incorporated in the base of a Si photocell is presented in [21]. The onset of the photocurrent was observed at 1.10 eV photon energy compared to 1.18 eV from a Si reference cell. This resulted in a slight increase in the short-circuit current. Unfortunately, the open-circuit voltage of the quantum dot photocell was reduced by about 90 mV compared to the reference. In [21] it is assumed that only an excited state in the Ge dots contributes to the current but the open-circuit voltage

Figure 10.14. TEM micrograph of $Si_{0.6}Ge_{0.4}$ quantum wells embedded in Si. The strain in the structure is relieved by forming a wavelike shape of the wells.

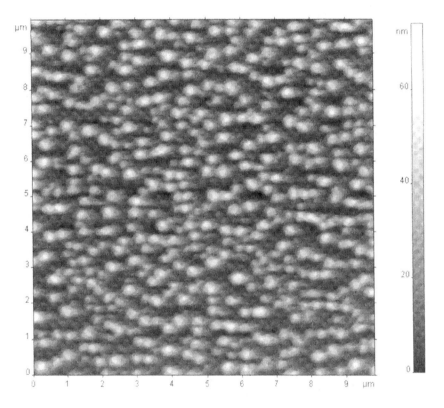

Figure 10.15. AFM image of Ge quantum dots on a Si surface.

is determined by the ground-state energy.

In conclusion, SiGe nanostructures offer interesting perspectives for the development of low bandgap TPV photocells based on Si technology but they are only just beginning to be developed. These cells might also be applied in solar photovoltaics for an optimum use of the solar spectrum. They may be used as infrared sensitive bottom cells in multiple junction devices based on Si.

10.6 Conclusion

TPV with Si solar cells and a selective Yb_2O_3 emitter is a suitable technology to be applied in residential heating systems. For a self-powered operation of the heater, an electrical power of about 120 W is sufficient and safety requirements as well as cost effectiveness seem to be fulfilled. The incandescent mantle structure of the emitter is very easy to fabricate and requires only about 10 g of Yb_2O_3. Its long-term thermal stability is still under investigation in the Hoval TPV furnace prototype. A clear disadvantage is its mechanical fragility. As a first alternative, ceramic foam were successfully tested. In the future, their thermal shock resistivity has to be improved and the emission below the silicon bandgap should be reduced.

The Si TPV technology might be replaced by more sophisticated technologies in the future. One possibility is the proposed novel thin-film TPV design which could drastically reduce the costs with an efficiency comparable to a Si TPV system. The photocell has a typical thickness of only about 1–3 μm, so that unhealthy constituents might be tolerable. Another possibility is the use of SiGe nanostructures to achieve a better match of the photocell to the selective emitter spectrum. Even though the first test devices have been fabricated, it seems too early to assess the widespread possiblities of this technology for TPV.

Acknowledgments

We thank Stefan Stutz for technical support with the FTIR spectrometry, Elisabeth Müller for the TEM microscopy and Oleg Stukalow for the AFM measurement. The work was supported by the Swiss Kommission für Technologie und Innovation (KTI), contract 5692.2 EBS.

References

[1] Coutts T J 2001 An overview of thermophotovoltaic generation of electricity *Solar Energy Mater. Solar Cells* **66** 441–52

[2] Fraas L M, Ballantyne R, Hui S, Ye S-Z, Gregory S, Keyes J, Avery J, Lamson D and Daniels B 1999 Commercial GsSb cell and circuit development for the midnight sun TPV stove *Thermophotovoltaic Generation of Electricity: Fourth NREL Conf. (AIP Conf. Proc. 460)* ed Coutts/Benner/Allman (New York: AIP) pp 480–7

[3] Fraas L M, Samaras J E, Huang H X, Minkin L M, Avery J E, Daniels W E and Hui S 2001 TPV generators using the radiant tube burner configuration *Proc. 17th European Photovoltaic Solar Energy Conf. Exh.* (Munich and Florence: WIP and ETA) pp 2308–11

[4] Wehrer R J, Wanlass M W, Wernsman B, Carapella J J, Ahrenkiel S P, Wilt D M and Murray C S 2002 0.74/0.55 eV GaInAs/InAsP monolithic, tandem, MIM TPV converters: design, growth, processing and performance *Proc. 29th IEEE Photovoltaic Specialists Conf.* (New York: IEEE) pp 884–7

[5] Andreev V M, Khvostikov V P, Khvostikova O V, Oliva E V, Rumyantsev V D and Shvarts M Z 2003 Low-bandgap Ge and InAsSbP/InAs-based TPV cells *Proc. 5th Conf. on Thermophotovoltaic Conversion of Electricity* (Woodbury, NY: AIP) pp 383–91

[6] Kushch A S, Skinner S M, Brennan R and Sarmiento P A 1997 Development of a cogenerating thermophotovoltaic powered combination hot water heater/hydronic boiler *Thermophotovoltaic Generation of Electricity: Third NREL Conf. (AIP Conf. Proc. 401)* ed Coutts/Allman/Benner (New York: AIP) pp 373–86

[7] Bitnar B, Mayor J-C, Durisch W, Meyer A, Palfinger G, von Roth F and Sigg H 2003 Record electricity-to-gas power efficiency of a silicon solar cell based TPV system *Proc. 5th Conf. on Thermophotovoltaic Conversion of Electricity* (Woodbury, NY: AIP) pp 18–28

[8] Bitnar B, Durisch W, Meyer A and Palfinger G 2003 New flexible photocell module for thermophotovoltaic applications *Proc. 5th Conf. on Thermophotovoltaic Conversion of Electricity* (Woodbury, NY: AIP) pp 465–72

[9] Palfinger G, Bitnar B, Durisch W, Mayor J-C, Grützmacher D and Gobrecht J 2003 Cost estimates of electricity from a TPV residential heating system *Proc. 5th Conf. on Thermophotovoltaic Conversion of Electricity* (Woodbury, NY: AIP) pp 29–37

[10] Bitnar B, Durisch W, Mayor J C, Sigg H and Tschudi H R 2002 Characterization of rare earth selective emitters for thermophotovoltaic applications *Solar Energy Mater. Solar Cells* **73** 221–34

[11] Durisch W, Bitnar B, Palfinger G and von Roth F Small 2003 TPV prototype systems *Proc. 5th Conf. on Thermophotovoltaic Conversion of Electricity* (Woodbury, NY: AIP) pp 71–8

[12] Vogt U, Herzog A and Thünemann M 2002 Porous ceramic materials produced by different methods *Proc. 6th Steinfurter Ceramic Seminar* (New York: IEEE) pp 1–6

[13] Hedström J and Ohlsén H 1993 ZnO/CdS/Cu(In,Ga)Se2 thin film solar cells with improved performance *Proc. 23rd IEEE Photovoltaic Specialists Conf.* pp 364–71

[14] Patent Thermophotovoltaik-System Int. patent no WO 03/017376A1

[15] Bitnar B, Durisch W, Grützmacher D, Mayor J-C, Müller C, von Roth F, Anna Selvan J A, Sigg H, Tschudi H R and Gobrecht J A 2000 TPV system with silicon photocells and a selective emitter *Proc. 28th IEEE Photovoltaic Specialists Conf.* (New York: IEEE) pp 1218–21

[16] Abbott P *et al* 2002 A comparative study of bulk InGaAs and InGaAs/InGaAs strain compensated quantum well cells for thermophotovoltaic applications *Proc. 29th IEEE Photovoltaic Specialists Conf.* (New York: IEEE) pp 1058–61

[17] Koschier L M, Wenham S R and Green M A 1998 Efficiency improvements in thin film silicon solar cells using SiGe alloys *Proc. 2nd World Conf. and Exh. on Photovoltaic Solar Energy Conversion* (Ispra: European Commission) pp 292–5

[18] People R and Bean J C Calculation of critical layer thickness versus lattice mismatch for Ge_xSi_{1-x}/Si strained-layer heterostructures *Appl. Phys. Lett.* **47** 322–4 and *Appl. Phys. Lett.* **49** 229

[19] Palfinger G, Bitnar B, Sigg H, Müller E, Stutz S and Grützmacher D 2003 Absorption measurement of strained SiGe nanostructures deposited by UHV-CVD *Physica* E **16** 481–8

[20] Konle J, Presting H, Kibbel H and Barnhart F 2002 Growth studies of Ge-islands for enhanced performance of thin film solar cells *Mater. Sci. Eng.* B **89** 160–5

[21] Konle J, Presting H and Kibbel H 2003 Self-assembled Ge-islands for photovoltaic applications *Physica* E **16** 596–601

Chapter 11

Solar cells for TPV converters

V M Andreev
Ioffe Physico-Technical Institute, 26 Polytechnicheskaya,
194021, St Petersburg, Russia

11.1 Introduction

The design and optimization of photovoltaic (PV) solar-energy-converting systems are strongly determined by the sunlight spectrum and by the fact that there is no back connection between a receiver and the Sun. However, there is a possibility of varying the operating concentration ratio (in other words, the operating current density of the pn-junction). In contrast to this, in thermophotovoltaic (TPV) systems, the optimization may imply choosing the emitter spectrum and returning the radiation from the receiver which has not been used back to the emitter surface supplying it with 'additional' power. Nevertheless, there are many common features in PV and TPV systems.

Investigations in the field of TPV started in the early 1960s [1] but the real advantage of the TPV approach was only demonstrated in the early 1990s [2]. Theoretical and semi-empirical modelling [3–11] has shown that the optimal semiconductor material bandgap has to be in the range of 0.4–0.6 eV in single-junction TPV cells designated for operation with blackbody (gray-body) emitters at temperatures of 1200–1500 °C. Germanium and silicon were the first materials to be suggested and applied to TPV conversion of radiation from fuel-fired emitters. However, the advantages of the first TPV systems based on these materials such as low cost and their commercial availability have not been realized.

Among the III–V compounds, gallium antimonide (GaSb) was the first material widely used for TPV device fabrication. InGaAsSb solid solutions ($E_G = 0.5$–0.6 eV) lattice matched to GaSb as well as InGaAs lattice matched ($E_G = 0.75$ eV) and mismatched ($E_G = 0.5$–0.6 eV) to InP substrates have been developed for TPV applications. Semiconductors with bandgaps wider than 0.75 eV, such as Si, can also be applied to TPV systems with high-temperature or selective emitters, in which the radiation maximum is shifted to the short-wavelength part of spectrum (to the photosensitivity region of these semiconductors).

TPVs in Russia may be a very promising field of activity for a number of natural features. Vast territories in Russia are characterized by a cold climate. Many people live far away from the electric power grid. At the same time, Russia possesses rich sources and deposits of natural gaseous and liquid fuel. For these reasons, the concept of TPV ensuring electricity and heat generation is of great interest to Russia.

The Photovoltaics Laboratory of the Ioffe Physico-Technical Institute has many years' experience in the field of devices based on III–V-compound, especially concentrator solar cells. AlGaAs/GaAs solar cells were first suggested and fabricated in the Ioffe Institute in the late 1960s [12]. During the last decade, low-bandgap PV cells based on InP/InGaAs, GaSb, GaSb/InGaAsSb and Ge were developed both for tandem solar cells and for TPV applications including the growth, treatment and packaging of narrow-gap cells as well as the design and manufacture of demonstration systems.

11.2   Predicted efficiency of TPV cells

One of the main common features of TPV and solar PV is that in both systems the energy source is characterized by a wide spectrum. This means that the most effective approach to improving solar system efficiency, which is the multi-junction approach, may be applied to improving the efficiency of the TPV system. A negative side of this approach in the TPV case is that the operating current density, which is limited by the emissivity properties of the emitter at given temperature, is reduced for both top and bottom junctions. For cells based on narrow-gap semiconductors, such a reduction in current density can cause a pronounced reduction in the photogenerated voltage, which reduces any possible efficiency improvement expected from the double-junction approach.

This section presents theoretical estimations of the potential of the double-junction approach for a TPV system when similar operating conditions for both single-junction and tandem cells with respect to the black-body (BB) spectrum and to various return efficiencies are considered. The limiting efficiencies were obtained by assuming an ideal pn-junction in a direct-band material and the ideal (lowest) saturation current densities (J_0) in the pn-junction limited only by the radiative recombination process [11, 12].

The BB spectrum was considered for a planar infinite 100%-efficient emitter. The absolute external quantum efficiency of the converter was assumed to be 0.95 for all photon energies greater than the bandgap $h\nu \geq E_G$.

Figure 11.1(a) shows the calculation results for single-junction cell efficiencies when the cell is kept at various temperatures. It is seen, for instance,

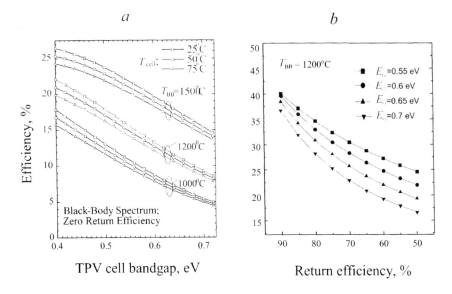

Figure 11.1. (*a*) Predicted maximum efficiencies of a TPV system based on single-junction cells for various temperatures of the blackbody emitter (T_{BB}). The cell temperature (T_{cell}) is 25, 50 or 75 °C. No photon recirculation is assumed. (*b*) Dependencies of a TPV system efficiencies on the return efficiency at different BB-emitter temperatures for single-junction TPV converters with bandgaps in the range $E_G = 0.55$–0.7 eV.

that at the BB temperature $T_{BB} = 1200$ °C, the efficiency of a GaSb-based cell ($E_G = 0.72$ eV) ranges near to 8–9%, whereas the efficiency of a narrow-gap cell ($E_G = 0.4$ eV) should achieve 20–22%.

One of the ways to improve TPV efficiency is to utilize the non-absorbed radiation in the cell part of the BB spectrum returning it to the BB surface where these low-energy photons can be transformed again into photons of the whole BB spectrum. The simulation results of this approach are shown in figure 11.1(*b*) for single-junction cells based on materials with $E_G = 0.55$–0.7 eV operating with BB emitters heated up to 1200 °C.

A second way is to combine two pn-junctions in a tandem, which is similar to the approach to improving solar cell efficiency. The simulation results of this approach are shown in figure 11.2 for the BB spectrum at $T_{BB} = 1200$ °C. Single- or double-junction TPV cells with various bandgaps are assumed to have temperature $T_{cell} = 25$ °C. Under zero return efficiency conditions (figure 11.2(*a*)), the estimated efficiency is about 26% in a tandem consisting of 0.55 eV top and 0.40 eV bottom sub-cells. One can compare these values with those of 22% and 16% for single-junction 0.40 eV and 0.55 eV-based cells, respectively (see figure 11.1(*a*), curve family $T_{BB} = 1200$ °C, $T_{cell} = 25$ °C).

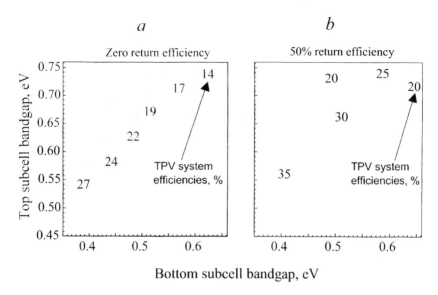

Figure 11.2. Predicted efficiencies of a TPV system based on two-junction cells at the black body temperature $T_{BB} = 1200\,°C$ and the cell temperature $T_{cell} = 25\,°C$: (*a*) zero return efficiency, (*b*) 50% return efficiency.

For an extremely high (90%) return efficiency, there is no necessity to apply a multi-junction approach because a wide-gap cell alone can operate with a very high efficiency, almost the same as the highest monochromatic efficiency of this cell at a given photocurrent density. Indeed, energy losses at thermalization of the photogenerated hot carriers are very low in this case as only photovoltaic conversion of light from the not so extended short-wavelength part of the BB spectrum takes place. Arranging of such a high return efficiency in a TPV system is not as realistic as it seems. The problem lies not only in the reflection coefficient of some spectrally selective mirrors but, also, in the view factor between the emitter and TPV receiver. This factor is assumed to be unity in our considerations. But the emitter cannot be entirely surrounded by the receiver in a practical TPV system and this means that there will always be additional radiation losses in the recycling process.

For a realistic 50% return efficiency and $T_{BB} = 1200\,°C$ (see figures 11.1(*b*) and 11.2(*b*), a double-junction approach gives a TPV system efficiency of 35% with tandem bandgaps of 0.55 and 0.4 eV for the top and bottom sub-cells, respectively. Single-junction cells based on semiconductors with bandgaps of 0.55 and 0.4 eV give efficiencies of 25% and 29%, respectively, for the same conditions. As a result, the improvement from the double-junction approach is impaired by a decrease in the return efficiency.

Based on theoretical simulation [3–11] and analytical models, the conversion

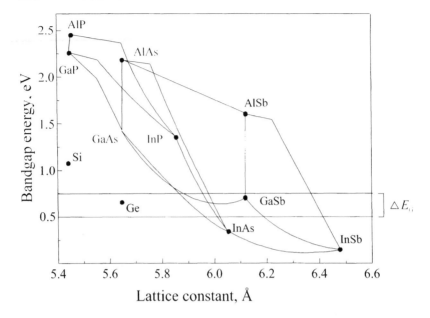

Figure 11.3. Dependence of the bandgap of Ge, Si and III–V compounds and their solid solutions on the lattice constant of these materials: ΔE_G, the range of single-junction cell bandgaps optimal for TPV applications.

efficiency of a TPV system can be estimated. An optical efficiency of about 20–25% is estimated for the system efficiency using the measured laboratory characteristics of available emitters, filters, high-bandgap cell spectrum and efficiencies, etc. The efficiency can be increased up to 30–35% using such advanced ideas as multi-layer matched selective emitters, multi-bandgap cells, back-reflecting PV cells and by optimizing the system performance instead of the performance of each component.

Figure 11.3 shows the dependence of the bandgaps on semiconductor lattice parameters. The bandgap range $\Delta E_G = 0.5$–0.75 eV, given in this diagram corresponds to Ge, GaSb, InGaAsSb/GaSb and InGaAs/InP semiconductors, which are the most suitable for TPV receiver fabrication.

Let us consider experimental methods for TPV cell efficiency measurements. In solar photovoltaics, the methods for solar cell testing are quite well developed, which allows a precise comparison of different cells and modules to be made. Using solar simulators (or tabulated AM0 or AM1.5 spectra) for the comparison of TPV cell efficiencies may give incorrect results because of the much higher short-wavelength (visible) radiation compared with the emitters used in TPV generators. This approach may be partly applied if the solar spectra are cut-off, for example, at $\lambda < 900$ nm. Another approach is the use of blackbody radiators (or tabulated BB spectra). Precise spectral response and $I–V$ curves measured at different illumination intensities corresponding to the conditions in TPV systems allow the efficiency of any emitter with a measured spectrum including the tabulated BB spectra to be calculated. Here, both these approaches will be applied to compare the performance of TPV cells.

11.3   Germanium-based TPV cells

Ge has a bandgap of 0.66 eV (300 K), which is much lower than the optimal value for effective sunlight conversion. Therefore, Ge was not used for single-junction solar cells. Nevertheless, it was the first material proposed and applied in the 1960s for manufacturing the first TPV systems [1]. However, TPV generators based on Ge PV cells were not effective because of the poor cell performance obtained in these cells. During the last decade, Ge-based PV cells have been intensively studied [13–23] because Ge had become the basic material (substrate and bottom cell) for tandem and triple-junction InGaP/GaAs/Ge solar cells. Already widely used for fabricating high-efficiency radiation-resistant space arrays, these cells began to be used for terrestrial concentrator photovoltaics.

In one of the first works [13] in this field, Ge and $Si_{0.07}Ge_{0.93}$ solar cells were developed for monolithic and mechanically stacked cascade application as the bottom cells in a tandem with 1.4–1.6 eV top cells. The Ge was grown from GeH_4. For SiGe alloy growth, simultaneous pyrolysis of GeH_4 and disilane was used. The cell structure consists of an n-base, p^+-emitter, n^+-back-surface field layer and a p^{++}-layer on the back surface, which acts as a minority-carrier mirror. The most successful passivation for the emitter of a Ge homojunction cell was found to be a p^+-$Al_{0.85}Ga_{0.15}As$ layer. A fill factor (FF) of 0.6 and open-circuit voltage (V_{OC}) of 0.2 V were measured under the AM0 spectrum. The lower V_{OC} value of 0.18 V obtained in SiGe cells may be due to defects at the SiGe/Ge hetero-interface and the influence of these defects on the dark current.

Monolithic GaAs/Ge [14, 15] and $Al_{0.08}Ga_{0.92}$/Ge [16] tandems with bottom Ge cells were developed for solar arrays. An increase in V_{OC} of up to 0.32 V was achieved in the GaAs/Ge cell under 100 suns (AM0). An efficiency of 3.5–4.6% was obtained [17, 18] under 10–170 suns (AM0) in the cells based on p-GaAs/p^{++}Ge/nGe. The p-GaAs top layer was epitaxially grown by MOCVD on n-Ge. This GaAs growth creates a pn-junction in Ge wafers through Ga and As in-diffusion. SIMS profiles show that the surface gallium-atom concentration in Ge is more than 10^{20} cm^{-3} and much higher than the arsenic-atom concentration (about 10^{19} cm^{-3}). The pn-junction depth in Ge was about 0.5 μm in the selected GaAs growth regimes.

A stable wet chemical growth process of oxide layers on Ge-substrates has been developed [19] for surface passivation on n^+pp^+Ge(As,In) diffused cells. A $V_{OC} = 0.24$ V and a short-circuit current density (J_{SC}) of 55.5 mA cm^{-2} in these cells were achieved at NASA Glenn RC under AM0, 25 °C conditions.

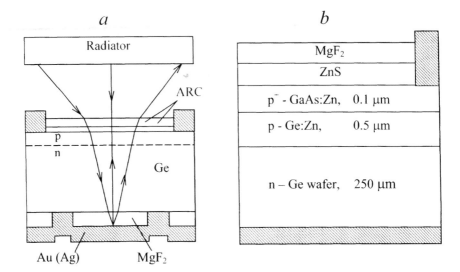

Figure 11.4. Schematic drawing of the sub-bandgap radiation reflection and reabsorption in radiator in a germanium TPV cell with a back-surface 'mirror' contact (*a*) and schematic cross section of p-GaAs/p-Ge/n-Ge TPV cell, fabricated by LPE growth of GaAs and Zn-diffusion (*b*) (from Andreev V M, An overview of TPV technologies, *Thermophotovoltaic Generation of Electricity*, Rome 2002. © 2003 AIP. Reprinted with permission).

Nowadays, Ge bottom cells are widely used for triple-junction GaInP/ GaAs/Ge [20–22] solar cells. Ge cells were fabricated at the National Renewable Energy Laboratory [21] by growing n-type GaInP epilayers on p-type Ge substrates by MOCVD. The resulting junctions are found to be diffused n-GaInP/n-Ge/p-Ge homojunctions. The open-circuit voltage in these cells increases from 0.25 V at $J_{SC} = 0.1$ A cm^{-2} up to 0.4 V at 30 A cm^{-2}.

Recently, Ge TPV cells with back-surface mirrors have been fabricated at the Ioffe Institute [23] by Zn diffusion into n-Ge-wafers and into p-GaAs-n-Ge heterostructures (figure 11.4). Back-surface reflection of non-absorbed sub-bandgap photons from a TPV cell allows the efficiency of a TPV system to be maximized owing to the possible re-absorption of these photons in the radiator. The back-surface mirror consists of MgF$_2$ and Au or Ag layers (figure 11.4). Back ohmic contacts of Au/Ge/Ni/Au were formed in the holes in the MgF$_2$ layer (with total hole-'shadowing' of 3%) and annealed at 400 °C. A front-grid contact with shadowing of 6% and ZnS/MgF$_2$ ARC was fabricated. The cell reflectance for these Ge cells at wavelengths longer than 1.9 μm is 82–87% (figure 11.5, curve 2). An external quantum yield as high as 0.9–0.95 (figure 11.5, curve 1) and the extremely high short-circuit current density of 31.6 mA cm^{-2} were measured in

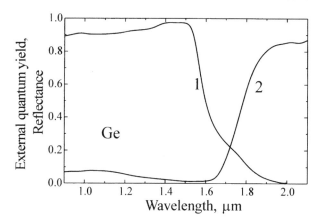

Figure 11.5. Spectral response (curve 1) and reflectance (curve 2) for Ge TPV cells with back contact fabricated through holes in MgF_2 layers (from Andreev V M, An overview of TPV technologies, *Thermophotovoltaic Generation of Electricity*, Rome 2002. © 2003 AIP. Reprinted with permission).

this Ge cell illuminated by sunlight with cut-off at $\lambda < 900$ nm AM0 spectrum. $J_{SC} = 50.3$ mA cm^{-2} was obtained under sunlight with cut-off at $\lambda < 580$ nm and the AM0 spectrum.

An efficiency of 3.8% was measured for the p-n Ge TPV cells with $V_{OC} = 0.39$ V and $FF = 0.64$ at $J_{SC} = 4.5$ A cm^{-2} without a back-surface mirror under the BB spectrum at $T_{BB} = 1473$ K. The calculated efficiency increases up to 11.8% in Ge cells with a back-surface mirror assuming an achievable return efficiency $R = 85\%$.

TPV cells based on p-GaAs/p-Ge/n-Ge heterostructures with a thin (0.1 μm) GaAs layer were fabricated [23] by low-temperature liquid-phase epitaxy and additional Zn diffusion. The high external quantum yield (85–90% in the spectral range of 900–1550 nm) in these cells demonstrates the low densities of recombination centres on the p-GaAs/p-Ge interface. In the fabricated GaAs/Ge cells, a photocurrent density of 3.8 A cm^{-2} has been estimated at BB irradiation with $T_{BB} = 1473$ K. Measured V_{OC} values as high as 0.4 V and a fill factor of 0.7 were used for efficiency calculations (see figure 11.6). A PV efficiency of 16% was estimated for the Ge cells with a back-surface mirror assuming $T_{BB} = 1473$ K and 100% returning photons with $\lambda > 1.8$ μm. For the case of an 85% return efficiency, the PV efficiency was estimated to be 9.8%. Slightly worse efficiencies were achieved for the Ge-based TPV cells than in the best GaSb TPV cells. Nevertheless, taking into account the lower cost of the Ge-based cells and the high sub-bandgap photon reflection, it can be concluded that these Ge cells are promising for TPV applications.

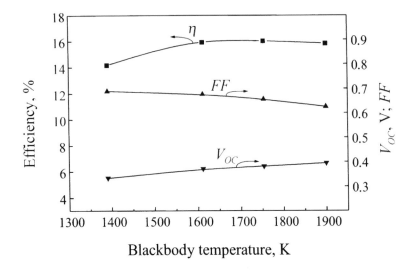

Figure 11.6. GaAs/Ge TPV cell efficiencies as a function of a blackbody IR-emitter temperature. The BB spectrum is cut-off at $\lambda > 1820$ nm. 100% reflection of sub-bandgap photons from the cell to the emitter is assumed (from Andreev V M, An overview of TPV technologies, *Thermophotovoltaic Generation of Electricity*, Rome 2002. © 2003 AIP. Reprinted with permission).

11.4   Silicon-based solar PV cells for TPV applications

TPV systems based on silicon PV cells have problems connected with the necessity to operate the emitters at very high temperature, at which all known materials are not reliable. There are similar reliability problems for the materials for selective emitters, which are based, for example, on rare-earth oxides.

The one-sun silicon solar cells produced commercially may be applied in TPV systems after solving the problem of ohmic loss. Conventional cells with a higher grid-coverage fraction may be used for a moderate illumination level. A cell of this type was fabricated [24] using n^+-p-p^+ structures. As a starting material, boron-doped silicon with low resistivity (0.1–0.2 Ω cm) and high minority-carrier lifetimes was used. A 25% efficiency was obtained under 100 suns (AM1.5) and a temperature of 28 °C.

Figure 11.7 shows two types of silicon solar cells specially developed for operating at high illumination intensities: (*a*) hybrid cells representing a combination of planar and vertical pn-junctions; and (*b*) a cell with back-point pn-junctions. The thickness of each sub-element in a cell shown in figure 11.7(*a*) is made many times smaller than the diffusion length of minority carriers to provide a high collection efficiency. In the long-wavelength spectral region, enhanced collection efficiency is achieved by reducing the distance the carriers have to travel

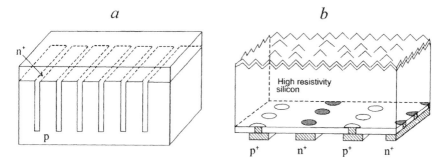

Figure 11.7. Design of silicon PV cells ensuring the high performance at increased output photocurrent densities of 1–10 A cm^{-2}: (*a*) cells with vertical pn-junctions; (*b*) with point contacts on the back side.

prior to being separated by the vertical pn-junctions. Reduction of the ohmic losses, essential for operation in TPV systems, is achieved by connecting the micro cells in parallel, thereby reducing the current density through the terminals of an individual cell in proportion with their number.

Multi-junction silicon cells [25] with point contacts of either polarity formed on the back surface are shown in figure 11.7(*b*) In these cells, the optical losses due to obscuration with the metallization grid are eliminated. Because the carrier separation occurs within pn-junctions at the backside of the structure, the base region may be made of high-resistivity float zone silicon with high minority-carrier lifetimes (over 3 ms). The front surface is passivated in order to lower the recombination losses. For the same purpose and, simultaneously, to lower optical losses, the structure thickness is reduced to about 100 μm. The long-wavelength light absorption efficiency in these structures is enhanced by texturing the top surface. Using this approach, an efficiency of 27.5% at 100 suns, AM1.5, has been obtained.

Commercial silicon solar cells have been applied in several TPV systems [26–28] with Yb$_2$O$_3$ selective emitters. The important Si cell parameters for use in a system with a Yb$_2$O$_3$ emitter are a high quantum efficiency up to 1100 nm wavelength and minimized reflection losses in the spectral range 800–1100 nm.

A maximum output power density up to 81.2 mW cm^{-2} was obtained [27] from commercially available silicon PV cells irradiated by a Yb$_2$O$_3$ selective emitter. A photoreceiver with a total area of 380 cm^2 ensured an electrical power of 21 W with 1.35 kW thermal input power. A larger PV receiver with an area of 2000 cm^2 generated 120 W electricity at 12 kW thermal input power. Recently [28] a new larger photocell generator consisting of high efficiency silicon cells provided a system efficiency of 2.8% with preheated (370 °C) combustion air.

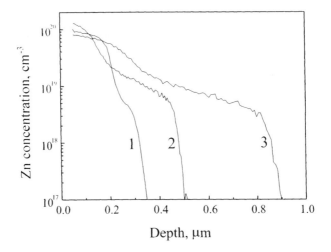

Figure 11.8. Profiles of Zn atoms detected by SIMS method after Zn diffusion in GaSb at 450 °C (curve 1), 500 °C (2) and 550 °C (3) during 40 min (from Andreev V M, TPV cell technologies, *The Path to Ultra-High Efficient Photovoltaics*, Ispra 2001. © 2002 European Commission. Reprinted with permission).

11.5   GaSb TPV cells

GaSb has a bandgap of 0.72 eV, which is by far the optimal single-junction cell for converting the solar spectrum. At first, GaSb cells were developed as bottom cells for concentrator mechanically stacked GaAs/GaSb tandems [29–34]. Recently the study of GaSb TPV devices has been considerably intensified [35–53]. It has also been shown [29, 30] that surface recombination rate for p-type GaSb is lower than that for p-type GaAs. For this reason GaSb TPV cells with high quantum yield were fabricated using a simple Zn-diffusion method without a lattice-matched AlGaAsSb window layer. The pseudo-closed-box diffusion technique was used. Mixtures of zinc and antimony, zinc and gallium or zinc-gallium-antimony were used as the diffusion sources.

It was found from the SIMS data that, near the surface, a Zn concentration of 10^{20} cm^{-3} is independent of the diffusion temperature at the selected zinc-antimony source and that a sharp Zn concentration gradient takes place in these structures (figure 11.8). The free-carrier (holes) distributions were recorded by Raman scattering spectroscopy accompanied by precise anodic oxidation of the structure [31, 33]. Comparison of SIMS and Raman profiles shows a good correspondence between the concentrations of free carriers and Zn atoms. This fact indicates that almost all diffused Zn atoms are electrically active.

JX Crystals Inc [2, 36, 37, 45] has organized the production of high-power TPV cells based on diffusion technology. The following performance has been achieved in flash-tested cells with a total area of 1.5 cm^2 [36]: $FF = 0.84$,

$V_{OC} = 0.52$ V, $I_{SC} = 6$ A, $P_{max} = 2.6$ W. Water-cooled cells in continuous operation in front of the heated SiC emitter operated at 1380 °C ensured the following output parameters: $FF = 0.67$, $V_{OC} = 0.44$ V, $P_{max} = 1.76$ W. Several TPV photoreceivers based on the developed cells have been fabricated [37, 45] for TPV generators. A water-cooled cylinder [37] contains 380 GaSb cells arrayed on the inside surface of the cylinder of 5.3" diameter. The cells were mounted on 20 circuits with 19 cells per circuit. Each circuit had a length of 8.5". Tested with a 1600 °C glow-bar, the cylinder generated 900 W with an electric power density of 2.5 W cm^{-2}. An advanced GaSb TPV single circuit was designed for the Midnight Sun stove panel [45]. This circuit utilizes a terraced AlSiC substrate with a thermal expansion coefficient close to that of GaSb. TPV cells were connected in series in a single pattern. The fabricated circuit had 22 cells in series and a total active area of 30 cm^2. This circuit produced electric power of up to 44 W.

To reduce the cost of a TPV system, GaSb cells based on a polycrystalline material have been developed [45, 46]. The achieved polycrystalline cell average performance is worse than in the single crystal case by about 10%. However, the potential cost savings are expected to dramatically out weigh this small performance drop.

The concept developed at the Fraunhofer ISE [43, 44, 47, 48] for GaSb TPV cells is to use strong built-in electric fields (up to 8 kV cm^{-1}) in the p-type emitter formed due to a specific steep Zn diffusion profile regulated through the Zn diffusion parameters and precise etching. This leads to an improvement in the device performance. Formation of strong built-in electrical fields ensures a decrease in the surface carrier recombination influence, which results in an increase in the generated current and voltage in the TPV cells. PV cells with an internal quantum yield exceeding 90% and an open-circuit voltage larger than 0.5 V at current densities >3 A cm^{-2} were fabricated [47,48] using this approach. Calculations [48] based on these cell measurements show that efficiencies of up to 30% could be achieved assuming a BB temperature of 1300–1500 K and a perfect band-edge filter.

MOCVD-grown AlGaAsSb/GaSb heterostructures with an AlGaAsSb (0.72–0.9 eV) window layer have also been fabricated at the Fraunhoffer ISE [61]. It was found that the p-GaSb/n-AlGaAsSb heterostructure ensures an excellent quantum efficiency as in corresponding homojunction devices, while the open-circuit voltage increased by approximately 20 mW due to a reduced dark current in the device.

A two-stage diffusion process as well as epitaxial GaSb cells have been developed at the Ioffe Institute [49–53]. In the first stage, GaSb wafers were exposed to Zn to form a shallow pn-junction in the photoactive area of a cell. During the second stage, a deep pn-junction (1–1.5 μm) was formed by an additional spatially selective diffusion process (figure 11.9) to reduce the current leakages under contact grid fingers. Then, anodic oxidation and selective etching were employed for precise thinning (to 0.2–0.3 μm) the photoactive diffused

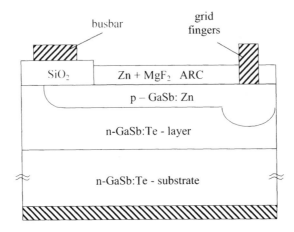

Figure 11.9. Cross section of an epitaxial GaSb TPV cell fabricated by the two-stage Zn-diffusion method (from Andreev V M, TPV cell technologies, *The Path to Ultra-High Efficient Photovoltaics*, Ispra 2001. © 2002 European Commission. Reprinted with permission).

p-GaSb layer. As seen from figure 11.10, the maximum photocurrent can be obtained from the cells with a p-region thickness in the range 0.15–0.3 μm. This explains the necessity for the second deep selective diffusion process. More precise investigations of photocurrent generated by GaSb cells with a pn-junction depth in this range allowed the optimal pn-junction depth of about 0.25 μm ensuring an increase in the photocurrent density up to 30 mA cm^{-2} under sunlight with AM0 spectrum cut-off at $\lambda < 900$ nm to be determined.

Te-doped GaSb layers were grown by liquid phase epitaxy (LPE) in attempts to prepare material of higher crystal quality. Ga-, Sb- and Pb-rich melts were used in the LPE processes. In the case of LPE growth from a Ga-rich melt, we employed the 'piston' boat technique, which enabled epilayers on 50 substrates 7 cm^2 in area to be grown in one run. Epilayers from Ga-rich melts were grown at a temperature of 450–400 °C. Figure 11.11 shows the spectral responses of the 'epitaxial' TPV cells. The maximum external quantum yield was obtained in cells which were prepared using epilayers grown from the Ga-rich melt. A photocurrent density as high as 54 mA cm^{-2} under the AM0 spectrum was obtained in GaSb cells and 29–32 mA cm^{-2} under the part of the AM0 spectrum cut-off at $\lambda < 900$ nm.

The cells 1 or 2 cm^2 in area have been developed for TPV generators. These cells were able to generate photocurrents up to 2–10 A (cw operation) with an I–V curve fill factor values of about 0.7. Figure 11.12 shows the histogram of photocurrent density for initial technology (left-hand side) and advanced technology (right-hand shaded part), characterized by a decreased p-GaSb front photoactive layer (to 0.25 μm thick) and an increased p-GaSb layer under the

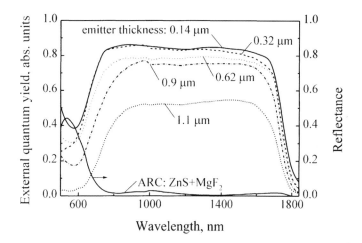

Figure 11.10. Spectra of photoresponse (on the designated illumination area including a contact grid with shadowing of 8%) for GaSb cells with different p-emitter thicknesses. The reflectivity spectrum in the case of $ZnS+MgF_2$ antireflection coating is shown as well (from Andreev V M, TPV cell technologies, *The Path to Ultra-High Efficient Photovoltaics*, Ispra 2001. © 2002 European Commission. Reprinted with permission).

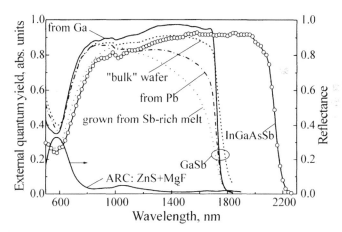

Figure 11.11. Spectral responses (on the photoactive area) of the TPV cells based on GaSb and lattice-matched InGaAsSb epitaxial layers. Reflectance from a cell with a $ZnS+MgF_2$ antireflection coating is shown as well (from Andreev V M, TPV cell technologies, *The Path to Ultra-High Efficient Photovoltaics*, Ispra 2001. © 2002 European Commission. Reprinted with permission).

contact grid fingers (up to 1 μm thick). V_{OC} and FF in the 1 cm^2 cells at $I_{SC} = $ 1 A had the following ranges: $V_{OC} = 0.42–0.45$ V and $FF = 0.70–0.72$. The

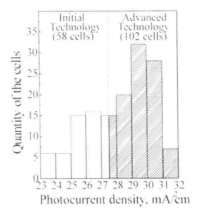

Figure 11.12. Histogram of photocurrent density under the part of AM0 spectrum with $\lambda > 900$ nm for 160 fabricated GaSb 1 cm $\times$ 1 cm^2 TPV cells.

cells 2 cm^2 in area generate photocurrents up to 9 A at $V_{OC} = 0.52$ V. Efficiencies of about 11% under the AM0 spectrum and 19% under part ($\lambda > 900$ nm) of the AM0 spectrum at a photocurrent density of 2–7 A cm^{-2} were achieved in these TPV cells.

The calculated efficiency exceeded 25% in the developed GaSb TPV cells assuming a BB emitter temperature of 1400–1600 K and a radiation spectrum cut-off at $\lambda > 1820$ nm (figure 11.13).

Several types of photoreceiver arrays have been fabricated from these GaSb cells for TPV generators. One of these receivers consists of 15 cells (1 cm $\times$ 2 cm) and provides an output power of 7 W (at a photocurrent of 2 A) illuminated by a propane-fuelled metallic emitter at a temperature of 1250 °C. The increase in gas inlets and emitter temperature allows the output power to be increased up to 20 W from this 15-cell receiver.

11.6   TPV cells based on InAs- and GaSb-related materials

This section surveys the TPV cells based on InAs- and GaSb-related alloys, mainly on InGaAsSb but also including InGaSb, GaAsSb, AlGaSb, AlGaAsSb and InAsSbP. The main feature of these materials is their narrow bandgap, which is very important for applications requiring a spectral response at up to 2.5 μm wavelength, for example, for TPV systems with low temperature (900–1100 °C) emitters. As a rule, these alloys are fabricated by epitaxial growth on GaSb substrates. However, InAsSbP thin-film structures are fabricated by growth on GaSb as well as on InAs substrates. There are some attempts to grow bulk ternary InGaSb and quaternary InGaAsSb crystals, which allow TPV cells to be prepared without epitaxial growth. AlGaSb and AlGaAsSb are used as wide-bandgap

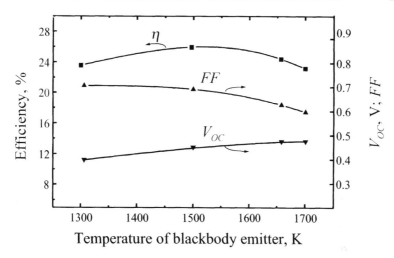

Figure 11.13. GaSb TPV cell efficiency (η), fill factor (FF) and open circuit voltage (V_{OC}) as a function of blackbody emitter temperature. Efficiencies were calculated for radiation spectra cut-off at $\lambda > 1820$ nm (from Andreev V M, An overview of TPV technologies, *Thermophotovoltaic Generation of Electricity*, Rome 2002. © 2003 AIP. Reprinted with permission).

layers in GaSb/InGaAsSb/AlGa(As)Sb heterostructures. TPV cells made of InAsSbP epitaxial structures allow the longer-wavelength spectral response to be extended up to 2.5–4 μm.

11.6.1 InGaAsSb/GaSb TPV cells

Lattice-matched InGaAsSb/GaSb heterostructures with $E_G = 0.5$–0.6 eV in the photoactive cell area were fabricated by LPE [41, 52–57], MBE [58, 59, 80, 81] and MOCVD [60, 75–79].

Double heterostructures with wide-bandgap, lattice-matched AlGaAsSb front-surface passivating 'window' layers and back-surface field cladding layers surrounding the InGaAsSb photoactive layer were grown by LPE in AstroPower, Inc [54, 55]. The absorption edge implied by the spectral response measurements ranged from 2250 to 2300 nm. At 2000 nm, an internal quantum efficiency as high as 95% has been measured in the fabricated TPV cells. Under an illumination intensity corresponding to a short-circuit current density of 1 A cm^{-2}, an open-circuit voltage of 0.26 V [54] has been measured.

As an alternative TPV structure, epitaxial layers of n-In$_{0.15}$Ga$_{0.85}$As$_{0.17}$Sb$_{0.83}$ (0.55 eV)/n-GaSb were diffused with zinc to form a pn-homojunction [55–57]. It was determined that, similar to GaSb cells, strong built-in electric fields near the surface lead to a reduction in the saturation value of the injection component of the dark current density of 4×10^{-6} A cm^{-2} and, hence, to an

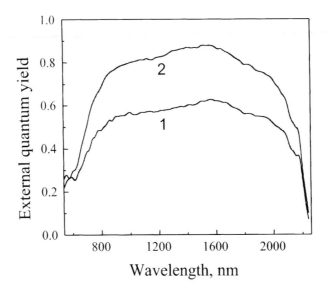

Figure 11.14. External quantum efficiencies of a 1 cm^2 InGaAsSb ($E_G = 0.55$ eV) TPV cell without anti-reflection coating (1) and with a ZnS/MgF$_2$ anti-reflection coating (2).

increase in the open-circuit voltage. Moreover, built-in electric fields as they reduce the influence of the surface carrier recombination mean good PV cell parameters without passivating the AlGaAsSb window layer. At an illumination intensity corresponding to $J_{SC} = 3$ A cm^{-2}, the open-circuit voltage was increased to 0.36 V.

A similar Zn-diffusion technique has been developed at the Ioffe Institute [39–41] for fabricating InGaAsSb/GaSb single-junction and tandem TPV cells. A high (80–90%) external quantum yield (see figures 11.11 and 11.14) was measured in the TPV cells fabricated by Zn diffusion into the n-GaSb/n-InGaAsSb/n-GaSb LPE structure with a thin (0.3 μm) GaSb front passivating layer. Quaternary In$_x$Ga$_{1-x}$As$_y$Sb$_{1-y}$ epilayers were grown by LPE from Sb-rich melts. The use of Sb-rich melts made it possible to decrease the concentration of stoichiometric defects, which are typical for GaSb-related solid solutions. The effect of the negative factor connected with the non-equilibrium interphase in GaSb (solid)–Ga-In-As-Sb (liquid) was also reduced. The crystal perfection of the grown solid solution is comparable with that of GaSb epilayers. A pn-junction was formed by both epitaxial growth and additional gas Zn-diffusion treatment. The external quantum yield of this cell is represented in figure 11.14. The photocurrent density was 54 mA cm^{-2} (AM0, 1 sun) in InGaAsSb cells without a prismatic cover and 58 mA cm^{-2} with such a cover.

Molecular beam epitaxy has also been applied [58, 59, 80, 81] to fabricate InGaAsSb/GaSb TPV cells. It has been shown [59] that the transport of photogenerated electrons in p-InGaAsSb is very efficient. Diodes based on the structure, which consists of an n-InGaAsSb base, a p-InGaAsSb thick emitter and a p^+-GaSb thin window layer, exhibit the electron diffusion length of 29 μm in an 8 μm thick emitter and have internal quantum efficiencies of 95% at a wavelength of 2000 nm, which are well suited to efficient TPV diode operation.

MOCVD technology is now widely used for large-scale production of Ge/GaAs/GaInP$_2$ solar cells for space solar arrays. The modern MOCVD installations are highly productive systems, which allow multilayer heterostructures to be grown on large-diameter (100 mm and higher) substrates. However, MOCVD of InGaAsSb TPV cells is usually realized on 50-mm GaSb substrates [77, 78].

Improved devices are obtained [60] when the optical absorption occurs in the p-layer due to the longer minority carrier diffusion length. It was shown that thin n-emitter/thick p-base devices are more promising, since surface passivation is less critical than for p-emitter devices.

By MOCVD, one can produce metastable alloys with narrower bandgaps than the miscibility regions of the phase equilibrium with a spectral response up to 2.5 μm [79].

11.6.2   Sub-bandgap photon reflection in InGaAsSb/GaSb TPV cells

Reflection of sub-bandgap photons can be an important effect in TPV cells based on InGaAsSb/GaSb as GaSb substrates are transparent for photons not absorbed in InGaAsSb structures. A good reflector can be formed by depositing gold films with an intervening thin layer of SiO$_2$ to prevent the formation of an absorbing alloy between gold and GaSb. It was found [73] that free-carrier absorption is about an order of magnitude higher in p-type GaSb than in n-type GaSb. Therefore, an n-type GaSb substrate thinned to about 100 μm was proposed to reduce optical losses in the substrate. In this case, a reflectance over 90% can be achieved for sub-bandgap photons. Another approach has been proposed [73] where a GaSb/InGaAsSb heterostructure was bonded to a semi-insulating GaAs wafer coated with an SiO$_2$/Ti/Au reflector and, after that, the GaSb substrate was then thinned by controlled etching. This structure can be used for fabricating monolithically interconnected modules, which are now well developed for InGaAs/InP heterostructures and which will be discussed later (in section 11.7).

11.6.3   Tandem GaSb/InGaAsSb TPV cells

The heterostructure of the monolithic tandem TPV cell (figure 11.15) consists of an n-GaSb (substrate); an n-p In$_x$Ga$_{1-x}$As$_y$Sb$_{1-y}$ ($E_G = 0.56$ eV, 1–3 μm thick n-layer and 0.2–0.5 μm thick p-layer) bottom cell; a p^{++}–n^{++}GaSb (0.8 μm total thickness) tunnel junction; an np GaSb (n-layer 3-5 μm thick and a p-layer 0.2–0.5 μm thick) top cell [41, 50]. These structures were fabricated using two-

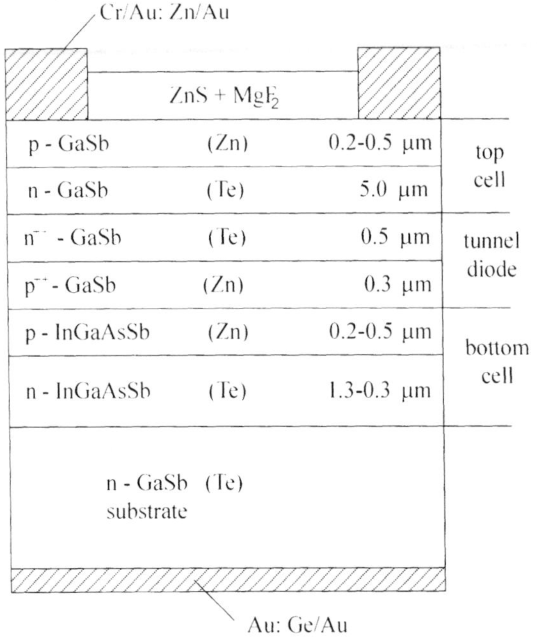

Figure 11.15. Cross section of the monolithic two-junction two-terminal TPV cell.

stage LPE growth and two Zn diffusions. The first step included LPE growth of an n-InGaAsSb layer and a GaSb cap layer from the Sb-rich melt on the GaSb substrate. Zn diffusion was carried out after the first LPE growth to form both the pn-junction in a quaternary solution and the first layer (p^{++} GaSb) of the tunnel junction in GaSb. Te-doped n^{++} GaSb-nGaSb layers (n^{++} GaSb as a second layer of the tunnel junction) were grown from Ga-rich melts. Finally, Zn diffusion was carried out to form a pn-junction in the GaSb top cell. The concentration of Te measured by SIMS was more than $5 \times 10^{19} \text{cm}^{-3}$ and the concentration of Zn was about 10^{19} cm^{-3} in the $p^{++}n^{++}$ tunnel junction region. After annealing at 600 °C for 20 min, the tunnel diodes exhibited ohmic characteristics under voltage biases of 10–20 mV corresponding to the current density of about 10 A cm^{-2}. $V_{OC} = 0.61$ V and $FF = 0.75$ at 0.7 A cm^{-2} were achieved with this tandem TPV cell.

11.6.4   TPV cells based on low-bandgap InAsSbP/InAs

Epitaxial InAsSbP/InAs heterostructures for TPV cells have been grown [55, 71–73] by the LPE method. Narrow-gap epitaxial InAsSbP (0.45–0.48 eV) cells were fabricated [73] from p-InAsSbP/n-InAsSbP/n-InAs heterostructures grown on (100) n-InAs substrates. Epitaxial growth of n-InAsSbP quaternary layers

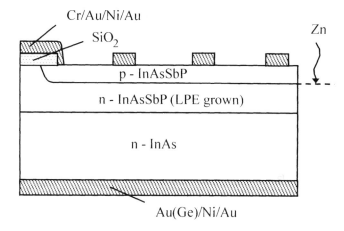

Figure 11.16. Cross section of a TPV cell based on the p-InAsSb/n-InAsSbP/n-InAs heterostructure.

lattice-matched to InAs was carried out by LPE from an Sb-rich melt. The lattice-mismatch ratio $\Delta a/a = 0.15\%$ was estimated from x-ray diffraction measurements. Liquid-solid phase equilibrium estimations were obtained using a simple or regular solution model. The epitaxy was performed at 570 °C during 10 min (step-cooling technique) under 10–11 °C super-saturation condition to avoid a composition gradient in the epitaxial layer. A conventional graphite sliding boat in a Pd-diffused hydrogen atmosphere was used in the LPE growth. The unintentionally doped layers were n-type with a background carrier concentration of $n \approx 10^{17}$ cm^{-3}.

The spectral response of TPV cells based on p-InAsSbP/n-InAsSbP/n-InAs structures was widened in the infrared range up to 2.5–3.4 μm by a combination of LPE growth and Zn diffusion. This means that InAsSbP-based cells can be considered as having excellent potential for applications in TPV generators with low-temperature (900–1000 °C) emitters. The low-temperature (340–350 °C) pseudo-closed box Zn diffusion process was applied [72] to form a pn-junction in a quaternary InAsSbP alloy. A patterned silicon nitride or SiO$_2$ dielectric layers on the top surface were used as a mask for Zn diffusion so that the junction was formed only within the designated illumination area, which allowed the current leakages to be reduced. The busbar was deposited on this dielectric layer and only contact grid fingers contacted to the junction region. A typical spacing between fingers was 100 and 200 μm with shadowing of 5–10%. A cross section of these InAsSbP-based devices is shown in figure 11.16. Figure 11.17, curve 1, represents the spectral response of p-InAsSbP/n-InAsSbP/n-InAs TPV cells. An insufficient diffusion length for the photogenerated holes in the epitaxial n-InAsSbP base layer probably causes a reduction in the photosensitivity in the long-wavelength

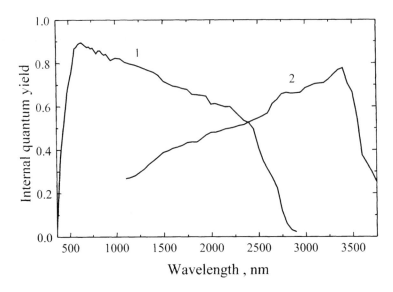

Figure 11.17. Spectral responses of the TPV cells based on: p-InAsSbP/n-InAsSbP/n-InAs heterostructure (curve 1), and p–n InAs fabricated by the Zn diffusion process (curve 2) (from Andreev V M, An overview of TPV technologies, *Thermophotovoltaic Generation of Electricity*. Rome 2002. © 2003 AIP. Reprinted with permission).

region. The V_{OC} value was 0.15 V in the InAsSbP cells with $E_G = 0.45$ eV at a photocurrent density of 3 A cm^{-2}.

Pn-InAs TPV cells were fabricated [72] using wafers prepared from ingots grown by the Czochralski technique, Te-doped. The InAs wafers were subjected to the Zn diffusion procedure by means of the 'pseudo-closed box' technique to form a pn-junction in photoactive area of the cells from a pure zinc source. figure 17, curve 2 represents the spectral response of a p-InAs/n-InAs TPV cell, which demonstrates rather high photosensitivity up to 3.5 μm at room temperature. The photosensitivity in the short-wavelength ($\lambda < 2.5$ μm) part of the spectrum can be increased by growing a wide-bandgap lattice-matched window InAsSbP layer on InAs.

11.7   TPV cells based on InGaAs/InP heterostructures

Lattice-matched In$_{0.53}$Ga$_{0.47}$As/InP heterostructures were developed initially for fabricating infrared photodetectors. Then, these diodes were used as laser power converters [64] and as bottom cells for monolithic InP(top)/InGaAs(bottom) [62] or mechanically stacked GaAs (top)/InGaAs(bottom) [31, 32] tandems. InGaAs layers were grown on InP substrates by LPE [31, 32, 63] and by MOCVD [62] methods. A 6% efficiency (AM0, 100 suns) was measured [31] in the LPE cells illuminated through the InP substrate and the GaAs IR transparent filter. TPV cells fabricated by simultaneous Zn and P diffusion into the LPE-grown InGaAs layer have demonstrated $V_{OC} = 0.42$–0.5 V and $FF = 0.75$–0.78 at $J_{SC} = 1$–15 A cm^{-2} [63].

Lattice-matched (or mismatched) InGaAs/InP monolithic interconnected modules (MIMs) have also been developed [64–67] for TPV applications. In this approach, the InGaAs structure was grown by MOCVD on a semi-insulating InP substrate. Then the structure was divided into several sub-cells, which were interconnected in series. This has the effect of reducing the current output of a single cell by a factor equal to the number of sub-cells, which reduces (in principle) the Joule losses and increases the power output at high illumination intensity. A MIM has other advantages over conventional one-junction cells, such as simplified array interconnections and heat sinking and a radiation recycling capability using a back-surface reflector.

Semi-insulating substrates are transparent to infrared radiation which means that extremely high sub-bandgap photon reflection can be obtained in the structures with a high-reflective back-surface reflector. A reflectivity of about 95% for photons with $\lambda > 1.7$ μm was obtained [65] from the back reflector made of Ag or Au deposited right to the back surface of semi-insulating InP. The reflectivity was enhanced up to 97–98% by interposing a MgF$_2$ layer between the InP and Au or Ag. The measured reflectivity from the MIM TPV cells was 80–85% in the spectral range 1.8–2.7 μm and increased at $\lambda > 2.7$ μm [66]. An external quantum efficiency of about 90% was achieved in the spectral range 0.6–1.5 μm that ensures $J_{SC} = 56$ mA cm^{-2} under the AM0 spectrum. $V_{OC} = 0.4$ V, $FF = 0.71$ and an efficiency of 11.7% (AM0) were achieved [66] in the developed MIM devices with a grid shadowing of 16%.

High-performance, lattice-mismatched InGaAs/InP MIMs with an InGaAs bandgap energy of 0.6 eV [68, 69] and 0.55 eV [70] were developed to make the spectral response of the TPV cells with narrower bandgaps better matched to the gray-body spectral power density profile for low-to-moderate emitter temperatures (i.e. 1000–1200 °C). The lattice-mismatched structures were grown by MOCVD using a compositionally step-graded InPAs buffer to accommodate a lattice mismatch of 1.1% between the active InGaAs cell structure and the InP substrate [68]. The internal quantum efficiency in 0.6 eV MIMs with a back-surface reflector was estimated to be near unity close to the band edge.

Good performance characteristics were demonstrated also for TPV converters made from a lattice-matched InAs$_{0.32}$P$_{0.68}$(0.96 eV)/Ga$_{0.32}$In$_{0.68}$As(0.6 eV)/InAs$_{0.32}$P$_{0.68}$ double-heterostructure, which was grown lattice-mismatched on an InP substrate, using the step graded region of InPAs [69]. The average V_{OC} value for the sub-cells is 0.39 V, which yields a voltage factor of 0.64 V/eV. High reflectance (80–90%) was achieved for $\lambda = 2$–12 μm, which indicates that sub-bandgap photon recirculation will be effective in a TPV system based on developed MIMs.

MIM devices based on InGaAs with a bandgap reduced to 0.55 eV were developed [70] as well. Novel buffering schemes improved cell architectures and a variety of interconnected schemes were investigated in order to maximize the device power density and efficiency. The external quantum efficiency of the device with no antireflective coating was found to be 0.7 and the dark current density was found to be 40 μA cm^{-2}. The 30-junction MIM had $V_{OC} = 8.3$ V and $FF = 0.61$ at the short-circuit current density of 1.15 A cm^{-2}. Such cells are very promising for TPV systems operating with low temperature (1000–1100 °C) emitters.

11.8   Summary

Gallium antimonide and narrow-bandgap ($E_G < 0.6$ eV) antimonide-based devices continue to make sufficient progress. Several teams have developed highly productive diffusion technologies for fabricating high efficiency GaSb-based TPV cells with good reproducible. The formation of a sharp Zn concentration gradient in the thin front diffused layers and a strong built-in electrical field, the use of epitaxial and two-stage diffusion technologies ensured improvement in the cell performance. Efficiencies exceeding 25% for in-band ($\lambda < 1800$ nm) blackbody (1200–1700 K) radiation were achieved in advanced GaSb TPV cells. To reduce the cost of the TPV system, polycrystalline GaSb was used to fabricate TPV cells. The potential cost saving compensates the small drop in performance in polycrystalline cells.

Using Zn diffusion into epitaxially-grown InGaAsSb (0.55 eV) lattice-matched to GaSb enabled the saturation value of the injection component of the dark current density to be reduced to 4×10^{-6} A cm^{-2} and the open-circuit voltage to be increased up to 0.36 V at an illumination intensity corresponding to $J_{SC} = 3$ A cm^{-2}.

High efficiency InGaAs TPV cells based on In$_{0.53}$Ga$_{0.47}$As/InP lattice-matched heterostructures, fabricated by both LPE and MOCVD, have been developed. MOCVD growth of monolithic interconnected modules based on both lattice-matched and mismatched InGaAs/InP structures have become more sophisticated in recent years. High sub-bandgap photon reflection has been realized in these structures with a back-surface reflector owing to the high transparency of semi-insulating InP substrates to infrared radiation. Good performance characteristics were demonstrated for TPV cells based on mismatched InGaAs (0.55–0.6 eV)/InP double-heterostructures, which were grown by MOCVD using the step graded region of InAsP to accommodate a lattice mismatch between the active InGaAs layers and the InP substrate.

Ge-based TPV cells ensured efficiencies higher than 16% were achieved under cut-off at $\lambda > 1800$ nm blackbody (1400–1900 K) spectrum at $J_{SC} = 3$–20 A cm^{-2} and a high sub-bandgap photon reflection of 85%.

The photosensitivity of TPV cells based on p-InAsSbP/n-InAsSbP/n-InAs and p-InAs/n-InAs structures was widened in the infrared range up to 2.5–3.4 μm. These PV low-bandgap cells are promising for applications in TPV generators with lower-temperature emitters.

Acknowledgments

The author would like to thank his colleagues at the Photovoltaics Laboratory of Ioffe Institute—V D Rumyantsev, V P Khvostikov, V A Grilikhes, V R Larionov, M Z Shvarts, L B Karlina, G V Il'menkov, N Kh Timoshina—for help, A Bett at the Fraunhofer ISE and M Mauk and O Sulima at AstroPower, Inc for useful discussions.

References

[1] Broman L 1995 *Prog. Photovolt.* **3** 65–74
[2] Fraas L M, Ballantyne R, Samaras J and Seal M 1994 *First World Conf. on Photovoltaic Energy Conversion (Hawaii)* (New York: IEEE) pp 1713–16
[3] Woolf L D 1985 *Conf. Record of the 18th IEEE PVSC* (New York: IEEE) pp 1731–32
[4] Wanlass M W, Ward J S, Emery K A and Coutts T J 1994 *First World Conf. on Photovoltaic Energy Conversion (Hawaii)* (New York: IEEE) pp 1685–91
[5] Gray J L and El-Husseini A 1996 *Second NREL Conf. on Thermophotovoltaic Generation of Electricity (AIP Conf. Proc. 358)* (New York: AIP) pp 3–15
[6] Iles P A, Chu C and Linder E 1996 *Second NREL Conf. on Thermophotovoltaic Generation of Electricity (AIP Conf. Proc. 358)* (New York: AIP) pp 446–57
[7] Coutts T J, Wanlass M W, Ward J S and Johnson S 1996 *Conf. Record of 25th IEEE PVSC (Washington, DC)* (New York: IEEE) pp 25–30
[8] Coutts T J 1999 *Renewable and Sustainable Energy Reviews* **3** 77–184
[9] Cody G D 1999 *Fourth NREL Conf. on Thermophotovoltaic Generation of Electricity (AIP Conf. Proc. 460)* (New York: IEEE) pp 58–67
[10] Coutts T J 1999 *Technical Digest of the International PVSEC (Sapporo)* (Tokyo: PVSEC) pp 137–40
[11] Andreev V M, Khvostikov V P, Larionov V R, Rumyantsev V D, Sorokina S V, Shvarts M Z, Vasil'ev V I and Vlasov A S 1997 *Conf. Record of 26th IEEE PVSC (Anaheim)* pp 935–8
[12] Andreev V M, Grilikhes V A and Rumyantsev V D 1997 *Photovoltaic Conversion of Concentrated Sunlight* (Chichester: Wiley)
[13] Venkatasubramanian R, Timmons M L, Pickett R T, Colpitts T S, Hills J S and Hutchby J A 1990 *Conf. Record of the 21th IEEE PVSC (Piscataway, NJ)* (New York: IEEE) pp 73–8
[14] Bullock J N 1989 *Trans. Electron. Devices* **36** 1238–43
[15] Tobin S P, Vernon S M, Bajgar C, Haven V E, Geoffroy L M, Sanfacon M M, Lillington D R, Hart R E, Emery K A and Matson R J 1988 *Conf. Record of the 20th IEEE PVSC (Las Vegas, NV)* (New York: IEEE) pp 405–10
[16] Timmons M L, Hutchby J A, Wagner D K and Tracy J M 1988 *Conf. Record of the 20th IEEE PVSC (Las Vegas, NV)* (New York: IEEE) pp 602–6

[17] Wojtczuk S J, Tobin S P, Keavney C J, Bajgar C, Sanfacon M M, Geoffroy L M, Dixon T M, Vernon S M, Scofield J D and Ruby D S 1990 *IEEE Trans. Electron. Devices* **37** 455–62

[18] Wojtczuk S, Tobin S, Sanfacon, M, Haven V, Geoffroy L and Vernon S 1991 *Conf. Record of the 22nd IEEE PVSC (Las Vegas, NV)* (New York: IEEE) pp 73–9

[19] Faur Mircea, Faur Maria, Bailey S G, Flood D J, Brinker D J, Wheeler D R, Alterovitz S A, Scheiman D, Mateescu G, Faulk J, Goradia C and Goradia M 1997 *Conf. Record of the 26th IEEE PVSC (Anaheim, CA)* (New York: IEEE) pp 847–51

[20] Karam N H, King R R, Cavicchi B T, Krut D D, Ermer J H, Haddad M, Cai L, Joslin D E, Takahashi M, Eldredge J W, Nishikawa W T, Lillington D R, Keyes B M and Ahrenkiel R K 1999 *IEEE Trans. Electron. Devices* **46** 2116–25

[21] Friedman D J, Olson J M, Ward S, Moriarty T, Emery K, Kurtz Sarah, Duda A, King R R, Cotal H L, Lillington D R, Ermer J H and Karam N H 2000 *Conf. Record of the 28th IEEE PVSC (Anchorage, AK)* (New York: IEEE) pp 965–7

[22] Cotal H L, Lillington D R, Ermer J H, King R R, Karam N H, Kurtz S R, Friedman D J, Olson J M, Ward J S, Duda A, Emery K A and Moriarty T 2000 *Conf. Record of the 28th IEEE PVSC (Anchorage, AK)* (New York: IEEE) pp 955–60

[23] Andreev V M, Khvostikov V P, Khvostikova O V, Oliva E V, Rumyantsev V D and Shvartz M Z 2002 *Fifth Conf. on Thermophotovoltaic Generation of Electricity (Rome) (AIP Conf. Proc. 653)* (Woodbury, NY: AIP) pp 383–91

[24] Green M A, Jianhua Z, Blakers A W, Taouk M and Narayanan S 1986 *IEEE Electron. Devices Lett.* **EDL-7** 583–5

[25] Sinton R A, Kwark Y, Gan J Y and Swanson R M 1986 *IEEE Electron. Devices Lett.* **EDL-7** 567–9

[26] Kushch A S, Skinner S M, Brennan R and Sarmiento P A 1997 *Third NREL Conf. on Thermophotovoltaic Generation of Electricity (AIP Conf. Proc. 401)* (New York: AIP) pp 373–86

[27] Bitnar B, Durisch W, Grutzmacher D, Mayor J.-C, Muller C, von'Roth F, Selvan J A A, Sigg H, Tschudi H R and Gobrecht J 2000 *Conf. Record of the 28th IEEE PVSC (Anchorage, AK)* (New York: IEEE) pp 1218–21

[28] Durish W, Bitnar B, Mayor J-C, Fritz von Roth, Sigg H, Tschudi H R and Palfinger G 2001 *Proc. 17th Photovoltaic Solar Energy Conf. and Exhibition (Munich)* (Munich and Florence: WIP and ETA) pp 2296–9

[29] Fraas L M, Girard G R, Avery J E, Arau B A, Sundaram V S and Thompson A G 1989 *J. Appl. Phys.* **66** 3866–70

[30] Fraas L M, Avery J E, Martin J, Sundaram V S, Girard G, Dinh V T, Davenport T M, Yerkes J W and O'Neill M J 1990 *IEEE Trans. Electron. Devices* **37** 443–9

[31] Andreev V M, Karlina L B, Kazantsev A B, Khvostikov V P, Rumyantsev V D, Sorokina S V and Shvarts M Z 1994 *First World Conf. on Photovoltaic Energy Conversion (Hawaii)* (New York: IEEE) pp 1721–4

[32] Andreev V M, Karlina L B, Khvostikov V P, Rumyantsev V D, Sorokina S V and Shvarts M Z 1995 *13th European Photovoltaic Solar Energy Conference (Nice)* (Bedford: H S Stephens & Associates) pp 329–32

[33] Sulima O V, Faleev N N, Kazantsev A B, Mintairov A M and Namazov A 1995 *Fourth European Space Power Conf. (Poitiers)* (ESA Publications Division) pp 641–3

[34] Andreev V M, Khvostikov V P, Paleeva E V, Rumyantsev V D, Sorokina S V,

Shvarts M Z and Vasil'ev V I 1996 *23rd Int. Symp. Compound Semiconductors (St Petersburg) (Inst. Phys. Conf. Ser. 155)* pp 425–8

[35] Dutta P S and Bhat H L 1997 *J. Appl. Phys.* **81** 5821–70

[36] Fraas L M, Huang H X, Ye S-Z, Hui S, Avery J and Ballantyne R 1997 *Third NREL Conf. on Thermophotovoltaic Generation of Electricity (AIP Conf. Proc. 401)* (Woodbury, NY: AIP) pp 33–40

[37] Fraas L, Samaras J, Huang H-X, Seal M and West E 1999 *Fourth NREL Conf. on Thermophotovoltaic Generation of Electricity (AIP Conf. Proc. 460)* (Woodbury, NY: AIP) pp 371–83

[38] Bett A W, Faleev N N, Mintairov A M, Namazov A, Sulima O V and Stollwerck G 1995 *13th European Photovoltaic Solar Energy Conf. (Nice)* (Bedford: H S Stephens & Associates) pp 88–91

[39] Andreev V M, Khvostikov V P, Paleeva E V, Sorokina S V and Shvarts M Z 1996 *Proc. 25th IEEE PVSC (Washington, DC)* (New York: IEEE) pp 143–6

[40] Andreev V M, Khvostikov V P, Sorokina S V, Shvarts M Z and Vasil'ev V I 1997 *Proc. 14th European Photovoltaic Solar Energy Conf. (Barcelona)* (Bedford: H S Stephens & Associates) pp 1763–6

[41] Andreev V M, Khvostikov V P, Larionov V R, Rumyantsev V D, Sorokina S V, Shvarts M Z, Vasil'ev V I and Vlasov A S 1997 *Conf. Record of the 26th IEEE PVSC (Anaheim, CA)* (New York: IEEE) pp 935–8

[42] Bett A W, Keser S, Stollwerck G and Sulima O V 1997 *Third NREL Conf. on Thermophotovoltaic Generation of Electricity (AIP Conf. Proc. 401)* (New York: AIP) pp 41–53

[43] Bett A W, Keser S, Stollwerck G and Sulima O V 1997 *14th European Photovoltaic Solar Energy Conf. (Barcelona)* (Bedford: H S Stephens & Associates) pp 993–8

[44] Bett A W, Keser S and Sulima O V 1997 *Cryst. Growth* **181** 9–16

[45] Fraas L M, Ballantyne R, Ye S-Z, Hui S, Gregory S, Keyes J, Avery J, Lamson D and Daniels B 1999 *Fourth NREL Conf. on Thermophotovoltaic Generation of Electricity (AIP Conf. Proc. 460)* (New York: AIP) pp 480–7

[46] Sulima O V, Bett A W, Dutta P S, Ehsani H and Gutmann R J 2000 *16th European Photovoltaic Solar Energy Conf. (Glasgow)* (London: James & James) pp 169–72

[47] Stollwerck G, Sulima O V and Bett A W 2000 *IEEE Trans. Electron. Devices* **47** 448–57

[48] Sulima O V and Bett A W 2000 *Technical Digest of 11th International Photovoltaic Science and Engineering Conf. (Sapporo)* (Tokyo: PVSEC) pp 441–2

[49] Rumyantsev V D, Khvostikov V P, Sorokina S V, Vasil'ev V I and Andreev V M 1999 *Fourth NREL Conf. on Thermophotovoltaic Generation of Electricity (AIP Conf. Proc. 460)* (New York: AIP) pp 384–91

[50] Andreev V M, Khvostikov V P, Larionov V R, Rumyantsev V D, Sorokina S V, Shvarts M Z, Vasil'ev V I and Vlasov A S 1998 *2nd World Conf. and Exhibition on Photovoltaic Solar Energy Conversion (Vienna)* (Ispra: European Commission) pp 330–3

[51] Khvostikov V P and Rumyantsev V D 1998 *2nd World Conf. and Exhibition on Photovoltaic Solar Energy Conversion (Vienna)* (Ispra: European Commission) pp 120–3

[52] Andreev V M, Vasil'ev V I, Khvostikov V P, Rumyantsev V D, Sorokina S V and Shvarts M Z 2000 *Technical Digest of 11th International Photovoltaic Science and Engineering Conf. (Sapporo)* (Tokyo: PVSEC) pp 443–4

[53] Andreev V M, Khvostikov V P, Rumyantsev V D, Sorokina S V and Shvarts M Z 2000 *Conf. Record of 28th IEEE PVSC (Anchorage, AK)* (New York: IEEE) pp 1265–8

[54] Shellenbarger Z A, Mauk M G, Cox J A, Gottfried M I, Sims P E, Lesko J D, McNeely J B and DiNetta L C 1997 *Third NREL Conf. on Thermophotovoltaic Generation of Electricity (AIP Conf. Proc. 401)* (New York: AIP) pp 117–28

[55] Mauk M G, Shellenbarger Z A, Cox J A, Sulima O V, Bett A W, Mueller R L, Sims P E, McNeely J B and DiNetta L C 2000 *J. Crystal Growth* **211** 189–93

[56] Bett A W, Ber B Y, Mauk M G, South J T and Sulima O V 1998 *Fourth NREL Conf. on Thermophotovoltaic Generation of Electricity (AIP Conf. Proc. 460)* (New York: AIP) pp 237–46

[57] Sulima O V, Beckert R, Bett A W, Cox J A and Mauk M G 2000 *IEE Proc.-Optoelectron.* **147** 199–204

[58] Wang C A, Choi H K, Turner G W, Spears D L and Manfra M J 1997 *Third NREL Conf. on Thermophotovoltaic Generation of Electricity (AIP Conf. Proc. 401)* (Woodbury, NY: AIP) pp 75–87

[59] Martinelli R U, Garbuzov D Z, Lee H, Morris N, Odubanjo T, Taylor G C and Connoly J C 1997 *Third NREL Conf. on Thermophotovoltaic Generation of Electricity (AIP Conf. Proc. 401)* (Woodbury, NY: AIP) pp 389–95

[60] Hitchcock C W, Gutmann R J, Borrego J M, Bhat I B and Charache G W 1999 *IEEE Trans. Electron. Devices* **46** 2154–61

[61] Agert C, Beckert R, Hinkov V, Sulima O V and Bett A 2001 *Proc. 17th European Photovoltaic Solar Energy Conf. and Exhibition (Munich)* (Munich and Florence: WIP and ETA) pp 372–5

[62] Wanlass M W, Ward J S, Emery K A, Gessert T A, Osterwald C R and Coutts T J 1991 *Solar Cells* **30** 363–71

[63] Karlina L B, Ber B A, Kulagina M M, Kovarsky A P, Vargas-Aburto C, Uribe R M, Brinker D and Scheiman D 2000 *Conf. Record of 28th IEEE PVSC (Anchorage, AK)* (New York: IEEE) pp 1230–3

[64] Wojtczuk S, Parados T and Walker G 1994 *Proc. XIII Space Photovoltaic Research and Technology Conf. (Cleveland, OH)* pp 363–71

[65] Fatemi N S, Wilt D M, Jenkins P P, Hoffman R W, Weizer V G, Murray C S and Riley D 1997 *Third NREL Conf. on the Thermophotovoltaic Generation of Electricity (AIP Conf. Proc. 401)* (New York: AIP) pp 249–62

[66] Wilt D M, Fatemi N S, Jenkins P P, Weizer V G, Hoffman R W, Jain R K, Murray C S and Riley D R 1997 *Third NREL Conf. on the Thermophotovoltaic Generation of Electricity (AIP Conf. Proc. 401)* (New York: AIP) pp 237–47

[67] Ward J S, Duda A, Wanlass M W, Carapella J J, Wu X, Matson R J, Coutts T J, Moriarty T, Murray C S and Riley D R 1997 *Third NREL Conf. on the Thermophotovoltaic Generation of Electricity (AIP Conf. Proc. 401)* (New York: AIP) pp 227–36

[68] Fatemi N S, Wilt D M, Hoffman R W, Stan M A, Weizer V G, Jenkins P P, Khan O S, Murray C S, Scheiman D and Brinker D 1999 *Fourth NREL Conf. of Thermophotovoltaic Generation of Electricity (AIP Conf. Proc. 460)* (Woodbury, NY: AIP) pp 121–31

[69] Wanlass M W, Carapella J J, Duda A, Emery K, Gedvilas L, Moriarty T, Ward S, Webb J D and Wu X 1999 *Fourth NREL Conf. of Thermophotovoltaic Generation of Electricity (AIP Conf. Proc. 460)* (Woodbury, NY: AIP) pp 132–41

[70] Murray C S, Fatemi N, Stan M, Wernsman B and Wehrer R J 2000 *Conf. Record of 28th IEEE PVSC (Anchorage, AK)* (New York: IEEE) pp 1238–41

[71] Mauk M G, Shellenbarger Z A, Cox J A, Tata A N, Warden T G and DiNetta L C 2000 *Proc. 28th IEEE PVSC (Anchorage, AK)* (New York: IEEE) pp 1028–31

[72] Khvostikov V P, Khvostikova O A, Oliva E V, Rumyantsev V D, Shvarts M Z, Tabarov T S and Andreev V M 2002 *Proc. 29th IEEE PVSC (New Orleans, LA)* (New York: IEEE) pp 943–46

[73] Charache G W, DePoy D M, Baldasaro P F and Campbell B C 1995 *The Second NREL Conf. on Thermophotovoltaic Generation of Electricity (AIP Conf. Proc. 358)* (Woodbury, NY: AIP) pp 339–50

[74] Kobayashi N and Horikoshi Y 1981 *Japan. J. Appl. Phys.* **20** 2301–5

[75] Bougnot G, Delannoy F, Foucaran A, Pascal F, Roumanille F, Grosse P and Bougnot J P 1988 *J. Electrochem. Soc.* **135** 1783–8

[76] Wang C A, Choi H K and Charache G W 2000 *IEE Proc. Optoelectron.* **147** 193–8

[77] Shellenbarger Z, Taylor G, Smeltzer R, Li J, Martinelli R and Palit K 2002 *The Fifth Conf. on Thermophotovoltaic Generation of Electricity (Rome)* (Woodbury, NY: AIP) pp 314–23

[78] Yu Y Z, Martinelli R U, Taylor G C, Shellenbarger Z, Smelzer R K, Li J, Burger S R, Cardines R P, Danielson L R, Wang C A and Conners M K 2002 *The Fifth Conf. on Thermophotovoltaic Generation of Electricity (Rome) (AIP Conf. Proc.)* (Woodbury, NY: AIP) pp 335–43

[79] Wang C A, Choi H K, Oakley D C and Charache G W 1998 *Thermophotovoltaic Generation of Electricity: Fourth NREL Conf. (AIP Conf. Proc. 460)* (Woodbury, NY: AIP) pp 256–64

[80] Uppal P N, Charache G, Baldasaro P, Campbell B, Loughin S, Svensson S and Gill D 1997 *J. Crystal Growth* **175–176** 877–82

[81] Griffin P, Ballard I, Barnham K, Nelson J, Zachariou A, Epler J, Hill G, Button C and Pate M 1998 *Solar Energy Mater. Solar Cells* **50** 213–19

Chapter 12

Wafer-bonding and film transfer for advanced PV cells

C Jaussaud[1], E Jalaguier[2] and D Mencaraglia[3]
[1] *CEA/DTEN*
[2] *CEA /DTS*
[3] *Supelec/LGEP*

12.1   Introduction

Wafer-bonding and film transfer have been developed in the microelectronic industry and these techniques are presently used to make silicon-on-insulator (SOI) structures. They are also of interest for photovoltaic cells and many studies have been done to develop thin-film cells based on film transfer [1–9]. They have also been studied to make multi-junction cells [10, 11]. In this chapter, we will describe recent developments in film transfer which could be of interest for multi-junction photovoltaic cells. Presently, multi-junction cells are made either by direct epitaxy onto a substrate or by mechanical stacking. In cells processed by epitaxy, the current is common to all cells and requires precise matching of the cells. Mechanically stacked multi-junction cells do need current matching between the cells and offer more flexibility for the choice of gaps (see figure 12.1).

However, mechanical stacking requires complex mounting. In this chapter, we will show the potential of wafer-bonding and thin-film transfer to make multi-junction cells that could combine the advantages of cells produced by epitaxy and of those produced by mechanical stacking.

12.2   Wafer-bonding and transfer application to SOI structures

Most of the examples given here have been obtained using the Smart-Cut® process. This process is a generic process which works on different materials

274

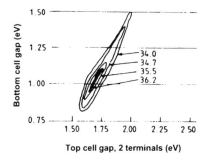

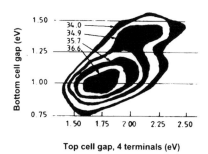

Figure 12.1. AM1 iso-efficiency for two cells one with two (left) and one with four terminals (right) (from [12]. © 1982 IEEE).

but up to now the most advanced studies have been carried out on silicon. It can be schematically described (figure 12.2) by the following steps to obtain SOI wafers:

- The atomic species (for instance hydrogen ions) are first implanted in a substrate A. This step induces the formation of an in-depth weakened layer.
- Then the substrate A is bonded onto a support B by wafer bonding.
- Next, the following splitting step, which takes place along the in-depth weakened layer, gives rise to the transfer of a thin layer from the substrate A onto the support B.
- Finally a treatment can be performed to remove the rough surface left after splitting. The asic mechanisms involved in splitting the silicon wafers have already extensively been reported (see, for instance, [13]).

Today, the major application of the Smart-Cut® process is the fabrication of SOI wafers. The process is well suited to obtaining SOI wafers, especially for very thin top silicon layers. This process allows a large flexibility in layer thickness with a very high silicon thickness homogeneity (better than ±5 nm over 200 mm wafers). While previously developed for a (0.1–0.2 μm) thick Si layer on top of (0.1–0.4 μm) thick buried oxides, new trends are to thin down the thickness of both the oxide and the Si film. Film thicknesses as thin as only few tens of nanometres are targeted today, for instance in ultra-low power consumption devices. The micro-roughness of as-finished SOI wafers is comparable to that of silicon bulk wafers. AFM (atomic force microscope) measurements are used to check this roughness over 0.5 μm × 0.5 μm up to 10 μm × 10 μm areas. In the case of 1 μm × 1 μm areas, typical values of ∼1.0 Å are measured. Furthermore, this technique enables very good crystalline quality of the top silicon layer to be obtained. For example, the density of HF defects (defects revealed by hydrofluoric acid) is today limited by the growth of the defects present in the initial silicon wafer used to obtain the top silicon layer. When SOIs are made from initial low

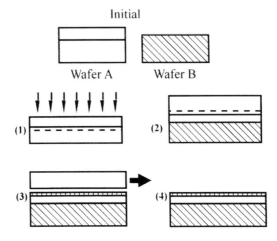

Figure 12.2. Schematic diagram of the Smart-Cut® process: (*a*) ion implantation (H, He) in wafer 1; (*b*) cleaning and bonding of wafers A and B; (*c*) splitting step; and (*d*) touch polishing.

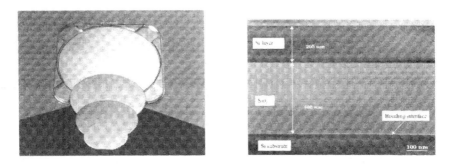

Figure 12.3. Left: UNIBOND® SOI wafers (100, 150, 200, 300 mm). Right: typical XTEM of the structure (by courtesy of SOITEC).

COP (crystal originated particles: defects that are seen during the counting of particles on the wafer, but which are small voids close to the silicon surface) wafers, the density of the HF defects can be lower than 0.1 cm^{-2}. In parallel, the results obtained on the new generation of SOIs indicate a density of Secco defects of only a few hundred defects cm^{-2} for a remaining silicon thickness after etching as thin as 30 nm [14]. All these results obtained on UNIBOND® SOI wafers indicate that the Smart-Cut® process allow very high quality SOI wafers to be achieved. Today, UNIBOND® SOI wafer production is currently running at a high volume and both the process and the facilities are well suited for up to 300 mm large wafers (figure 12.3).

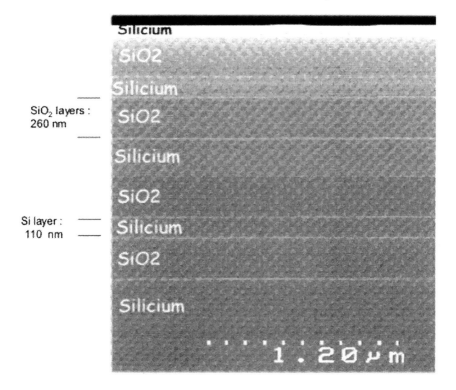

SiO₂ layers :
260 nm

Si layer :
110 nm

Figure 12.4. SEM view of a multiple SOI layer stack (by courtesy of SOITEC).

The possibility of transferring thin silicon films using Si_3N_4 layers has also been demonstrated [13]. Using Si_3N_4 instead of SiO_2 is favourable for heat dissipation since the thermal conductivity of Si_3N_4 is 30 times higher than that of SiO_2. The possibility of bonding an Si_3N_4 layer onto an SiO_2 layer has also been demonstrated, giving rise to new SOI structures.

The flexibility of the process can be used to realize complex structure [13]. For example, this layer transfer can be repeated several times on the same structure. Figure 12.4 shows the stacking of four SOI layers. In this example, the Si layers are 110 nm thick and the SiO_2 layers are 260 nm thick. These multiple SOI layers can be achieved by repeating a single layer transfer several times onto the same wafer or by transferring SOI layers stacks already formed onto a single substrate or SOI wafer.

12.3   Other transfer processes

The transfer of thin films with patterned structures onto different substrates is very attractive for applications such as thin-film transistors on glass or quartz for TFT-

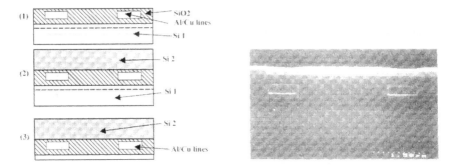

Figure 12.5. Transfer of patterned structures. Left: (1) SiO$_2$ deposition + Al/Cu deposition and patterning + SiO$_2$ deposition and planarization + H$^+$ implantation; (2) Bonding on Si 2; (3) Splitting at implanted layer. Right: SEM micrograph of one of the manufactured structures.

LCDs, intelligent sensors and actuators or smart power. For example, metal lines embedded in an oxide have been transferred onto a Si substrate [15].

In this example (see figure 12.5), the process was the following: 380 nm thick oxide growth on a silicon substrate, 340 nm thick aluminium-copper and 40 nm titanium nitride deposition and patterning, CVD deposition of 1.7 μm thick oxide. This structure is then planarized, hydrogen is implanted through it and silicon wafer 1 is bonded to silicon wafer 2. Heat treatment at 450 °C has been chosen here to split this structure at the implanted layer.

The Smart-Cut$^\circledR$ process has been applied to the transfer of SiC [16], AsGa [17], InP [18] and germanium films [10].

In the case of SiC, AsGa and InP, the transfers have been done via an oxide. In the case of germanium, the transfer indicated in [19] was done via direct Ge/Si bonding. Transfers of Ge onto silicon can also be done via an oxide. Figure 12.6 shows examples of transfers done via an oxide at CEA/LETI.

For some applications (power devices, laser diodes, solar cells), a conductive bonding is interesting. So, the Smart-Cut$^\circledR$ process has been applied to wafers bonded via metal layers [19]. The metal chosen for bonding is palladium. This metal was chosen because it forms alloys with silicon, GaAs and InP. The process consists of the following steps: H$^+$ implantation into one of the wafers, Pd deposition on both wafers, bonding at room temperature by contacting the two wafers, 500 °C annealing for consolidation of the bonding (formation of Pd$_2$Si). The implanted layer splits during the 500 °C annealing. After splitting, a 560 nm thick GaAs layer is transferred onto the Si wafer, and an ohmic contact is formed between the two wafers. This technique has also been applied to transfer a 470 nm InP film onto a Si wafer. Figure 12.7(a) shows XTEM of the bonding interfaces and figure 7(b) a photograph of a silicon film transferred onto a 100 mm silicon

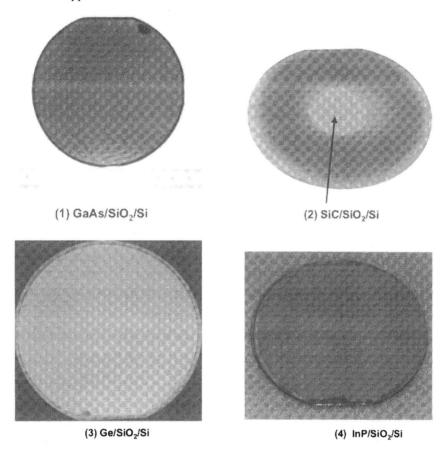

Figure 12.6. Transfer of (1) Ge, (2) InP (3) GaAs and (4) SiC onto silicon.

wafer. In the case of GaAs and InP, only small areas (typically 1 cm²) are transferred and more work is required to obtain the transfer of films on a whole 100 mm wafer.

12.4   Application of film transfer to III–V structures and PV cells

The techniques of film transfer have already been used for transistors and solar cells applications. We give here some examples of structures and devices realized using bonding and transfer.

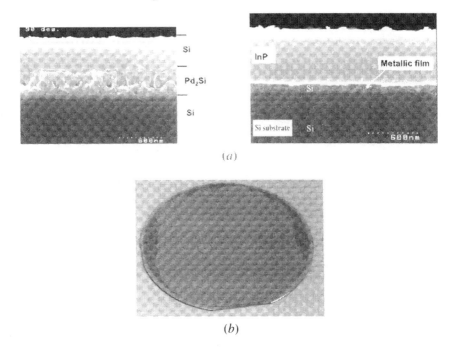

Figure 12.7. (*a*) XTEM of films transferred onto silicon with metallic bonding. Left: Si transfer on Si. Right: InP transfer on Si. (*b*) Photograph of films transferred onto silicon with metallic bonding [19] (Si film transferred onto a 100 mm Si wafer).

12.4.1 HEMT InAlAs/InGaAs transistors on films transferred onto Si

Transistors made on III–V films transferred onto silicon have been realized [20]. Lattice-matched InAlAs/InGaAs layers were grown in the reverse order compared to conventional HEMT structures on a 2 in InP substrate with etch stop layers between the substrate and the InAlAs/InGaAs. The InP wafer with the grown layers was bonded onto a 2 in FZ silicon wafer by means of SiO_2–SiO_2 bonding. After the InP substrate and etch stop layers were removed, the HEMT structure was realized (figure 12.8). These devices showed cut-off frequencies close to those of conventional HEMT transistors, indicating the good quality of the transferred layers. In this example, the InP substrate was etched but, in parallel, the possibility of transferring a thin InP film onto a silicon substrate by the Smart-Cut(R) process and growing the layers to process HEMT on top of this structure was demonstrated [20]. Physical characterization of the HEMT layers indicate very good crystalline characteristics.

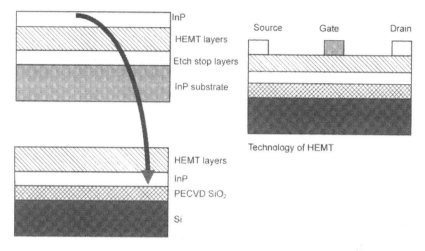

Figure 12.8. HEMT processed on III–V films transferred onto silicon [20].

12.4.2 Multi-junction photovoltaic cells with wafer bonding using metals

Experiments in view of developing a four-junction cells using metallic wafer bonding have been conducted [11]. The most difficult point is the interconnection between the cells. It must meet three requirements: it must be optically transparent to allow non-absorbed light to be transmitted from one cell to the one under; it must be electrically conductive; and mechanically rugged.

Bonding with metallic layers has been obtained. However, the properties of the contacts (optical transmission of 60% for the Sn layer used for bonding, and ohmic contact with contact resistivity of 2.5 Ω cm^2) must be improved to apply this technique to multi-junction cell fabrication.

12.4.3 Germanium layer transfer for photovoltaic applications

Multi-junctions solar cells can be processed on germanium substrates. However, these substrates are very expensive and the possibility of replacing them with germanium layers bonded on silicon substrates has been studied [10]. The process used consists in implanting H$^+$ ions into germanium, to bond the implanted germanium wafer to a virgin silicon wafer and splitting the germanium layer. In this experiment, direct bonding of germanium and silicon was used, in order to obtain an ohmic contact without altering the transmission by a metallic layer. A specific contact resistance of 400 Ω cm^2 was obtained after annealing at 350 °C. This value of resistance is too high for solar cells applications but it could be improved by annealing at higher temperature. Triple-junction solar cell structures with photoluminescence intensity and decay lifetimes comparable to those of solar cells grown on bulk germanium were grown on these Ge/Si hetero-structures. The transferred surfaces are limited to 1 cm^2, due to the difficulty of

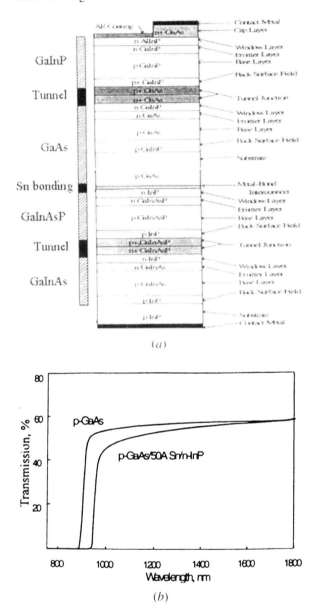

Figure 12.9. (*a*) Four-junction cell manufactured using metallic wafer bonding. (*b*) Sn layer optical transmission (from [11]. © 1997 IEEE).

direct bonding: while bonding via an oxide occurs spontaneously when the two wafers are put in contact, direct bonding must be done under pressure.

12.5 Conclusion

Wafer-bonding can be used to fabricate multi-junction cells. The main advantages are that the stacking is not limited by lattice parameter adaptation and the cells can be electrically independent, giving more freedom in the choice of gaps. Three bonding techniques can be used: direct bonding, bonding via a metallic layer and bonding via an oxide. Direct bonding is attractive since there is no light absorption at the interfaces. However, this technique is difficult and requires very good surface preparation (very low roughness and very clean surfaces, free of oxides) and must be done under pressure. Bonding with metallic layers is easier but the bonding layer must be transparent and so very thin, making the bonding difficult. However, bonding via an oxide is a very well controlled technique, used in the semiconductor industry to make SOI structures on 300 mm wafers and the oxide used for the bonding has very good optical transmission. Since the layers are isolated from each other, the technique can be used to make four (or more) terminal multi-junction cells, which offer more flexibility in the choice of the gaps.

References

[1] Sakaguchi K and Yonehara T 2000 *Solid State Technol.* **43** 88–92
[2] Tayanaka H, Yamauchi K and Matsushita T (Sony Corporation) 1998 *2nd World Conf. on PVSEC (Vienna)* (Ispra: European Commission) pp 1272–7
[3] Bilyalov R R, Solanki C S and Poortmans J 2001 *17th European PVSEC* (Munich and Florence: WIP and ETA)
[4] Berge C, Bergmann R B, Rinke T J, Schmidt J and Werner J H 2001 *17th European PVSEC* (Munich and Florence: WIP and ETA)
[5] Bergmann R B, Rinke T J, Wagner T A and Werner J H *PVSEC 11 (Sapporo)* (Tokyo: PVSEC) pp 541–6
[6] McCann M J, Catchpole K R, Weber K J and Blakers A W 2001 *Solar Energy Mater. Solar Cells* **68** 173–215
[7] Bergmann R B and Werner J H 2002 *Thin Solid Films* **403–404** 162–9
[8] Brendel R 2001 *J. Appl. Phys.* **40** 4431–9
[9] Bergmann R B and Rinke T J 2000 *Prog. Photovolt. Res. Appl.* **8** 451–64
[10] Zahler J M, Ahn C G, Zaghi S, Atwater H A, Chu G and Iles P 2002 *Thin Solid Films* **403–404** 558–62
[11] Sharp P R *et al* 1997 *26th PVSC (Anaheim, CA)* (New York: IEEE) p 895
[12] Fan J C C, Tsaur B Y and Palm B J 1982 *Proc. 16th IEEE Photovoltaic Spec. Conf.* (New York: IEEE) pp 692–701
[13] Aspar B *et al* 2001 *J. Electron. Mater.* **30** 834
[14] Rouviere J L, Rousseau K, Fournel F and Moriceau H 2000 *Appl. Phys. Lett.* **77** 1135
[15] Aspar B, Bruel M, Zussy M and Cartier A M 1996 *Electron. Lett.* **32** 1985–6
[16] Di Cioiccio L, Le Tiec Y, Letertre F, Jaussaud C and Bruel M 1996 *Electron. Lett.* **32** 1144–5
[17] Jalaguier E, Aspar B, Pocas S, Michaud J F, Zussy M, Papon A M and Bruel M 1998 *Electron. Lett.* **34** 408–9

[18] Jalaguier E, Aspar B, Pocas S, Michaud J F, Papon A M and Bruel M 1999 *11th Conf. On Indium Phosphide and Related Materials (Davos)* (New York: IEEE)

[19] Aspar B, Jalaguier E, Mas A, Loca telli C, Rayssac O, Moriceau H, Pocas S, Papon A M, Michaud J F and Bruel M 1999 *Electron. Lett.* **35** 1–2

[20] Bollaert S, Wallaert X, Lepilliet S, Cappy A, Jalaguier E, Pocas S and Aspar B 2002 *IEEE Electron. Devices Lett.* **23** 73

Chapter 13

Concentrator optics for the next-generation photovoltaics

P Benítez and J C Miñano
Instituto de Energía Solar—Universidad Politécnica de Madrid
ETSI Telecomunicación Ciudad Universitaria s/n—28040
Madrid, Spain

13.1 Introduction

Next-generation photovoltaic (PV) converters aim to be ultra-high efficiency devices. In order to be so efficient, it is expected that the cost of these converters will also be very high per unit area. Although it has been claimed [1] that the next-generation PV approaches should aim at high efficiency but also low-cost per unit area using thin-film technologies, perhaps it will be more probable that we first find a highly efficient next-generation solution with high cost per unit area as there will be fewer restrictions.

As an example, let us imagine that next-generation devices achieving 70% efficiency were to be invented in the near future. The electricity cost based on one-sun modules made from these devices would not be economic if their cost were to be higher than about seven times the cost of currently available commercial 15% efficient cells [2]. Such a device has not yet been found and, thus, cost analyses are obviously risky! However, the present general feeling is that, unless a new breakthrough approach appears, the relative cost per unit area of a 70% efficient device is likely to be much higher than seven times the present cost per unit area of commercial silicon cells (in fact, the present next-generation approach with the most advanced degree of development, the multi-junction tandem cell, has proven efficiencies around twice those of commercial silicon crystalline cells but their medium-term estimated cost is about 300 times higher [1]). This scenario would imply that one-sun modules based on this 70% efficient next-generation device will not be competitive with conventional modules, except in applications

285

where other factors (like efficiency) were much more important than the cost of electricity.

Nevertheless, next-generation PV can aim to achieve the goal of competing in cost with conventional non-renewable energies, if the device cost per unit area were to be reduced by concentrating the sunlight. In addition, the potential increase in the efficiency due to the concentration also helps them to compete with conventional sources. Solar concentrators have been studied since PV cells were first applied for terrestrial applications. During these three decades, expectations of the concentrator's capacity for cost reduction have been high but no commercial concentration product has so far proved this. The reasons for such a lack of success are multiple, as pointed out recently [2]. Let us review the following reasons:

(1) Until now, the PV market has not been driven by the PV electricity cost. Moreover, most customers have been public institutions or subsidized private ones. Furthermore, even high-cost renewable energies are gaining acceptance because sensitivity to the environmental aspects is increasing.

(2) Concentrators have not found a niche market as flat modules did. For instance, in devices integrated into buildings, tracking concentrators have had no applicability. Since tracking concentrators seem to be less reliable than non-tracking flat modules, they have not yet been accepted for automatic applications, where the maintenance cost is more important than the module cost.

The present rapidly growing PV market seems to be leading to a situation in which the subsidies will be reduced [4] and then it is likely that the lowest cost technology will start to dominate the PV market. Besides that, when the market becomes big enough, niche applications for market penetration seem to be easier to find. In this situation, concentrators and next-generation cells are expected to play an important role.

13.1.1   Desired characteristics of PV concentrators

The next-generation concepts need concentrators for which the features are, in many cases, not yet fully defined. It seems that most of the present next-generation devices will need the same features as those required for the present PV concentration cells and, thus, we will refer to them here.

The PV concentrator constitutes a specific optical design problem, with features that make it very different from other optical systems. Even designs for other optical concentration applications (like solar thermal energy, wireless optical communications, high-sensitivity sensors, etc) may not be suitable for PV. Generally speaking, PV concentrators must be

(1) efficient,
(2) suitable for mass production (repetitive and inexpensive),

(3) capable of producing high concentration ($>1000\times$) for a high cell cost,
(4) noticeably insensitive to manufacturing and mounting inaccuracies and
(5) capable of producing sufficiently uniform illumination of the cell.

In order to be efficient and inexpensive, only a few optical surfaces can be used in the design. The present trend towards high concentration systems arises from the high cost of the present III–V solar cells (which can only be offset with a high concentration) and the promising efficiency results of these cells under concentrations as high as $1000\times$: over 30% with two-junction tandem cells [3] and over 26% with single-junction cells [4].

The insensitivity to inaccuracies may be important to achieve successful concentration products. This is related to the angular aperture of the concentrator, usually called the concentrator acceptance (half) angle α. Achieving a high acceptance angle (let say, $\alpha > 1°$, which is close to four times the sun's angular radius $\alpha_s = 0.265°$ relaxes the accuracy of the optical surface profiles, module assembling, installation, supporting structure, etc. Due to the inverse relation between acceptance angle and concentration (see section 13.1.2), a higher concentration for a given (sufficient) acceptance angle means a more problematic design. The present experience in concentrator systems is still too brief to understand the dependence of the electricity cost on the acceptance angle. This dependence seems to be much stronger in PV applications than in solar thermal concentration systems, as discussed in section 13.1.4, and, therefore, the longer experience with solar thermal installations is no help. Our opinion is that for mass production and low-cost installation, this parameter may be critical.

The illumination uniformity is probably the most singular aspect of PV concentrators with respect to other fields of application. However, getting uniform illumination is a problem in other classical optical designs (like the condenser design in projection optics) but in these problems no other features are necessary: cost is usually high, complexity is allowed and the optics exit aperture is oversized (which would imply a small acceptance angle α in the PV framework).

13.1.2 Concentration and acceptance angle

Concentrators can be accurately analysed in the framework of geometrical optics, where the light can be modelled with the ray concept. This is deduced from statistical optics [5], in which geometrical optics appears as the asymptotic limit of the electromagnetic radiation phenomenon when the fields can be approximated as globally incoherent (also called quasi-homogeneous). In this approximation sources and receivers are non-punctual (usually called extended sources and receivers) and the diffraction effects are negligible.

Assume that a certain PV concentrator, which points in one nominal direction given by the unit vector, v, ensures that all the light rays impinging on the concentrator entry aperture forming angles smaller than α with v are transmitted onto the cell with incidence angles smaller than β. Let us suppose that

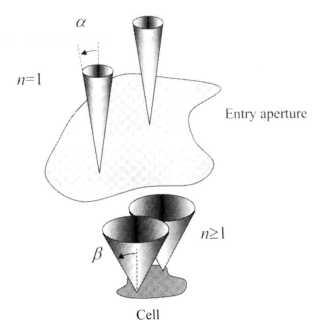

Figure 13.1. Description of concentrator geometry for the application of the étendue conservation theorem.

the refractive index of the medium surrounding the cell is n. From the étendue conservation theorem of geometrical optics [7], it is deduced that the following inequality is verified:

$$C_g \sin^2 \alpha \leq n^2 \sin^2 \beta \tag{13.1}$$

where the geometrical concentration C_g is the ratio A_E/A_R, A_E being the area of the projection of the entry aperture onto a plane normal to v and A_R the receiver area. Equation (13.1) is valid for any contour of areas A_E and A_R, and v is called the *normal incidence* direction vector. This equation is usually referred to as the *acceptance–concentration product bound*. The bound $n^2 \sin^2 \beta$ is a maximum for $\beta = 90°$, i.e. when the receiver is illuminated isotropically. This situation is usually referred to as the thermodynamic limit of the concentration because surpassing this maximum bound would imply the possibility of heating a black receiver with a concentrator to a temperature higher than that of the sun, which contradicts the second principle.

13.1.3   Definitions of geometrical concentration and optical efficiency

Several definitions of both the geometrical concentration C_g and optical efficiency η_{opt} of PV concentrators are commonly used in the literature. Of course, the definitions do not affect the device performance and cost but this situation may be confusing for the interested reader when trying to compare different concentrators. We will briefly discuss about these definitions next and we will adopt one criterion for this chapter.

13.1.3.1 Geometrical concentration

The different possible definitions of the geometrical concentration come from the definitions of both the entry aperture area and cell area.

With respect to the concentrator entry aperture area, sometimes the fully occupied area $A_{E,full}$ (for instance, the one defined to tessellate concentrator units) is used, while other times, if a portion of the aperture is clearly inactive by design (for instance, if there is a gap), it is excluded from the aperture area, leading to $A_{E,act}$. It should be noted that the concentrator topology may affect the decision of how to define the aperture area: inactive portions in linear systems (e.g. the central gap in parabolic trough technologies) have commonly been excluded [6], [7], while in rotational optics the full aperture area has been used [9]. Besides, when the inactive area is more distributed (as in the shadow of the nearly-vertical facets of Fresnel lenses), it seems illogical to exclude it.

For the cell area, sometimes the optically active cell area $A_{C,act}$ (i.e. that to be illuminated) is used to define C_g, while other times the full cell area $A_{C,full}$ is computed. The most common case is that of a squared cell whose active area is circular (i.e. the area's ratio is 1.27). Also in this case, some cell technologies claim to exclude the distributed shading area below the grid-lines from the active cell area.

Let us point out that including the inactive area of the cell does not affect the optical efficiency definition (see next section), while including the inactive areas of the entry aperture will.

Therefore, according to the two options for both A_E and A_C, four definitions of C_g are possible. Let us highlight two: $C_{g,1} = A_{E,full}/A_{C,full}$ and $C_{g,2} = A_{E,act}/A_{C,act}$. The first definition $C_{g,1}$ is the only useful for estimating material costs, while the second definition $C_{g,2}$ is suitable for optical and electrical calculations (in particular, it applies in (13.1)).

As an example, consider the cells in a flat module. In this case $C_{g,2} = 1$ but $C_{g,1} < 1$, due to the gaps between the cells and the typical non-squared cell contour (if the cell were circular, $C_{g,1}$ would equal 0.78, in today's typical modules $C_{g,1} \approx 0.95$–0.98).

Here, we will assume, in all cases, that C_g is $C_{g,2}$ (and, hence, the active areas are $A_{E,act}$ and $A_{C,act}$) but we will not exclude either the nearly vertical inactive facets in Fresnel-type structures nor the cell-shading grid-lines from the active areas.

13.1.3.2   *Optical efficiency*

A first definition of the optical efficiency [6, 8], $\eta_{opt,1}$, is the light power transmission efficiency through the concentrator up to the cell surface for light rays impinging at the entry aperture from a given nominal direction v (usually refer to as the *normal incidence*). This definition is wavelength dependent and cell independent.

A second definition [1] $\eta_{opt,2}$ is obtained when the nominal parallel incident rays produce one-sun irradiance with the solar spectrum at the entry aperture and it is given by the ratio of the concentrator photocurrent to the product of the geometrical concentration and the cell photocurrent when illuminated at one sun with the solar spectrum. For the one-sun photocurrent, the cell is assumed to receive the light at normal incidence inside the same refractive index medium as the one that surrounds the cell inside the concentrator. The solar spectrum classically used for this definition is AM1.5d but its suitability for characterizing concentrators is presently under discussion. When compared to $\eta_{opt,1}$, this second definition considers the spectral response of both the concentrator and the cell and includes the possible angular dependence of the external quantum efficiency of the cell (which may not be negligible [1, 9]).

A third definition [7] $\eta_{opt,3}$ uses the photocurrent ratio, as in the second definition $\eta_{opt,2}$ but with two differences: (1) the rays incident on the concentrator have the same angular extension as the sun; and (2) the cell photocurrent at one sun is obtained by encapsulating it as a flat module. This third definition is obviously suitable for measurements with real sunlight. The first difference is often not relevant because the concentrator photocurrent is usually the same for any parallel rights impinging from inside the sun disk when the sun is centred at the nominal position. However, the second difference is definitely not negligible, because the Fresnel and absorption losses on the glass cover of the flat module are around 4–5%. Therefore, this third definition can be seen as the ratio of the concentrator to the flat-module optical efficiencies of type $\eta_{opt,2}$, and then it is a relative optical efficiency.

Here, the optical efficiency η_{opt} will mean $\eta_{opt,2}$.

13.1.4   The effective acceptance angle

According to the definition given in section 13.1.2 for the acceptance angle α, 100% of the rays within the acceptance angle cone are collected. This acceptance angle lacks practical interest for two reasons. First, because an optical efficiency better than that with which the geometric rays are collected must be considered. Therefore, the discussion must be done in terms of the *directional optical efficiency*, $\eta_{opt}(u)$ i.e. the optical efficiency of the designed concentrator as a function of the direction unit vector u of the incoming light. Second, because it is obvious that if a few percent of the rays are not collected at a certain incidence angle, the concentrator is still useful at that incidence angle.

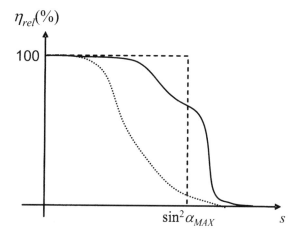

Figure 13.2. Examples of relative directional optical efficiencies. The dotted curve corresponds to a concentrator which does not achieve the isotropic illumination, while the full and broken curves do.

We will discuss in this section how an *effective acceptance angle* within which the concentrator can be considered to be useful should be defined and how critical the value of this effective acceptance angle may be. This topic is usually ignored, as proven by the absence of any comment on the acceptance angle in most publications about concentrating systems.

Let us assume that the geometrical concentration ratio C_g of the concentrator and the refractive index n of the medium surrounding the cell are fixed, and that the maximum value η_{MAX} of the function $\eta_{opt}(\boldsymbol{u})$ is reached at normal incidence $\boldsymbol{v}$ for which 100% of the light rays are collected.

We will call the *relative directional optical efficiency* $\eta_{rel}(\boldsymbol{u}) = \eta_{opt}(\boldsymbol{u})/\eta_{MAX}$. Again from the étendue conservation theorem, it can be deduced that $\eta_{rel}(\boldsymbol{u})$ fulfils:

$$\int_{\boldsymbol{u} \in H} \eta_{rel}(\boldsymbol{u})(\boldsymbol{u} \cdot \boldsymbol{v}) \, d^2\boldsymbol{u} \le \frac{\pi n^2}{C_g} \qquad (13.2)$$

where · denotes the scalar product (thus $\boldsymbol{u} \cdot \boldsymbol{v} = \cos\theta$ is the cosine factor), $d^2\boldsymbol{u}$ is the differential solid angle around $\boldsymbol{u}$ and H is the hemisphere of directions where the concentrator receives the light (i.e. in H, $\boldsymbol{u} \cdot \boldsymbol{v} > 0$). As an example, if $\eta_{rel}(\boldsymbol{u})$ is rotationally symmetric with respect to the axis $\boldsymbol{v}$, equation 2 becomes:

$$\int_{s=0}^{s=1} \eta_{rel}(s) \, ds \le \frac{n^2}{C_g} \qquad (13.3)$$

where $s = \sin^2\theta$. Equation (13.3) implies that the area below the curve $\eta_{rel}(s)$ is bounded and that the equality (in both equations (13.2) and (13.3)) is achieved only when the cell is isotropically illuminated and when the (non-geometrical) optical losses are the same for all values of θ.

Figure 13.2 shows three examples of the function $\eta_{rel}(s)$. The broken line is a pill-box-type function stepped at $\sin^2 \alpha_{MAX}$, where $\alpha_{MAX} = \sin^{-1}(n^2/C_g)$ is obtained from the equality in equation 1 with $\beta = 90°$. It is straightforward to check that this broken line fulfils the equality in equation (13.3). The full curve corresponds to a concentrator that also achieves the equality in equation (13.3) (and thus encloses the same area in figure 13.2) but it is not stepped and the dotted curve corresponds to a concentrator which does not achieve the isotropic illumination, enclosing a smaller area.

The following question arises: What percentage of optical loss can be allowed at each direction u? Or, in other words, what is the best directional optical efficiency, $\eta_{opt}(u)$?

Probably, the best (and obvious) answer to that question is: 'The one that provides the lowest cost per unit of produced energy'. However, this answer recalls the trade-off between efficiency and cost, always present in PV and that involves too many parameters. Let us reduce the problem to maximizing the energy production, which is more adequate for the discussion at the optical design stage.

Let us consider the case in which we have a concentrator array of N concentrator units, each one illuminating a solar cell. To discuss which directional optical efficiency will provide the most energy, many different statistical phenomena will be involved:

- the sun's time-dependent radiance;
- postioning errors caused by static inaccuracies, e.g. the positioning of the concentrator units when they are grouped in modules, installation of these modules;
- positioning errors caused by time-dependent inaccuracies, e.g. sun-tracking, deformations of the supporting structure by the system weight or wind loads;
- dispersion of the function $\eta_{opt}(u)$ or the different concentrator units, due to the dispersion in the optics and cell manufacturing; and
- time-dependent dispersion of the function $\eta_{opt}(u)$ for the different concentrator units, due to (non-homogeneous) dust accumulation.

Of course, all this information cannot be known in advance and will be site specific. Only with long experience and detailed analysis of a given mass-produced technology could this information be available for specific sites. Of course, all these errors could be modified by changing their cost (again, the efficiency-cost trade-off appears). The key information is how the cost is affected by the accuracy requirements and it is clear that experience is needed to provide the necessary feedback.

For a complete analysis, not only would a statistical description of each static random variable and of each time-dependent stochastic process be necessary but also their correlation functions. For instance, the concentrator unit positioning errors may be uncorrelated but wind loads are highly correlated with the different concentrator units in a module. Also, in many sites there may be a strong correlation between the sun radiance and wind (i.e. higher winds on cloudier days).

Moreover, the time-dependent errors are usually averaged over time due to the ergodicity of the stochastic processes but a finer analysis should take into account that time-averaging will be correct for the high-frequency components of the stochastic processes (e.g. turbulent winds), because the maximum-power-point tracker cannot follow them, but it will not be correct at low frequencies (e.g. steady winds).

Let us consider a simplified situation in which the dominant inaccuracy is the static positioning errors of the concentrator units. Although these are specific, they will illustrate the main issues arising in the selection of the concentrator function $\eta_{\mathrm{opt}}(\boldsymbol{u})$. To focus the discussion on the angular dependence of the optical efficiency, we will also assume that η_{MAX} is fixed, and then we can talk in terms of $\eta_{\mathrm{rel}}(\boldsymbol{u})$ or $\eta_{\mathrm{opt}}(\boldsymbol{u})$ indeterminately.

The analysis of the static positioning errors is obviously equivalent to the supposition that all the concentrator units are equally positioned but each one sees the sun in a different position. Then, we need to know the probability of the centre of the sun being located in a certain direction $\boldsymbol{u}_{\mathrm{c}}$. Let us assume that we know $g(\boldsymbol{u}_{\mathrm{c}})$, the probability density function of the sun's centre direction. The power cast by one concentrator unit with the sun centered in the direction $\boldsymbol{u}_{\mathrm{c}}$ is proportional to the average value of $\eta_{\mathrm{rel}}(\boldsymbol{u})$ within the sun disk. Then the function

$$\eta_{\mathrm{rel}}^{\mathrm{sun}}(\boldsymbol{u}_{\mathrm{c}}) = \frac{1}{\pi \sin^2 \alpha_s} \int_{|\boldsymbol{u} - \boldsymbol{u}_{\mathrm{c}}| < \sin \alpha_s} \eta_{\mathrm{rel}}(\boldsymbol{u})(\boldsymbol{u} \cdot \boldsymbol{v})\, \mathrm{d}^2 \boldsymbol{u} \qquad (13.4)$$

indicates the relative directional optical efficiency for the sun centred at $\boldsymbol{u}_{\mathrm{c}}$. Neglecting the cosine factor in equation (13.4), this averaging can also be seen as the convolution between $\eta_{\mathrm{rel}}(\boldsymbol{u})$ and the pill-box-type radiance distribution $R(\boldsymbol{u})$ of the sun (normalized by $\sin^2 \alpha_s$ to enclose a unit volume). The function $\eta_{\mathrm{rel}}^{\mathrm{sun}}(\boldsymbol{u}_{\mathrm{c}})$ is obviously less abrupt than $\eta_{\mathrm{rel}}(\boldsymbol{u}_{\mathrm{c}})$.

In order to illustrate the effect of selecting directional optical efficiency $\eta_{\mathrm{opt}}(\boldsymbol{u})$ let us consider the three rotational examples in figure 13.2, whose sun-averaged relative optical efficiencies $\eta_{\mathrm{rel}}^{\mathrm{sun}}(s_{\mathrm{c}})$ are shown in figure 13.3(a). For this discussion, we will consider that $\alpha_{\mathrm{MAX}} = 1°$.

Of course, if the pointing errors are very small relative to the width of the function $\eta_{\mathrm{rel}}^{\mathrm{sun}}(s_{\mathrm{c}})$, i.e. $g(s_{\mathrm{c}})$ is very narrow around $s_{\mathrm{c}} = 0$, any of the three functions will fulfil $\eta_{\mathrm{rel}}^{\mathrm{sun}}(s_{\mathrm{c}}) \approx 1$ where $g(s_{\mathrm{c}}) \neq 0$ and, thus, the three will produce the same power output. However, in contrast to the dotted curve, the full and broken curves allow for the accuracies in the system to be relaxed, which potentially can reduce the cost.

Therefore, let us discard the dotted curve and assume that the pointing tolerances $g(s_{\mathrm{c}})$ have been adjusted to the width of the functions $\eta_{\mathrm{rel}}^{\mathrm{sun}}(s_{\mathrm{c}})$, as shown in figure 13.3(b). In this figure, as an illustrative example, the vertical

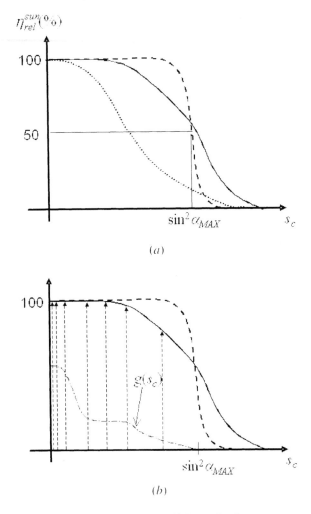

Figure 13.3. (*a*) Relative directional optical efficiency for the sun corresponding to the examples shown in (*b*). The probability density function $g(s_c)$ of the sun's apparent position is crucial for comparing the different directional optical efficiencies.

arrows indicate the sun's position for an array of $N = 7$ concentrator units according to the probability density $g(s_c)$.

The energy produced by the seven concentrator units will depend on their electrical connections. If they are parallel connected, the array performs approximately as if the seven units are equally illuminated with an optical efficiency that coincides with the average of the values of η_{rel}^{sun} for the seven sun positions. Thus, both the full and broken curves will produce a similar power output, because only one unit is in the lower efficiency portion of the full curve (because g is minimum in that range of s_c values).

This averaging process between the concentrator units can be exchanged with the prior averaging done between $\eta_{rel}(u)$ and $R(u)$ which leads to $\eta_{rel}^{sun}(u_c)$. Then, instead of talking about the different apparent positions of the sun's centre, the discussion uses the function $\eta_{rel}(u)$ illuminated by an apparent radiance distribution $R_{eff}(u) = R(u) * g(u)$ (the $*$ denotes the convolution operator), which is usually called the radiance distribution of the *effective sun* [10]. This concept is very common for solar thermal concentrators but, although it has also been used in PVs [11], it must be used carefully because it is only valid for parallel connected concentrator arrays (or single concentrator slope errors and scattering) but not for the series connection case, as shown next.

If the seven concentrators units in figure 13.3(*b*) were series connected, the array performs approximately as if the seven units are equally illuminated with an optical efficiency coinciding not with the average but with the minimum η_{rel}^{sun} of the seven sun positions. Although this is a pessimistic estimation [12], it illustrates the fact that the full curve will, in this case, produce significantly less energy than the broken curve, because the most misaligned unit limits the current. In practice, by-pass diodes are placed to partially relieve this situation (especially if only a few cells are very poorly illuminated). However, in any case it is evident that, in contrast to the parallel array, the low efficiency portions of the function η_{rel}^{sun} will never contribute effectively to a series array at its maximum power point.

Of course, mixed series–parallel configurations produce results in between the all-parallel and all-series cases. Nevertheless, the series connection limitation is the reason why only the values of the relative optical efficiency $\eta_{rel}(u)$ over a certain threshold, usually fixed at 90%, are considered to be useful. Note that this resulting threshold is very different from those of other applications, for instance, optical communications receivers, where the sensitivity threshold is usually defined at the 50% level.

The area of directions given by the isoline at 90% of the function $\eta_{rel}(u)$ is called the *effective acceptance area* and the system must be designed to guarantee that the inaccuracies are within the effective acceptance area (or at least that being beyond it is very unlikely).

It is usual to define the effective acceptance angle, α_{90}, as half the length of the smallest angular thread passing through *v* lying on the isoline of the effective acceptance area. Note that if the effective acceptance area is fixed, the circular contour of this area is the one that maximizes the effective acceptance angle.

Therefore, all the information in the directional optical efficiency $\eta_{opt}(u)$ is usually summarized in two numbers: the optical efficiency (at normal incidence, as defined in section 13.1.3.2) and the effective acceptance angle α_{90}. Obviously, much information is lost doing this and for more detailed models the whole function $\eta_{opt}(u)$ should be considered.

The answer to the question about which is the best directional optical efficiency is not yet closed. It seems that for high concentration systems, achieving stepped and circular effective acceptance areas is important, because the theoretical limit for the acceptance angle is short. However, the reflectivity of non-textured cells is high for glazing angles (even if antireflection-coated) and, thus, illuminating such cells nearly isotropically reduces the efficiency and the cost benefit of the higher tolerances is useless. Another example of the always-present trade-off between efficiency and cost comes from the case in which additional secondary optics are introduced increasing the acceptance angle but reducing the optical efficiency.

Moreover, there must be a value of α_{90} beyond which there will be no advantages in cost reduction for relaxing the tolerances further. This could be the case in low concentration systems.

13.1.5   Non-uniform irradiance on the solar cell: How critical is it?

It is well known that non-uniform illumination of solar cells under high concentration may decrease the power output significantly. For the concentrator designer, the key question is what local concentration distribution can be considered good enough for a given cell.

Usually, the only information available for concentrator designers is the cell efficiency under uniform illumination for different concentration factors C_U. With this information, the common rule of thumb is that the cell efficiency under a certain local concentration distribution $C(x, y)$ of average concentration $\langle C \rangle$ and peak concentration C_{MAX} will be between the cell efficiencies under uniform illuminations of $C_U = \langle C \rangle$ and $C_U = C_{MAX}$. This simple rule of thumb does not indicate to which cell efficiency it will be closer, because more parameters appear (how much area of the cell is illuminated with each concentration level, how important is the series resistance power loss at $\langle C \rangle$ and at C_{MAX}, etc). The concentrator designer usually takes C_{MAX} as the merit function to minimize (without evaluating the probability of any other concentration value), although its correlation with the best performance is uncertain.

In order to improve the merit function, cells under non-uniform illumination need to be modelled. One possible model consists in using equivalent circuits of the discretized cell and solving the resulting system of nonlinear equations using circuit simulators (e.g. PSPICE) [13]. Figure 13.4 shows an example of such modelling for a selected GaAs cell with crossed-grid front metallization.

The numerical resolution with this discretized circuit model provides the concentrator designer with an accurate merit function (the cell efficiency at the maximum power point) with which to optimize the design. However, such simulations are time consuming and do not indicate how to modify the concentration distribution on the cell to produce significant improvements in the merit functions. Therefore, this optimization process, without additional information, is likely to be very inefficient.

The concentrator designer would find faster tools and a deeper knowledge to orientate, speed-up and terminate the optimization of the optical design useful.

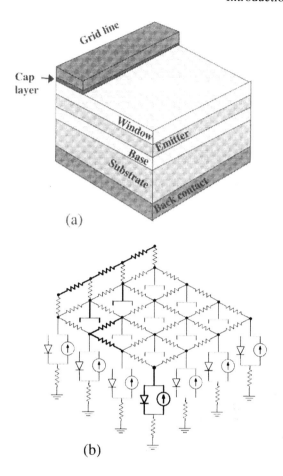

(a)

(b)

Figure 13.4. The unit cell (*a*) of the GaAs solar cell was discretized as shown in (*b*) to be analysed with PSIPICE under selected non-uniform pill-box-type illuminations with $\langle C \rangle = 1000$ suns. (*c*) The model provides the maximum power output, which can be used as the merit function.

Assuming that the cell parameters and average concentration $\langle C \rangle$ on it are fixed and known, the objective is then to have a cell model able to answer the following questions:

(a) How far is a given local concentration distribution $C(x, y)$ from the optimum, measured in terms of loss of cell efficiency?
(b) What modification in $C(x, y)$ will approach the optimum most rapidly?
(c) If a given distribution $C(x, y)$ has a specific feature (for instance, a certain irradiance peak), is it limiting the cell performance?

Question (a) implies the determination of the optimum irradiance

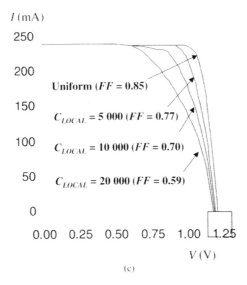

Figure 13.4. (Continued.)

distribution. Such an optimum should be defined with constraints, taking into account the fact that concentrator designs have limited capabilities in practice. For instance, it is obvious that, for a given cell, the optimum must not illuminate the grid-lines in front-contacted cells. However, if the use of micro-optics is excluded, the optimum does not seem to be attainable with low-cost optics and, thus, this optimum will not have any practical interest for the designer. In this case, it would be more practical to define the optimum with the practical constraint that the irradiance will not vary significantly along distances greater than the grid-line thickness or even greater than the grid-line pitch in high concentration cells (this can formally be stated in terms of the correlation length of the possible functions $C(x, y)$). Then, with this constraint, the shadow factor of the grid-lines will not be avoided and it will be approximately independent of $C(x, y)$.

It is well known that if the grid-lines' contribution to the series resistance is negligible (compared to the summations of the front-contact, emitter, base and rear contact contributions), this practical optimum is the uniform illumination (see the appendix). Note that if we add the obvious additional constraint to $C(x, y)$ i.e. that it must vanish outside the active cell area, the uniform illumination is not attainable either, because the first constraint does not allow for an abrupt transition near the edge.

It is also well known that in the case of a non-negligible contribution by the grid-lines to the series resistance, uniform illumination is not the optimum: the cell efficiency will improve if non-uniformity reduces the average path of the current along the grid-lines. Although the efficiency increase with respect to uniformity is slight, the optimum distribution $C(x, y)$ can be very different from the uniform one if the grid-line series resistance dominates [14]. This indicates to the concentrator designer where and how high local concentrations levels can be allowed for in these cells without degrading their performance.

We will focus the discussion here on the first case (negligible contribution of the grid-lines to the series resistance), and we will consider that the average concentration $\langle C \rangle$ is fixed. This cell can be modelled as a set of parallel connected unit cells of differential area $dx\,dy$. We will assume that the single exponential model for the dark current is valid, that the photocurrent density is proportional to the local concentration factor and that the different series resistance contributions can be modelled as a single constant parameter and the specific series resistance r_s, is independent of the concentration factor (i.e. high injection and current crowding effects at the emitter are neglected). Then, if $J(C, V)$ is the photocurrent density though a differential unit cell illuminated with a local concentration C and with an applied voltage V, the following equation is fulfilled:

$$J(C, V) = C J_{L,1sun} - J_0 \exp\left(\frac{V + J(C, V)r_s}{V_T}\right)$$ (13.5)

where $J_{L,1sun}$ is the photocurrent density at one-sun illumination, J_0 the saturation current density, $V_T = kT/e$ and V the cell voltage (common for all unit cells).

The cell current I under an illumination $C(x, y)$ is obtained by integrating the photocurrent density J in the cell area A_C. From equation (13.5):

$$I(V) = \langle C \rangle I_{L,1sun} - I_0 \exp\left(\frac{V}{V_T}\right)\left[\frac{1}{A_C}\int_{A_C} \exp\left(\frac{J(C(x, y), V)r_s}{V_T}\right) dx\,dy\right]$$ (13.6)

where the function $J(C, V)$ is (implicity) given by equation (13.5).

Let us define the *exponential concentration, EC*, as the function

$$EC(x, y, V) = \exp\left(\frac{J(C(x, y), V)r_s}{V_T}\right).$$ (13.7)

With this function, equation (13.6) can be written as

$$I(V) = \langle C \rangle I_{L,1sun} - \langle EC(V) \rangle I_0 \exp\left(\frac{V}{V_T}\right).$$ (13.8)

Equation (13.8) is the $I-V$ characteristic of the non-uniformly illuminated cell. Assuming that the cell parameters and $C(x, y)$ are known, the maximum power point (V_m, I_m) can be calculated from equation (13.8). This calculation can be performed numerically with a simple fast program (or even with a spreadsheet) in a computer.

Note that this calculation answers question (a), giving the concentrator designer a quick tool with which to evaluate the cell efficiency, which is the best merit function for the optimization.

The effect of $langle EC(V) \rangle$ in equation (13.8) can be interpreted as an apparent (voltage-dependent) increase in I_0. If an apparent voltage drop is preferred for the analysis, the function

$$V_x(V) = V_T \ln(\langle EC(V) \rangle) \qquad (13.9)$$

should be defined and used.

The minimum $\langle EC(V) \rangle$ for each V is obtained at the optimum case, i.e. under uniform illumination (see appendix). Consider now a non-uniform local concentration distribution $C(x, y)$ with the same average concentration $\langle C \rangle$. Recalling question (b): What modification in $C(x, y)$ will approach the optimum most rapidly?

Let us imagine that the designer can freely select two equal areas $(dxdy)_A$ and $(dxdy)_B$ of the cell surface illuminated with local concentrations C_A and C_B, respectively, and that the designer can modify the concentrator to decrease the local concentration on $(dxdy)_A$ to $C_A - dC$ and correspondingly increase it on $(dxdy)_B$ to $C_B + dC$ (note that the average concentration $\langle C \rangle$ is maintained). Then, for a given voltage V, the variation in $\langle EC(V) \rangle$ with the concentrator modification can be immediately calculated from equation (13.6):

$$d(\langle EC(V) \rangle) = dC[EC(C_B, V) - EC(C_A, V)]. \qquad (13.10)$$

The fastest decrease in $\langle EC(V) \rangle$ towards the minimum is obtained when the right-hand side in equation (13.10) is the most negative. Due to the monotonically increasing dependence of EC on C (which can be easily deduced from equations (13.5) and (13.7)), the modification to $C(x, y)$ which approaches the optimum most rapidly will be produced if the concentrator designer selects $(dxdy)_A$ and $(dxdy)_B$ such that $C_A = C_{MAX}$ and $C_B = C_{MIN}$.

This result was clearly expected: we must reduce the local concentration where it is at a maximum to increase where it is at a minimum. However, equation (13.10) gives more information than that. To show this, let us point out that in practice, the concentrator designer does not have such wide degrees of freedom in the design and the fastest modification rate of the concentration distribution is, in general, not attainable. However, it is not unusual for the designer to be able to select between two alternatives. Let us simplify them as follows: the designer can move the concentration dC either from $(dx\,dy)_{A1}$ to $(dx\,dy)_{B1}$ or from $(dx\,dy)_{A2}$ to $(dx\,dy)_{B2}$. Which to choose? According to equation (13.10), the best option is not that for which the concentration difference $C_{A,k} - C_{B,k}$ is the greatest but that for which the exponential concentration difference $EC_{A,k} - EC_{B,k}$ (at the maximum power point before modification) is the greatest.

Therefore, for the cells considered in this model, it can be concluded that it is not the concentration distribution but the exponential concentration distribution that plays the key role. Then, the concentrator designer when ray-tracing should display the function $EC(x, y, V_m)$ over the cell area rather than $C(x, y)$. Note that the exponential will cause a rather abrupt dependence of EC on C. This means that if there is even a small area of high local concentration (containing even a small power!), it may dominate the series resistance effect, because it dominates the value of $\langle EC(V_m)\rangle$ over the rest of the lower local concentration levels. Note that if such a high concentration peak on a small area is not dominant, the cell efficiency gain by eliminating that peak will be unimportant (note that this does not contradict the fact that the fastest optimization will need to reduce that peak).

Determining which concentrator levels are dominantcan be deduced by integrating $EC(V_m)$ within the area between isotopes of the function $C(x, y)$. However, this is much easier to do if the exponential concentration $EC(x, y, V_m)$ is expressed as a function of the local concentration C, instead of the spatial variables (x, y), as shown next. This will make the information about the distributions more precise, due to the reduction to a single dimension.

Let us consider the distribution of values of the local concentration factor C, which is studied in terms of the probability density function $f(C)$. The differential of area of the cell that is illuminated with local concentration in the interval $(C, C + dC)$ is given by $f(C)\,dC$. The function $f(C)$ can be calculated by standard statistical methods considering $C(x, y)$ as a function of the two-dimensional uniform random variable (X, Y) on the cell area. In practice, during the optical design, it can be calculated directly from the ray trace results.

Note that the nomenclature in equation (13.8) is general, independent of the change invariables. However, equation 6 must be rewritten in terms of $f(C)$ as

$$I(V) = \langle C\rangle I_{L,1sun} - I_0 \exp\left(\frac{V}{V_T}\right)\int_0^\infty \exp\left(\frac{J(C, V)r_s}{V_T}\right) f(C)\,dC. \quad (13.11)$$

Recalling the integrand in equation (13.11) as

$$w(C, v) = \exp\left(\frac{J(C, V)r_s}{V_T}\right) f(C) = EC(C, V)f(C). \quad (13.12)$$

Thus:

$$I(V) = \langle C\rangle I_{L,1sun} - I_0 \exp\left(\frac{V}{V_t}\right)\int_0^\infty w(C, V)\,dC \quad (13.13)$$

The function $w(C, V)$ gives us the measure at a given voltage V of the relative importance of the different local concentration intervals $(C, C + dC)$ on the cell performance degradation due to the series resistance losses. And obviously $\langle EC(V_m)\rangle$ is given by the area enclosed by $w(C, V_m)$.

Consider, as an example, the cell parameters $r_s = 2.1$ mΩ cm^2, $V_T = 25$ mV, $J_{L,1sun} = 25$ mA cm^{-2}. These parameters are close to those of high-concentration GaAs cells. Assume also that a given concentrator produces an irradiance distribution on the cell with the probability density function $f(C)$ shown in figure 13.5, whose average is $\langle C\rangle = 1490$ suns. This function $f(C)$ will illustrate the concepts presented, although it does not fulfil the smoothness constraint mentioned for the definition of the optimum distribution $C(x, y)$.

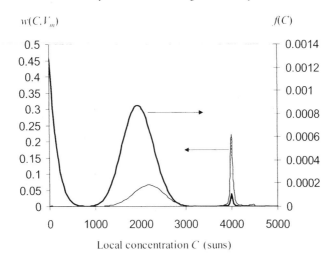

Figure 13.5. Example of probability density function $f(C)$ of the local concentration level and the contribution of each concentration interval $(C, C + dC)$ at low voltages given by the function $w(C, V_m)$.

From equation (13.13), which is the I–V characteristic of the whole cell, the maximum power point has been calculated, leading to $V_m = 1.02$ V and a cell efficiency $\eta_C = 24.8\%$ (to be compared with the uniform illumination of $\langle C \rangle = 1490$ suns, $\eta_C = 25.4\%$, the maximum achievable).

In figure 13.5 the function $w(C, V_m)$ is also shown and the amplification at high concentrations caused by the exponential function is noticeable. The maximum concentration is $C_{MAX} = 4500$ suns. However, the maximum contribution per concentration interval $(C, C + dC)$ occurs at 4000 suns. Let us compare this contribution to that around other local maximum, at 2200 suns: Since $f(4000)/f(2200) = 0.162$ and $w(4000, V_m)/w(2200, V_m) = 3.42$, the portion of cell area that is illuminated with concentration levels in (4000, $4000 + dC$) is 16.2% of the corresponding illuminated area for (2200, $2200 + dC$) but the contribution to $\langle EC(V_m) \rangle$ is 3.42 times greater around 4000.

Talking about the contribution of wider concentration intervals, there are two lobes in the function $w(C, V_m)$: one above and the other below 3500 suns. The lobe above 3500 only corresponds to 2.1% of the incident light power (calculated by integrating $Cf(C)$). However, its relative contribution to $\langle EC(V_m) \rangle$, which is calculated by integrating $w(C, V_m) = EC(V_m)f(C)$, is much higher: 23.6%. Therefore, the relative contributions of the two lobes are of the same order of magnitude, and neither of them dominate $\langle EC(V_m) \rangle$ (this comparison in terms of orders of magnitude results from the fact that cell performance is affected only by the changes of order of magnitude of $\langle EC(V_m) \rangle$, as occurs with I_0).

For a deeper understanding of the analysis, it is interesting to make a

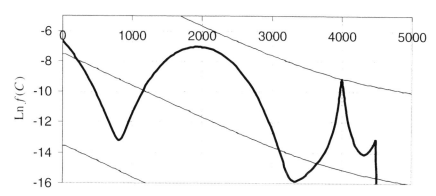

Local concentration C (suns)

Figure 13.6. Graphical analysis of the relative contributions of the local concentration factors to the degradation by non-uniform illumination. The tilted nearly-straight lines are isotopes with an equal contribution to $\langle EC(V_m)\rangle$. The ratio of the contribution values of adjacent isotopes is $\exp(-6) = 0.25\%$.

graphical representation of the one-parametric family of functions $f(C)$ defined as

$$w(C, v) = \exp\left(\frac{J(C, V)r_s}{V_T}\right) f(C) = B \tag{13.14}$$

where B is a positive parameter. For a given value of B, this function $f(C)$ is the loci of the points of the plane $f(C) - -C$ that contribute equally to $\langle EC(V_m)\rangle$ per concentration interval $(C, C + dC)$. For $V = V_m$, the one-parametric family of functions is:

$$f(C) = B \exp\left(-\frac{J(C, V_m)r_s}{V_T}\right) \Leftrightarrow \ln f(C) = \ln B - J(C, V_m) \tag{13.15}$$

Figure 13.6 shows some curves of this family in the plane $\ln f(C)-C$. The selected values for parameter B fulfil $B_{k+1} = \exp(-6)B_k$, where $\exp(-6) \approx 0.25\%$.

The representation has been done in the plane $\ln f(C)-C$ because the curves of this family are close to straight lines. In fact, since $J(C, V_m)r_s \ll V_T$ and $J \approx C J_{L.1sun}$ for the lower concentration values, the function in equation (13.15) coincides, for those values, with the straight line:

$$f(C) = B \exp(-C/C_0) \Leftrightarrow \ln f(C) = \ln B - \frac{C}{C_0} \quad \text{where } C_0 = \frac{V_T}{J_{L.1sun}r_s}. \tag{13.16}$$

For the cell parameters in this example, $C_0 = 475$.

Equation (13.16) suggests using the approximation $J \approx C J_{L.1sun}$ for evaluating the exponential concentration $EC(C, V)$ for any concentration C. This

Table 13.1. Comparison for the selected example of the simulation at the maximum power point for calculations done with the models presented and uniform illuminations

Illumination	V_m (mV)	η_C (%)
Uniform $C = 1490$	1043	25.4
Uniform $C = 2000$	1017	24.9
Uniform $C = 4000$	945	22.9
Uniform $C = 4500$	926	22.3
$f(C)$ in figure 13.5, exact EC	1014	24.8
$f(C)$ in figure 13.5, approx EC	1001	24.5

leads to very simple expressions. For instance:

$$EC(C,V) = \exp\left(\frac{J(C,V)r_s}{V_T}\right) \approx \exp(C/C_0) \qquad (13.17)$$

which is independent of V. This approximation is accurate for low C and low V, and its effect in the I–V characteristic of the cell (equation (13.8)) is that it is accurate close to short-circuit operation and very inaccurate close to open-circuit operation.

Near the maximum power point, since the function $J(C,V)$ will be intralinear with respect to C, this approximation is pessimistic in the sense $J \approx CJ_{L,1sun}$ will overestimate EC. This overestimation can be also seen in figure 13.6, where the isotopes of this approximation would be the straight lines given by equation (13.16), which are the tangents lines, at low concentration, to the isotopes shown in that figure.

As $\langle EC(C,V)\rangle$ will also be independent of V, using the apparent voltage drop defined in equation (13.9), the series resistance effect with this approximation is equivalent to a constant voltage drop of

$$V_s = V_T \ln\left(\int_0^\infty \exp(C/C_0)f(C)\,dC\right). \qquad (13.18)$$

For the example under analysis, $V_s = 114$ mV.

The accuracy of this approximation for calculating the maximum efficiency has not been studied in general but only for the selected example. Table 13.1 shows the comparison between the model with and without the approximation of EC, along with the calculation for several uniform illuminations with remarkable values in the local concentration distribution ($\langle C\rangle$ and the two peaks in $f(C)$ in figure 13.5).

From the comparison in table 13.1, it can be deduced that the illumination in figure 13.5 is similar to a uniform illumination of 2000 suns. This implies that a concentration over 3500 suns, which has only 2.1% of the incident power, a dominant concentration at 4000 and a maximum of 4500 suns does not degrade the cell performance. The question is: how much area or concentration in a certain area is needed to dominate the series resistance degradation?

Let us finalize this section answering this question with a simple analysis, using the approximated model of equation (13.17). Consider a concentration distribution which only takes two values, C_{MAX} and C_{MIN}, each one occupying the areas A_{MAX} and A_{MIN}, respectively. Then, the highest concentration level will dominate when $EC_{MAX}A_{MAX} \gg EC_{MIN}A_{MIN}$. For instance, the highest concentration will produce 90% of the value of $\langle EC \rangle$ when

$$\frac{EC(C_{max})A_{max}}{EC(C_{min})A_{min}} \geq \frac{90}{10} \tag{13.19}$$

which leads to

$$C_{max} \geq C_{min} + C_0 \ln \left(\frac{9A_{min}}{A_{max}} \right) \qquad \text{where } C_0 = \frac{V_T}{J_{L,1sun}r_s}. \tag{13.20}$$

For $C_0 = 475$, $C_{MIN} = 1490$, $A_{MAX} = 0.1A_C$ and $A_{MIN} = 0.9A_C$, we get $C_{MAX} > 3580$. Again, note that the light power impinging on the area A_{MAX} is only $0.1*3580/(0.1*3580 + 0.9*1490) = 21.1\%$ of the total incident power on the cell. As the approximation is pessimistic, the exact calculation with equation (13.10) would show that the threshold is greater than 3580 suns.

Note that all the results in this section are independent of the function $J(x, y)$, depending only on $f(C)$. This means that the performance of the cell will depend on the local concentration values but not on the spatial position where each value is produced. The reason for this property is clear: this cell model is completely symmetric with respect to the differential unit cells. As an example, if the irradiance distribution is translated inside the cell active area, the cell's I–V characteristic will remain unchanged. Of course, this does not happen in cells with non-negligible grid-line series resistance, which have been excluded from this model.

13.1.6 The PV design challenge

Equation (13.1) showed that if a high concentration is needed to reduce the receiver cost and a sufficiently high acceptance angle α is needed to make the system practical, the illumination angles β on the receiver must, of necessity, be high. A first consequence of illuminating the cell with wide angles β is that such an illumination will make the cell design (especially antireflection coatings, thickness of tandem cells, etc) more difficult and, in general, for the same cost, the final cell performance will be worse than for the case of a low illumination angle β.

The higher the angles α and β are the more difficult the design problem for achieving good illumination uniformity will be, which can be critical as we have just seen in section 13.1.5. The difficulty that corresponds to the acceptance angle α can be illustrated as follows: if α were equal to the sun radius α_s (which probably makes the system impractical because of the low tolerances), then any simple point-focus imaging concentrator (i.e. one that images the sun on the cell) would provide a nearly-uniformly illuminated disk on the cell (the sun image). When α is bigger, for instance, $\alpha = 4\alpha_s$, the imaging concentrator will produce a disk with a 16 times higher concentration with an area 16 times smaller on the cell, and this sun image will move around the cell when the sun (due to tracking errors) moves within the acceptance area. For high concentration cells (usually with $J_L r_s > V_T$), such local concentration levels will produce a dramatic reduction in the conversion efficiency and, thus, the imaging concentrator is useless for this high α.

Therefore, a practical concentrator design cannot perform as the imaging concentrator, it must be different to obtain good illumination uniformity without reducing the acceptance angle α too much. It should be noted that concentrator efficiency records are usually obtained with imaging concentrators (Fresnel lenses) with a small acceptance angle to have good illumination uniformity and low β to maximize the cell performance. Such records are obviously good for showing the limiting performance but are not representative of the achievable efficiency in practical systems. Of course, this is not exclusive to concentration systems and also happens in some one-sun records: there are record cells that achieve a very high efficiency for normal incidence but significantly lower efficiency for non-normal rays, so they are not practical for static one-sun modules (since their daily average efficiency will not be that high).

Designing for good illumination uniformity without reducing the acceptance angle so that it is close to that of the sun is a difficult task. There are two methods in classical optics that potentially can achieve this, if conveniently adapted to the PV requirements. The two methods specifically used for condenser designs in projection optics are the light-pipe homogenizer and the Kohler illuminator (commonly called an *integrator*) [15].

The light-pipe homogenizer uses the kaleidoscopic effect created by multiple reflections inside a light pipe, which can be hollow with metallic reflection or solid with total internal reflections (TIR). This strategy has been proposed several times in PVs [11, 16–18] essentially by attaching the cell to the light-pipe exit and making the light-pipe entry the receiver of a conventional concentrator. It can potentially achieve (with the appropriate design for the pipe walls and length) good illumination on a squared light-pipe exit with the sun in any position within the acceptance angle. For achieving high illumination angles β, the design can include a final concentration stage by reducing the light-pipe cross section near the exit. However, this approach has not yet been proven to lead to practical PV systems (neither has any company commercialized it as a product yet).

However, the integrating concentrator consists of two imaging optical elements (primary and secondary) with positive focal length (i.e. producing a real image of an object at infinity, as a magnifying glass does). The secondary element is placed at the focal plane of the primary, and the secondary images

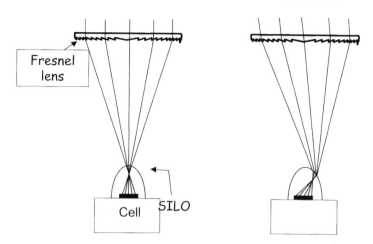

Figure 13.7. The 'Sandia concept 90' proved that the cell could be nearly uniformly illuminated on a square (with a squared Fresnel lens) for any position of the sun in the acceptance angle: (*a*) normal incidence; (*b*) incidence near the acceptance angle.

the primary on the cell. This configuration ensures that the primary images the sun on the secondary aperture and, thus, the secondary contour defines the acceptance angle of the concentrator. As the primary is uniformly illuminated by the sun, the irradiance distribution is also uniform and the illuminated area will have the contour of the primary, which will remain unchanged when the sun moves within the acceptance angle (equivalently when the sun image moves within the secondary aperture). If the primary is tailored to be square, the cells will be uniformly illuminated in a squared area. A squared aperture is usually the preferred contour to tessellate the plane when making the modules, while a squared illuminated area on the cell is also usually preferred as it fits the cell's shape.

Integrator optics for PV was first proposed [19] by Sandia Labs in the late 1980s and it was commercialized later by Alpha Solarco. This approach uses a Fresnel lens as the primary and a single-surface imaging lens (called SILO, from SIngLe Optical surface) that encapsulates the cell as secondary, as illustrated in figure 13.7.

This simple configuration is excellent for getting a sufficient acceptance angle α and highly uniform illumination but it is limited to low concentrations because it cannot get high β. Imaging secondaries achieving high β (high numerical aperture, in the imaging nomenclature) have, up to now, proved impractical. Classical solutions, which would be similar to high power microscopes objectives, need many lenses and would achieve $\beta \approx 60°$. A simpler solution that nearly achieves $\beta = 90°$ is the RX concentrator [20, 21] (see figure 13.8). Although the Lens+RX integrator is still not practical, it is

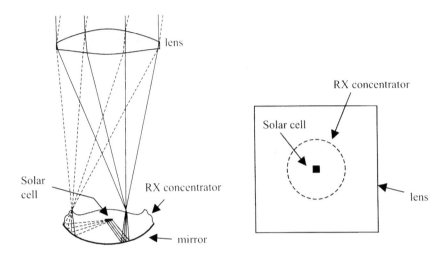

Figure 13.8. The use of an RX concentrator as the imaging secondary in the integrator (not shown to scale on the left) with a double aspheric imaging primary shows that it is theoretically possible to come very close to optimum PV concentrator performance.

theoretically interesting, because it encourages the definition of an optimum PV concentrator performance and shows that it is (at least) nearly attainable.

Let us define the *optimum PV concentrator performance* as that provided by a squared aperture concentrator collecting all rays within the acceptance angle α (this being several times the sun radius α_s) and achieving isotropic illumination of the cell ($\beta = 90°$), producing a squared uniform irradiance covering the whole cell, independently of the sun position within the acceptance area. Note that since all rays reaching the cell come from the rays within a cone of angular radius α no rays outside this cone can be collected by the optimum concentrator.

As an example, for $\alpha = 1°$, this optimum performance concentrator will have a geometrical concentration $C_g = 7387\times$, which is the thermodynamic concentration limit for that acceptance angle.

This definition of optimum PV concentrator performance is just a definition and it does not try to be general. For instance, as already mentioned, the high reflectivity of non-textured cells for glazing angles may make the isotropical illumination useless. As another example, illuminating a squared area inside the cell is perfect for back-contacted solar cells but it may be not so perfect for front-contacted cells, for which an inactive area is needed to make the front contacts (breaking the squared shape active area restriction). Finally, for medium concentration systems (say, $C_g = 100$) the aforementioned optimum performance would imply an ultra-wide acceptance angle $\alpha = 8.6°$. As pointed out in section 13.1.4, it seems logical that over a certain acceptance angle, there must be no cost benefits due to the relaxation of accuracies (and $8.6°$ seems to be above

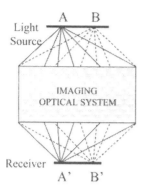 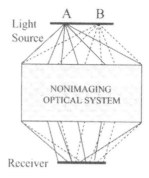

Figure 13.9. The difference between imaging and non-imaging optical systems is that in the former a specific point-to-point correspondence between the source and the receiver is required.

such a threshold). If this is the case, coming close to the optimum performance seem to be unnecessary for this medium concentration level.

The present challenge in optical designs for high-concentration PV systems is to design concentrators that approach the optimum performance and, at the same time, are efficient and suitable for low-cost mass production.

13.1.7 Non-imaging optics: the best framework for concentrator design

Non-imaging optics (also called anidolic optics) is the branch of optics that deals with maximum efficiency power transfer from a light source to a receiver [], and, thus, it is the best framework for PV concentrator design. The term *non-imaging* comes from the fact that for achieving high efficiency the image formation condition is not required (but neither is it excluded, as the RX showed!), and then, in contrast to imaging optical systems, the ray-to-ray correspondence will not be restricted (see figure 13.9). This idea emerged in the mid-1960s when the first non-imaging device, the compound parabolic concentrator (CPC), was invented, and it was possible to attain the thermodynamic limit of concentration in two dimensions with a very simple device that did not have any imaging properties.

Concentrator designs in non-imaging optics are carried out within the framework of geometrical optics. Then the concentrators act as transformers of extended ray bundles. In the PV framework, the bundle of rays impinging on the surface of the entry aperture of the concentrator within a cone of acceptance angle α is called the input bundle and is denoted by M_i. The bundle of rays that links the surface of the exit aperture of the concentrator with the cell is the exit bundle M_o. Collected bundle M_c is the name given to the set made up of the rays common to M_i and M_o, connected to one another by means of the concentrator. The exit bundle M_o is a subset of M_{MAX}, the bundle formed by all the rays that can impinge on the cell (i.e. M_{MAX} is the ray bundle that illuminates the cell isotropically).

There are two main groups of design problems in non-imaging optics. Although both groups have been usually treated separately in the non-imaging literature, PV concentrators constitute an example of a non-imaging design that belongs to both groups.

In the first group, which we will refer to as 'bundle-coupling', the design problem consists in specifying the bundles M_i and M_o, and the objective is to design the concentrator to couple the two bundles, i.e. making $M_i = M_o = M_c$. When $M_o = M_{MAX}$, the maximum concentration condition is achieved. This design problem is arises, for instance, in solar thermal concentration or in point-to-point IR wireless links (for both emitter and receiver sets).

For the second group of design problems, 'prescribed-irradiance', it is only specified that one bundle must be included in the other, for example, M_i in M_o (so that M_i and M_c will coincide), with the additional condition that the bundle M_c produces a certain prescribed irradiance distribution on one target surface at the output side. As M_c is not fully specified, this problem is less restrictive than the bundle-coupling one. These designs are useful in automotive lighting, the light source being a light bulb or an LED and the target surface is the far-field, where the intensity distribution is prescribed. It is also interesting for wide-angle ceiling IR receivers in indoor wireless communications, where the receiver sensitivity is prescribed to compensate for the different link distances for multiple emitters on the desks (in this case, M_o is included in M_i, and the irradiance distribution is prescribed at the input side, at the plane of the desks).

The design problem of the perfect PV concentrator performance, as defined in section 13.1.6, can be stated with the following two design conditions: (1) coupling M_i (defined with the acceptance angle α and the squared entry aperture) and M_{MAX} (defined with the refractive index around the cell and the squared cell active area); and (2) defining that every bundle M_i', defined as the cone with arbitrary axis direction and the sun's radius α_s, if contained in M_i, must produce the same relative irradiance distribution, which is prescribed to be uniform on the squared cell's active area. Therefore, (1) is a bundle-coupling problem at maximum concentration, while (2) is a prescribed-irradiance problem for *every* subset M_i'. Due to the clear difficulty of this design, traditionally only partial solutions have been found, some aim at condition (1) while others aim at condition (2).

The two groups of designs are carried out with the help of the edge-rays. If M is a ray bundle, the topological boundary of M as a set of the phase space is called the edge-ray bundle of M (and denoted by δM). For the two-dimensional example in figure 13.10, the rays of bundle M_i are all the rays linking the light source with the entry aperture, while its edge-rays δM_i are only the rays passing either by the source edges or the entry aperture edges. The same applies for M_o and δM_o. For a more detailed discussion about the definition of edge rays and the application of the theorem, see [22].

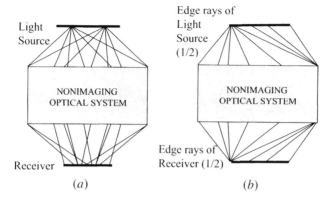

Figure 13.10. The edge rays of the input and output ray bundles in (*a*) are those passing through the edges of the source, the receiver, the entry aperture or the exit aperture (half of them are shown in (*b*)).

For the first group of designs, the *edge-ray theorem* states that in order to couple two bundles M_i and M_o, it is sufficient to couple the edge rays δM_i and δM_o. This theorem is very useful, since it is sufficient to design for edge rays, which form a ray bundle with fewer rays (it has one dimension less). Also in the prescribed-irradiance problem, only edge rays need to be considered for the design.

Several types of optical surfaces have been typically used in non-imaging design: first, conventional refractive surfaces and mirrors, which deflect the ray bundles as in classical imaging optical systems (lenses, mirror telescopes, etc); and second, a special mirror type, called the flow-line mirror, because it coincides with a flow line of the bundle. In 2D geometry, the flow line is defined as the line which is tangent to the bisector of the ray bundle at each point. This line indicates the direction of the energy flow. In a flow-line mirror, the ray bundle is guided by the mirror which acts like a light funnel. The main practical difference between conventional and flow-line mirrors is that to achieve a high concentration–acceptance angle product, the flow-line mirrors need to touch (or be very close) to the cell. This is usually considered inconvenient, especially if the cell is small.

The previously mentioned optical surfaces can be either continuous or microstructured. Examples of microstructured surfaces are those of Fresnel or TIR lenses. Microstructured surfaces with small facets can be studied as an apparent surface (obtained by enveloping the facet vertices) with a deflection law which differs from reflection or refraction laws. Usually, microstructured surfaces are not ideal in the sense that a fraction of the microstructure output is inactive, which implies that the transmitted ray bundle will not fill the cell bundle M_{MAX} totally. As a consequence, in these microstructures the concentration-acceptance angle product will necessarily be lower than the bound of equation (13.1). As an example, the microstructure of a Fresnel lens concentrator is not ideal: the cell is not illuminated from the inactive nearly-vertical facets.

The non-imaging designs are usually done in 2D geometry, and the 3D device is generated by linear or by rotational symmetry. This means that only part of the real 3D rays are designed: typically, the meridian rays in the rotational concentrators and the normal rays in the linear ones. There are essentially four design methods in 2D geometry:

(1) The Welford-Winston method (also called the TERC method) [23–27]. This is the method that was used to invent the CPC and also other designs later applied to PVs, such as the DTIRC (which is a dielectric-filled CPC with a prefixed circular aperture).
(2) The flow-line method [28].
(3) The Poisson bracket method [29].
(4) The simultaneous multiple surface (SMS) design method [30].

The SMS method has proven to be, in our opinion, the most versatile and practical of the four methods. It consists in the design of at least two aspheric optical surfaces (refractive or reflective), which are calculated simultaneously and point-by-point, to produced the desired edge-ray transformation. Not only does the SMS method produce designs that can work close to the thermodynamic limit but the resulting devices also have such excellent practical features as high compactness or simplicity for manufacturing. Besides that, in contrast to the other three design methods, the SMS method does not need to use flow-line mirrors (however, flow-line mirrors can also be designed in the SMS [22]). In section 13.2.2, a description of SMS concentrators designed for high concentration systems is given.

13.2   Concentrator optics overview

13.2.1   Classical concentrators

The progress in solar cells development since its invention has been huge. The investment needed has also been important. Present cells have become much more efficient and the technology much more advanced. The next-generation approaches discussed in this book try to further cell development, increasing cell sophistication if necessary.

However, PV concentrator development has not been accompanied by such progress. This can be seen in the fact that most PV concentrator systems have been based either in the parabola or in the Fresnel lens (see figure 13.11): the parabola has been known since Menaechmus (380–320 BC) and its optical properties since Apolonius (262–190 BC), and Galileo (1564–1642) used lenses for making the first refractive telescope. The thinning of such lenses to avoid

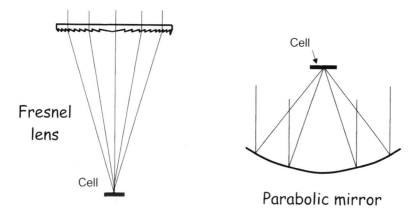

Figure 13.11. Fresnel lenses and parabolic reflectors are the classical concentrators in PVs.

absorption was proposed by Georges de Buffon in 1748 (and later optimized by Fresnel in 1822).

The parabola and Fresnel lens perform far from the theoretical concentration–acceptance angle limits. The question is: Are they good enough or do we need to come closer to the limit? As discussed in section 13.1.4, we guess that they are not good enough, especially due to the present high concentration trend.

Non-imaging optics came into the PV field as an option for low (static) concentration [31] or as a CPC-type secondary optical element for higher concentration PVs [32]. The first full non-imaging device for non-static concentration was a non-imaging linear Fresnel lens with a curved aperture [33].

Among all these non-imaging systems, only the addition of the CPC-type secondary to a classical primary could work for very high concentration systems ($>1000\times$) with a sufficient acceptance angle ($\alpha_{90} > 1°$) [26]. However, apart from the lack of illumination homogeneity and the practical problems related to flow-line mirrors, this solution is necessarily non-compact (depth H to entry aperture diameter D ratio (H/D) ≈ 1.5 for the Fresnel lens and $H/D \approx 1$ for the parabolic mirror). This is due to the design approach, in which the secondary and primary are designed separately and, thus, the secondary does not consider the actual rays exiting the primary but considers the primary as a Lambertian source [34]. The smaller the H/D ratio is, the lower accuracy of this approximation will be and, thus, the concentration–acceptance angle product will be reduced.

13.2.2   The SMS PV concentrators

Using the SMS method to design concentrators has produced new families during the last decade [35]. Among them, several families (RR [30], RR-RRI$_F$ [30, 36], XR [30], XX$_F$ [37]], SMTS [38], DSMTS [6], RXI [39], RXI$_F$ [40], RXI-RX [40], TIR-R [41]) have been suggested for PV applications.

The nomenclature used for referring to the different designs (excluding SMTS and DSMTS) is as follows: each concentrator is named with a succession of letters indicating the order and type of incidence of the optical surfaces that the sun-ray encounters on its way to the cell. The following symbols are used: R=refraction, X=reflection, I=total internal reflection. The sub-index F is added to X and I of these mirrors coincidence with the flow line).

In the case of high concentration systems (i.e. RRI$_F$, XR, RXI and TIR-R), these devices have been designed with rotational symmetry and conceived for small cells as *miniconcentrators* (concentrator entry aperture diameter = 30–85 mm). In order to make modules, the concentrators are truncated as squares for tessellation. In these designs, the acceptance angle remains unchanged after truncation, and then the acceptance angle- concentration product is reduced by a factor $2/\pi = 64\%$. This reduction would be smaller if hexagonal concentrators were used but square tessellation is usually considered more aesthetic because it is free from edge effects.

The interest in miniconcentrators comes from the fact that small cell technology is very close to that of the LEDs, which is very well developed and highly automated. As an example, highly efficient 1 mm^2 GaAs cells for 1000-sun operation have been demonstrated [4]. The power to be dissipated is around 0.7 W. However, the Luxeon LED of Lumileds has a high-flux blue chip, also 1 mm^2, and it was designed to dissipate 1.3 W.

We will review the aforementioned high-concentration SMS devices, assuming that the aperture is square and the cell's active area is circular (unless otherwise specified). As will be seen later, apart from the TIR-R, these designs have aimed to provide excellent acceptance angle–concentration products but the uniformity of the irradiance distribution of the cell is not good. This is because their design focused on it as a bundle-coupling problem and not as a prescribed irradiance problem (see section 13.1.7). The TIR-R is the first SMS designed which has approached both problems.

The RR concentrator for PVs is formed by one primary lens with a flat entry surface and a continuous or Fresnel exit surface, and a refractive secondary that encapsulates the solar cell (see figure 13.12(*a*)). In contrast to the conventional Fresnel lens and non-imaging secondary concentrators, the primary and secondary are designed simultaneously, leading to a better concentration-acceptance angle product without compromising the compactness. It achieves $\beta = 45°$ with good optical efficiency (around 85%), and for $\alpha_{90} = 1°$, it can get about $C_g = 1500\times$ for a square aperture and circular cell, although the irradiance homogeneity is very poor. A modification of this device has been proposed in the framework of the Hisicon EU project for front-contacted silicon solar cells [42], designed and manufactured by LETI, which is the project leader. This silicon cell concept performs well for much higher concentration levels than the back-contacted cells (and, of course, better than the two-side contacted cells). The grid-lines in the Hisicon cells are aluminum prisms (which contact the p^+ and n^+ emitters, alternatively), acting as a linear cone concentrator that concentrates $C_g = 1.52\times$ in the cross-sectional dimension of the prisms (see figure 13.13). The modified RR device consists in a squared primary with a secondary element which has a refractive rotational symmetric top surface that is crossed with two linear flow-line TIR mirror (see figure 13.12(b)). Then, in the cross section normal to the prisms, the secondary coincides with an RR concentrator with 2D concentration of $C_g = 12\times$, while in the cross section parallel to the prisms it coincides with an RRI$_F$ concentrator. The flow-line mirrors I$_F$ have linear symmetry perpendicular to the grid-lines providing the RR with an additional concentration $C_g = 2.08\times$ as the grid-lines in this dimension. Therefore, the cell is rectangular (1:2.08 aspect ratio), the grid-lines being parallel to the shorter rectangle side. The geometrical concentration (defined according to section 13.1.3.1 using the cell's active area, i.e. that of the naked silicon) is $(12 \times 2.08) \times (12 \times 1.52) = 455\times$ for the square aperture and rectangular cell, and achieves a design acceptance angle of $\alpha_{90} = 1.8°$.

The XR concentrator is composed by a (non-parabolic) aspheric primary mirror and a refractive secondary that encapsulates the solar cell (see figure 13.14). As in the RR design, and in contrast to the classical parabola plus non-imaging secondary configuration, the simultaneous design of the two optical surfaces of the XR enables it to be very compact ($H/D \approx 0.25$, compared to $H/D \approx 1$–1.5 for classical systems), without sacrificing performance. The XR can even get isotropic illumination of the receiver ($\beta = 90°$), if necessary. For $\beta = 70°$, $n = 1.41$, a squared entry aperture and round cell, and $\alpha_{90} = 1°$, the XR can get $C_g = 3100\times$ for a square aperture and circular cell. Silver mirrors (reflectivity $\approx 93\%$) would lead to an optical efficiency around 87–88%. The illumination homogeneity is still not good (although better than in the RR).

The XR has been suggested for space applications with a glass secondary, where the increase in glass thickness with respect to conventional space modules improves the cell shielding. Note that, as the glass diameter is small, the weight of the glass (which is critical) is not necessarily increased.

The RXI concentrator, shown in figure 13.15, was used for PV applications with single-junction GaAs cells in the framework of the Hercules EU project [8]. This device can be also very compact ($H/D \approx 0.3$) and has the feature that the front surface is used twice: once as refractive one and the other as a (totally internally) reflective one. The prototypes in the Hercules project were manufactured by low-cost injection moulding, and silver evaporation. The geometrical characteristics are a circular entry aperture (40 mm diameter) and a 1 mm^2 square cell active area. The measurements showed that $\alpha_{90} = 1.6°$ and $\eta_{opt} = 86\%$ at $C_g = 1197\times$. The measured photocurrent density of the

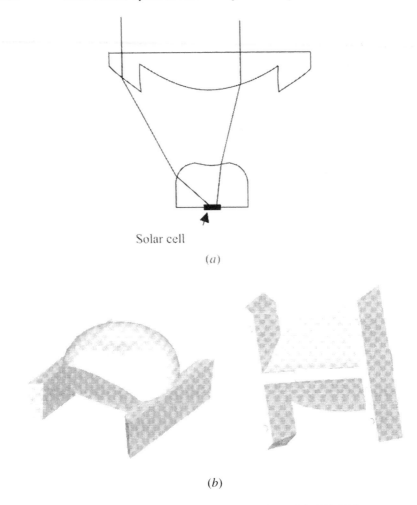

Solar cell

(*a*)

(*b*)

Figure 13.12. (*a*) The RR concentrator. (*b*) The secondary of the RR-RRI$_F$ concentrator (supporting, not optically active, elements are also shown).

concentrator was 20.6 mA cm^{-2}, and the cell was being illuminated up to $\beta = 70°$ (as indicated in 13.1.3.1, all these numbers exclude the 4.7% inactive area of the front mirror). The illumination homogeneity was poor, and the cell fill factor was only $FF = 0.77$ (with uniform flash illumination, $FF = 0.85$ was obtained).

The RXI needs mirror coatings. The combination of low cost, high reflectivity and durability of the mirror coating is difficult to achieve, and has still to be proved. In this respect, the new mirror technology marketed by 3M [43], which consists in a dielectric interferential multiplayer structure based on giant birefringent optics (GBO) is very promising. This product is sold as films and

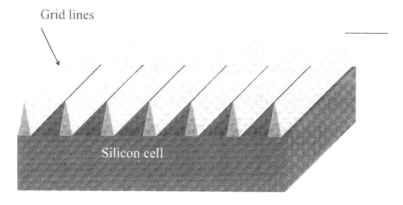

Figure 13.13. Concentrating grid-lines in the front-contacted solar cells of the Hisicon EU project

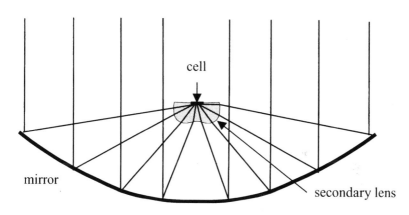

Figure 13.14. The XR concentrator.

is (surprisingly) low cost, ultra-highly reflective (above 98% in the GaAs-useful spectrum has been proved) and potentially durable (due to the absence of metals). The availability of new technology for concentrator development may encourage designers to focus on mirror-based solutions.

The RXI also has some additional practical problems. First, the front surface is not flat, so it could possibly accumulate dust, hence impeding easy alignment by gluing the concentrator units to flat glass (which would also provide good outdoor resistance and UV filtering). Second, the TIR is not protected, and its reflectivity (theoretically 100%) decreases significantly when absorbing particles (like dust) accumulates when exposed outdoors.

The TIR-R concentrator, shown in figure 13.16, solves the problems of the RXI and also does not need mirror coatings. This device is composed of a TIR

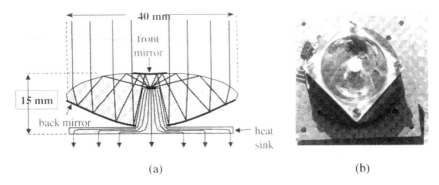

Figure 13.15. The RXI concentrator: (*a*) cross-sectional view, (*b*) example of one injection-moulded prototype in the HERCULES project.

lens primary and a refractive secondary that encapsulates the solar cell. The TIR lens was invented by Fresnel [44] as a revolutionary optics for lighthouses at the beginning of the 19th century. More recently, during the last decades, TIR lenses have proven to be excellent devices for illumination applications [45], although the method used for their design was not suitable for getting good PV performance. The use of the SMS method on the TIR-R configuration has permitted to achieve high concentration–acceptance angle product to be achieved, good illumination uniformity and also sufficiently big aspheric facets (this is important for keeping a high efficiency, due to possible vertex rounding when manufactured).

The SMS TIR-R can achieve a geometrical concentration around $C_g = 2300\times$, for $\alpha_{90} = 1°$, squared entry aperture and circular cell active area. The cell is illuminated up to $\beta = 70°$. The compactness is $H/D \approx 0.4$–0.5. The optical efficiency can be in the $\eta_{opt} = 80$–83% range, if the draft angle of the nearly-vertical facets is $2°$ and the radius of the vertices is around $20\ \mu$m.

The irradiance uniformity of the TIR-R designs is significantly improved over that of previously presented designs. It is because the edge-ray assignation was used to solve the bundle-coupling problem and the prescribed-irradiance problem for normal incidence rays. This strategy could be also applied to any of the SMS designs presented. As an example, a design for $C_g = 1600\times$ (squared entry aperture and circular cell active area) achieves $\alpha_{90} = 1.2°$, and peak irradiance for the sun at normal incidence about 1900 suns (assuming 850 W m^{-2} for the irradiance at the entry aperture), and when the sun is off centre, the irradiance peak increases further, up to about 3000 suns.

The TIR-R concentrator is being developed for $C_g = 1250\times$ on GaAs cells in the framework of the Inflatcon and Hamlet EU projects leaded by Isofotón, and for $C_g = 300\times$ back-point-contacted silicon solar cells by Sunpower Corporation.

In the case of Isofotón's development, the module will consist on

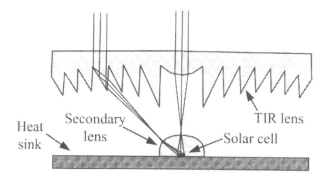

Figure 13.16. The TIR-R concentrator.

miniconcentrator units with a squared aperture about 10 cm^2 (i.e. around 1000 concentrators per m^2), illuminating III–V cells with an acceptance angle over 1.2°. The manufacturing process will be highly automated. The most surprising feature is that the module thickness is around 2 cm, similar to a framed flat module. Isofotón aims to get over 20% efficient modules with GaAs cells and over 25% efficient modules with tandem cells, and to reach the goal of total cost of 2.5 €/W$_p$ in medium-large generation plants.

13.3   Advanced research in non-imaging optics

The recent progresses in non-imaging optics research are focused on to designs in 3D geometry. The interest in 3D designs for PVs lies in the potential capability of 3D methods to obtain practical solutions close to the optimum PV concentrator performance, defined in section 13.1.6.

Not only do 2D designs not control most of the 3D rays, but the rotational or linear symmetric devices are theoretically unable to solve some design problems due to its symmetry. For instance, linear symmetric concentrators cannot achieve isotropic illumination of cell surrounded by an optically dense medium ($n > 1$) [46], and rotational concentrators cannot achieve maximum concentration on a spherical receiver [47].

Designing in 3D is more difficult than in 2D, due to the greater number of rays to be controlled. For instance, while the edge rays in 2D constitute a one-parameter family of rays, they are three-parameter ones in 3D (because the dimensions increase in both the spatial and angular coordinates).

In general, the 3D design deals with the design of free-form optical surfaces (i.e. surface without rotational or linear symmetry). Note that these surfaces can presently be manufactured with optical precision (even for imaging applications) thanks to the development of multiple–axis high-accuracy diamond-turning machines. Its cost is not essentially higher for mass production, because in any mass-production replication technique it only affects the fixed cost of the master.

Some 3D design methods were developed in the past aiming to solve the bundle-coupling problem (see section 13.1.7) in 3D geometry, as extensions of two of the methods in 2D (the Flow-Line method [28, 48, 49] and the Poisson bracket method [50]). One device obtained with Poisson bracket method in 3D, which proved the existence of theoretically exact solutions for coupling a 3D bundle of acceptance angle α with the maximum concentration bundle ($\beta = 90°$) on a flat receiver of arbitrary contour is remarkable.

However, these methods do not lead to practical PV devices yet: some of them use inhomogeneous refractive index media and all of them use flow-line metallic mirrors. New methods are now under development, which do not have these practical limitations:

(1) A single free-form refractive or reflective surface [51–53] to solve the prescribed-irradiance problem based on the approximation of infinitesimal source. This strategy (usually called point-to-point mapping [15]) is well known for rotational optics, where its solution simply involves the integration of a nonlinear ordinary differential equation [54], but designing one free-form surface needs a nonlinear partial differential equation of the Monge–Ampère type to be solved. This design could provide, if desired, a squared aperture and sufficiently good illumination uniformity on a squared area for the real (finite radius) sun. However, the design strategy provides very small illumination angles β and, consequently, the acceptance angle seems insufficient (essentially, the suns radius), which may make it impractical.

(2) The SMS method in 3D [55, 56]. With this approach, at least, two free-form surfaces are designed, transforming a selected subset of edge rays at each point of the entry aperture of the input bundle M_i into edge-rays of the output bundle M_o. It can also been applied either for bundle-coupling or for prescribed-irradiance problems. 3D-SMS can be applied to the configuration of any of the 2D-SMS devices presented here.

These advanced techniques may be applied to PV concentrator designs in the near future. In our opinion, the 3D-SMS method has the potential of finding the first practical device approaching the optimum PV performance.

13.4 Summary

The combination of concentration and next-generation PV devices is expected to play an important role in future PV electricity generation. Due to the present high cost of the most developed next-generation approach (the tandem cells), there is a trend towards achieving practical devices working at high concentration levels.

The desired characteristics of PV concentrators (efficient, mass-producible, insensitive to errors, good illumination uniformity, capable for high concentration, tessellatable) make their design a challenging problem for optical designers. These characteristics lead to the definition of an optimum PV concentrator performance, which can serve as the goal for the design. Approaching that goal with practical devices is still an open problem, which may not be a need for low concentration systems.

The importance of designing for a sufficient acceptance angle has highlighted one of the ignored issues in PV concentration, which may be one key feature for the future success of high-concentration next-generation systems. The acceptance angle is a measure of the allowable tolerances of the system (especially module manufacturing and installation) and, although the dependence of the system cost with the acceptance angle is uncertain at present, it may be critical. The nonlinear effects of the series connection of cells lead to a more restrictive definition for the acceptance angle other than in non-PV applications, making that the useful incidence angles are those for which the concentrator optical efficiency is high.

The problem of defining the degree of non-uniformity that can be allowed for a given cell has been discussed, specifically for front-contacted cells with low series resistance on the grid-lines. The concept of exponential concentration seems to play an important role, and the tools for the calculation of the cell efficiency, the strategies for modifying the irradiance distribution to produce the fastest approach to the optimum and the analysis of the contribution of the different concentration levels are given.

Non-imaging optics is the best framework for optical concentrator design. Classical optics (based on parabolic mirrors and Fresnel lenses) seems to be limited to obtaining high-concentration devices achieving good illumination homogeneity and sufficient acceptance angle. The non-imaging SMS design method has proven to be a versatile tool for designing concentrators that achieve high acceptance–angle concentrator products (and recently sufficiently uniform illuminations) and, at the same time, have nice practical features (simplicity, compactness). The use of the SMS techniques in 3D geometry may be a good path to follow for improving the present designs towards higher concentration levels and better illumination uniformity, keeping the tolerances at a practical level.

Acknowledgments

This work has been under the EU contract ERN6-CT2001-00548.

Appendix: Uniform distribution as the optimum illumination

We will formally prove with the model presented in section 13.1.5 that for cells whose grid-line series resistance is negligible, when the local concentration distribution is assumed to vary slowly between the grid-lines, the distribution providing maximum cell efficiency is the uniform one.

Let us consider the cell model presented in section 13.1.5. In the trivial case in which the parameter r_s is negligible (i.e. $J(C, V)r_s \ll V_T$ for any $C < C_{MAX}$ and any $V < V_{oc}$), any local concentration distribution $f(C)$ produces the same efficiency (equation (13.11) does not depend on $f(C)$).

Assuming now that r_s is not negligible, consider two concentration distributions with the same average concentration $\langle C \rangle$: one uniform distribution, i.e. $f_U(C) = \delta(C - \langle C \rangle)$, where δ denotes the Dirac-delta, and another non-uniform one, with probability density function $f_{NU}(C)$. From equations (13.5) and (13.11) we can write:

$$I_U(V) = \langle C \rangle I_{L,1sun} - I_0 \exp\left(\frac{V + J_U(V)r_s}{V_T}\right) \quad \text{with } I_U = A_C J_U \quad (13.21)$$

$$I_{NU}(V) = \langle C \rangle I_{L,1sun} - I_0 \exp\left(\frac{V}{V_T}\right) \int_0^\infty \exp\left(\frac{J_{NU}(C, V)r_s}{V_T}\right) f_{NU}(C)\, dC \quad (13.22)$$

and:

$$I_{NU}(V) = \int_0^\infty J_{NU}(C, V) f_{NU}(C)\, dC = A_C < J_{NU}(V) \rangle. \quad (13.23)$$

We shall prove that for every value of V, $I_U(V) > I_{NU}(V)$ and, thus, $P_U(V) = V I_U(V) \rangle V I_{NU}(V) = P_{NU}(V)$.

From the properties of the exponential function, we can write

$$\exp(x) \geq \exp(x_0)(1 + (x - x_0)) \quad (13.24)$$

for any value of x, and the equality is only fulfilled at $x = x_0$. Let us apply this property to $x = J_{NU}r_s/V_T$ and $x_0 = J_U r_s/V_T$, we can obtain, first, from equation (13.22):

$$I_{NU}(V) \leq I_U(V) - \frac{r_s I_0}{V_T A_C} \exp\left(\frac{V - J_U(V)r_s}{V_T}\right)(I_{NU}(V) - I_U(V)). \quad (13.25)$$

From here we deduce:

$$[I_U(V) - I_{NU}(V)]\left[1 + \frac{r_s I_0}{V_T A_C} \exp\left(\frac{V - J_U(V)r_s}{V_T}\right)\right] \geq 0. \quad (13.26)$$

As the second factor is positive, we get $I_U(V) > I_{NU}(V)$.

References

[1] Algora C 2004 *Next Generation Photovoltaics* (Bristol: Institute of Physics Publishing) ch 6
[2] Swanson R 2000 The promise of concentrators *Prog. Photovolt. Res. Appl.* **8** 93–111

[3] Bett A, Dimroth F, Lange G, Meusel M, Beckert R, Hein M, Riesen S V and Schubert U 2000 '0% monolithic tandem concentrator solar cells for concentrations exceeding 1000 suns *Proc. 28th IEEE PVSC* (New York: IEEE) pp 961–4

[4] Algora C, Ortiz E, Rey-Stolle I, Díaz V, Peña R, Andreev V, Khvostikow V and Rumyantsev V 2001 A GaAs solar cell with an efficiency of 26.2% at 1000 suns and 25.0% at 2000 suns *IEEE Trans. Electron. Devices* **48** 840–4

[5] Mandel L and Wolf E 1995 *Optical Coherence and Quantum Optics* (New York: Cambridge University Press)

[6] Benítez P, Mohedano R and Miñano J C 1997 DSMTS: a novel linear PV concentrator *Proc. 26th IEEE Photovoltaic Specialists Conf.* (New York: IEEE) pp 1145–8

[7] Luque A *et al* 1997 Some results of the Euclides photovoltaic concentrador prototype *Prog. Photovolt. Res. Appl.* **5** 195–212

[8] Álvarez J L, Hernandez M, Benítez P and Miñano J C 1998 Experimental measurements of RXI concentrator for photovoltaic applications *2nd World Conf. and Exhibition on Photovoltaic Solar Energy Conversion* pp 2233–6

[9] Terao A *et al* New developments on the flat-plate micro-concentrator module *3rd World Conf. on Photovoltaic Energy Conversion* to be published

[10] Rabl A 1985 *Active Solar Collectors and their Applications* (Oxford: Oxford University Press)

[11] Feuermann D and Gordon J M 2001 High-concentration photovoltaic designs based on miniature parabolic dishes *Non-imaging Optics: Maximum Efficiency Light Transfer VI (Proc. SPIE 4446)* ed R Winston (SPIE) pp 43–51

[12] This approximation is equivalent to assume that fill factor of the array is equal to that of the isolated units, but in fact the fill factor is better in the series array. See Luque A, Lorenzo E and Ruiz J M 1980 Connection losses in photovoltaic arrays *Solar Energy* **25** 171–8

[13] Maroto J C and Araújo G L 1995 Three-dimensional circuit analysis applied to solar cell modeling *Proc. 13th European Photovoltaic Solar Energy Conf.* (Bedford: H S Stephens & Associates) pp 1246–9

[14] Benítez P and Mohedano R 1999 Optimum irradiance distribution of concentrated sunlight for photovoltaic energy conversion *Appl. Phys. Lett.* **74** 2543–5

[15] Cassarly W 2001 Non-imaging optics: concentration and illumination *Handbook of Optics* 2nd edn (New York: McGraw-Hill) pp 2.23–2.42

[16] Ries H, Gordon J M and Laxen M 1997 High-flux photovltaic solar concentrators with kaleidoscope-based optical designs *Solar Energy* **60** 11–16

[17] O'Ghallagher J J and Winston R 2001 Nonimaigng solar concentrato with near-uniform irradiance for photovoltaic arrays *Non-imaging Optics: Maximum Efficiency Light Transfer VI (Proc. SPIE 4446)* ed R Winston (SPIE) pp 60–4

[18] Jenkings D G 2001 High-uniformity solar concentrators for photovoltaic systems *Non-imaging Optics: Maximum Efficiency Light Transfer VI (Proc. SPIE 4446)* ed R Winston (SPIE) pp 52–9

[19] James L W 1989 Use of imaging refractive secondaries in photovoltaic concentrators SAND89-7029 Alburquerque, NM

[20] Miñano J C, Benítez P and González J C 1995 RX: a non-imaging concentrator *Appl. Opt.* **34** 2226–35

[21] Benítez P and Miñano J C 1997 Ultrahigh-numerical-aperture imaging concentrator *J. Opt. Soc. Am.* A **14** 1988–97

[22] Benítez P, Mohedano R, Miñano J C, García R and González J C 1997 Design of

CPC-like reflectors within the simultaneous multiple surface design method *Non-imaging Optics: Maximum Efficiency Light Transfer IV (Proc. SPIE 3139)* ed R Winston (SPIE) pp 19–28

[23] Hinterberger H and Winston R 1968 Efficient light coupler for threshold Cerenkov counters *Rev. Sci. Instrum.* **39** 419–20

[24] Winston R and Welford W T 1978 Two-dimensional concentrators for inhomogeneous media *J. Opt. Soc. Am.* **68** 289–91

[25] Miñano J C, Ruíz J M and Luque A 1983 Design of optimal and ideal 2-D concentrators with the collector immersed in a dielectric tube *Appl. Opt* **22** 3960–5

[26] Ning X, Winston R and O'Gallager J 1987 Dielectric totally internal reflecting concentrators *Appl. Opt.* **26** 300–5

[27] Winston R and Ries H 1993 Nonimaing reflectors as functionals of the desired irradiance *J. Opt. Soc. Am.* **11** 1902–8

[28] Winston R and Welford W T 1979 Geometrical vector flux and some new non-imaging concentrators *J. Opt. Soc. Am.* **69** 532–6

[29] Miñano J C 1985 Two-dimensional non-imaging concentrators with inhomogeneous media: a new look *J. Opt. Soc. Am. A* **2** 1826–31

[30] Miñano J C and González J C 1992 New method of design of non-imaging concentrators *Appl. Opt.* **31** 3051–60

[31] Winston R 1974 Principles of solar concentrators of a novel design *Solar Energy* **16** 89–94

[32] Rabl A and Winston R 1976 Ideal concentrators for finite sources and restricted exit angles *Appl. Opt.* **15** 2880–3

[33] Kritchman E M *et al* 1979 Efficient Fresnel lens for solar concentrator *Solar Energy* **22** 119–23

[34] Welford W T and Winston R 1980 Design of non-imaging concetrators as second stges in tandem with image forming first-stage concentrators *Appl. Opt.* **19** 347–51

[35] Miñano J C, Benitez P, Gonzalez J C Falicoff W and Caulfield H J *High Efficiency Non-Imaging Optics* US and International Patents Pending: US Patent Publication no US 2003/0016539, International Publication no Wo 01/69300

[36] This device is similar to the XRI$_F$ proposed in [22], but changing the first reflective surface by a refractive surface

[37] Friedman R P, Gordon J M and Ries H 1995 Compact high-flux two-stage solar collectors based on tailored edge-ray concentrators *Non-imaging Optics: Maximum Efficiency Light Transfer III (Proc. SPIE 2538)* ed R Winston (SPIE) pp 30–41

[38] Benítez P, Miñano J C and González J C 1995 Single-mirror two-stage concentrator with high acceptance angle for one axis tracking PV systems *Proc. 13th European Photovoltaic Solar Energy Conf.* (Bedford: H S Stephens & Associates) pp 2406–9

[39] Miñano J C, González J C and Benítez P 1995 A high-gain, compact, non-imaging concentrator: RXI *Appl. Opt.* **34** 7850–6

[40] Benítez P, Hernández M, Miñano J C and Muñoz F 2001 New non-imaging static concentrators for bifacial photovoltaic cells *Non-imaging Optics: Maximum Efficiency Light Transfer VI (Proc. SPIE 3781)* ed R Winston (SPIE) pp 22–9

[41] Álvarez J L, Hernández M, Benítez P and Miñano J C 2001 TIR-R concentrator: a

new compact high-gain SMS design *Non-imaging Optics: Maximum Efficiency Light Transfer VI (Proc. SPIE 4446)* ed R Winston (SPIE) pp 32–42

[42] Luque A 1989 Back point contact silicon solar cell *Solar Cells and Optics for Photovoltaic Concentration* (Bristol: Adam Hilger) chs 10–13

[43] www.3m.com

[44] Stevenson A 1835 *Report to the Committee of the Commissioners of the Northern Lights* appointed to take into consideration the subject of illuminating the lighthouses by means of lenses (Elton Engineering Books)

[45] Medvedev V, Parkyn W A and Pelka D G 1997 Uniform high-efficiency condenser for projection systems *Non-imaging Optics: Maximum Efficiency Light Transfer IV (Proc. SPIE 3139)* ed R Winston (SPIE) pp 122–34

[46] Miñano J C 1984 Application of the conservation of the étendue theorem for 2D-subdomains of the phase space in non-imaging optics *Appl. Opt.* **23** 2021–5

[47] Ries H, Shatz N E and Bortz J C 1997 Consequences of skewness conservation for rotationally symmetric non-imaging devices *Non-imaging Optics: Maximum Efficiency Light Transfer IV (Proc. SPIE 3139)* ed R Winston (SPIE) pp 47–58

[48] Gutiérrez M, Miñano J C, Vega C and Benítez P 1996 Application of Lorentz geometry to non-imaging optics: new 3D ideal concentrators *J. Opt. Soc. Am.* A **13** 532–40

[49] Benítez P and Miñano J C 2001 Elliptics bundles in homogeneous refractive index media: towards the general solution *Non-imaging Optics: Maximum Efficiency Light Transfer VI (Proc. SPIE 4446)* ed R Winston (SPIE) pp 1–10

[50] Miñano J C 1986 Design of three-dimensional non-imaging concentrators with inhomogeneous media *J. Opt. Soc. Am.* A **3** 1345–53

[51] Cafarelli L A and Oliker V I 1994 Weak solutions of one inverse problem in geometrical optics *(private communication)*

[52] Kochengin S A, Oliker V I and von Tempski O 1998 On the design of reflectors with prespecified distribution of virtual sources and intensities *Inverse Problems* **14** 661–78

[53] Ries H and Muschaweck J A 2001 Tailoring freeform lenses for illuminations *Novel Optical Systems Design and Optimization IV (Proc. SPIE 4442)* (SPIE) pp 43–50

[54] Elmer W B 1980 *The Optical Design of Reflectors* 2nd edn (New York: Wiley) ch 4.4

[55] Benítez P, Mohedano R and Miñano J C 1999 Design in 3D geometry with the simultaneous multiple surface design method of non-imaging optics *Non-imaging Optics: Maximum Efficiency Light Transfer V (Proc. SPIE 3781)* ed R Winston (SPIE) pp 12–21

[56] Miñano J C and Benítez P SMS design method 3D: Calculating multiple free form surfaces from the optical prescription *Frontiers in Optics, the 87th OSA Annual Meeting* to be published

Appendix: Conclusions of the Third-generation PV workshop for high efficiency through full spectrum utilization

As a result of the presentations during the workshop and the subsequent discussion, the participants have agreed the following conclusions:

- Present solar cells are not likely to reach a cost that will allow penetration of the PV electricity market because, in their present form—with poor utilization of the solar spectrum—they make ineffective use of the solar resource that—although immense—comes with moderate densities.
- A number of options were presented that, in principle, may ensure better use of the solar resource.
- Of such options, multi-junction solar cells seem to be the one closer to practical exploitation. It was generally agreed that high concentration was needed to render them cost effective. Operation of the cells in high concentration has also been proven in many cases. Additional R&D including the use of nanotechnology (superlattices) might enable this promising line of research to reach its full potential.
- Alternative options based on better utilization of the low-energy photons could be achieved with intermediate-band or -level solar cells. In this option, two lower-energy photons are used to raise the free energy of the electrons to be delivered to the external circuit to a high level. Some theoretical work has been done recently on this topic and initial experimental work is based on nanotechnology, mainly quantum dot arrays. The concept, although apparently a long-term one, is worth exploration. Practical ideas to separate the up-converting function from solar cell fabrication, using optical coupling, have been presented in this workshop for the first time. Concentration seems to be one way of making the complex devices based on this principle cost effective.
- However, thin films based on this concept may also be a way to develop them in a low cost way, possibly with a poorer performance. Quantum dots based on a highly porous material grown in cheap ways or even organic

semiconductors operating with the two-photon principle may become one of the most attractive ways to realize cost-effective high-efficiency solar cells.

- Using the high energy of the electrons excited by high-energy photons, before they thermalize with the lattice, may allow for higher efficiency devices. This can, in principle, be achieved by ionization of a second electron–hole pair or by extracting the electrons while they are still at a high temperature. Concepts for both solutions have been presented although they still seem to be far from demonstration. Arrays of quantum dots seem to be important in all these technologies and mastering this new field of applied science seems to be the key for possible exploitation of such concepts.

- Thermophotovoltaics and thermophotonics, which is extracting electricity by a solar cell illuminated by a heated emitter, i.e. a heated LED, presents several theoretical advantages. In principle, such devices could reach the Carnot efficiency if the cell and LED were ideal. Practical thermophotovoltaic devices operating with fuel-heated emitters already exist. Further research on this concept and ways of heating the emitter with solar energy are worth further study. This might become a medium-term option to compete with the other options studied.

- There is a general agreement that many of the solutions studied in this workshop may be cost effective only under highly concentrated sunlight. It was stressed that while many technological aspects of solar cells have advanced greatly, the optical concepts used for the concentrator are still based on concepts that were already known in the Ancient World. However, such devices were not the best that could be achieved and a need to exhaust the theoretical potential of concentrators was felt necessary to reach cost-effective solutions. New synthesis methods are available and further research on them seems justified to render cost-effective concentrators for novel PV cells.

- Novel technological aspects, some based on micromechanics, appear to be of utility for the purpose of this workshop. Direct bonding techniques were presented as an alternative to monolithic multi-junction cells in which some of the constraints of their fabrication as well as many of the drawbacks of conventional stacking disappear. This option is worth further exploration because it might lead to unexpected and interesting results.

Index

A3B5 (see III–V compounds)
absorber, 51–52, 56–61, 166–169, 178, 196–197
absorption efficiency, 51–52,
acceptance angle, 85, 108, 117, 120, 125, 287, 290–297, 305–321
AlAs, 22, 27, 100,
AlGaAs, 22–28, 36–39, 44, 47, 68–77, 88, 92–93, 100, 114, 156, 202–204, 209, 247, 251, 261
AlGaAsP, 26, 93
AlGaAsSb, 77, 93, 260, 261
AlInAsP, 26, 93
AlInP, 30, 93
anidolic (see non-imagining)
antireflection coating, 26, 27, 70–71, 77 80, 113, 116, 121, 127, 224, 238, 252, 258–259, 296, 305
aperture, 82, 84, 287–290, 307–315, 318–320
ARC, see antireflection coating
Auger, 149, 166, 198–200, 207–208, 210–212, 216–217

bio-fuels, 2–3, 45
Bloch, 153
Bragg reflectors, 25–27, 98–99

carrier collection limit, 167, 175, 179–193
cascade (see multi-junction)

cascade solar cells (see multi-junction solar cells)
CCL (carrier collection limit)
CdTe, 64,
charge current, 57–59
chemical
    energy current, 54
    potential, 50–57, 147
CIGS, CIS, 64, 235–240
climatic change, 4–6, 18
CO_2, 6–8
coal, 2, 8, 11
concentration (of solar radiation), concentrators, 34–44, 51, 56, 59, 61, 64–90, 108–138, 147–148, 246, 285–326
concentrator RX, 307–309, 317
concentrator CPC, 309, 312–313
concentrator RR, XR, SMTS, DSMTS, 314
concentrator RXI, 118–119, 314–319
concentrator TIR, 117–118, 306, 311, 315, 317–319
concentrator, design methods, 312, 320
contact resistance, 69, 113, 281
costs (of photovoltaics) (see also module, multi-junction and TPV), 1–17, 22, 24, 30–31, 43, 108–109, 121–136, 176, 196, 298, 326–327
$Cu(InGa)S_2$, see CIGS

defects, 25, 32, 167–172, 178–182, 193, 251, 261, 275–276
down-converters, 61–62, 141
dual-junction (*see* multi-junction)

effective mass equation, 151
efficiency (*see also* multi-junction)
efficiency
  Carnot, 34–35, 51–52, 54, 60, 62, 327
  limits, 35, 45, 50–64, 66, 67, 91, 120–121, 135, 140, 144, 148–150, 160, 196–197, 218
  (of solar cells and related systems), 16, 20–41, 64, 70–84, 98, 112, 121–136, 196, 226–230, 235–238, 247–254, 260, 266–267, 287, 302
  optical, 41–42, 80, 288–296, 314–315
electron affinity, 57
electron localization and delocalization, 152, 154, 157, 161, 203, 205
emitter, TPV, 45–46, 93–94, 99–101, 103, 223–243, 246–256, 260, 264, 267–268, 327
energy current, 50–55
entropy, 50–56, 143–144
EUCLIDES, 129

Fermi, quasi-Fermi levels, 53, 58–59, 140, 141–147, 158–161, 165–166, 173, 184, 203
fossil-fuels, 3, 5, 8, 19, 51, 99
Fresnel lens, 37–44, 71–72, 75, 80–84, 87, 289–290, 306, 311–314, 318, 321

GaAs, 21–30, 35–39, 44, 47, 64–80, 88, 92–101, 109–125,
130–136, 142, 155–156, 173, 202–210, 247, 251–256, 263, 278–279, 296–298, 314–319
GaAsTi, 155
GaInNAs, 31, 74
GaInP, 29–30, 44, 66, 73–77, 80, 92–93, 96, 98, 101, 114, 121–122, 125, 135, 251–252, 263
GaP, 44, 93
gas (energy source), 2–3, 8, 224, 225, 247
GaSb, 42, 45–46, 67–72, 77, 88, 93, 101, 121, 135, 224, 246–265
Ge, 20, 27–31, 35, 65, 70, 73–74, 77, 86, 98, 100–101, 122–123, 127, 130, 135, 225, 240, 241–243, 246–247, 250–254, 263, 268, 278–281
germanium (*see* Ge)
greenhouse effect, 3–4, 19

H_2 (*see* hydrogen)
heat dissipation, 37, 40–41, 82, 126, 132
heteroface (heterojunction, heterostructure, hetero-interface), 1, 21–27, 30–35, 45–47, 68–70, 114, 134, 169–178, 181–182, 216, 251–253, 257, 261–268, 281
hot carrier solar cell, 60, 196–221
hot carriers, 168, 249
hydrogen, 8, 57, 69, 264, 275, 278
  fuel, 15, 17

IBSC (*see* intermediate band solar cell)
III–Vs (*see also* AlAs, AlGaAs, AlGaAsP, AlGaAsSb, AlInP, AlInAsP, InAsP,

GaAs, InAlAs, GaInNAs,
GaInP, GaP, GaSb, InGaAs,
InGaAsSb, InGaSb,
InGaAsP, AnAsSbP, gallium
arsenide solar cells, AlAs,
AlGaAs, GaInP, GaInNAs,
GaP, InGaAs, InGaAsSb)
III–V compounds, 1, 20–36, 43–45,
64–68, 73–74, 86– 87, 93,
108, 110, 114, 120, 135,
156, 209, 214, 246–247,
250, 279, 281, 287, 319
impact ionization (*see also* Auger),
197–199, 207–209,
213–214, 216–217
InAlAs, 279
InAsP, 225
InAsSbP, 260
InGaAs, 45, 73–77, 80, 93–95,
99–102, 121, 125, 135, 156,
210, 225, 241, 246–247,
263–268, 279
InGaAsP, 93–95, 100,
InGaAsSb, 45, 93, 246–247, 250,
256–265
InGaP (*see* GaInP)
InGaSb, 261
InP, 92–93, 99, 247, 263–268,
278–279
intermediate band solar cell, 34–35,
108, 120, 135, 140–164, 326
intermediate level solar cell, 61–62,
326
ion implantation, 167, 168–193,
276, 278,

lattice matching, 25–31, 66–68,
73–77, 86, 92–97, 100–103,
156, 171, 209, 241, 246,
250, 256, 259, 261,
264–268, 279, 281
learning curve, 10–17, 65, 127–134,
136

light trapping, 97–99, 102, 150,
158, 169
liquid phase epitaxy (*see* LPE)
LPE, 23, 25, 27, 67–70, 138, 252 –
268

matrix elements, 152–153
MBE, 24, 26, 34, 156, 241,
261–262
membrane, for electrons and holes
57–60, 166–167, 174
metal grid, 74, 113–114, 117, 119
metal-organic chemical vapour
deposition (*see also*
MOCVD)
metal-organic vapour phase epitaxy
(*see also* MOVPE)
MIND (*see* multi-interface novel
devices)
MOCVD, 24, 26–28, 30–31,
251–252, 257, 261–262,
266–268
module
concentrator, 35–44, 64–65,
71–72, 80–87, 318–319
FLATCON, 82–85
in TPV, 225, 238–240, 250,
263–268
learning rate, 128
sales, market, costs, 9–16, 36,
64–66, 82–86, 128–133,
285–293, 296, 305–309,
315–321
space, 44
molecular beam epitaxy (*see* MBE)
monochromatic energy conversion,
29, 46, 55–56, 59, 62, 166,
224, 249
MOVPE (*see also* MOCVD), 73,
77, 86, 127–128
MQW (*see* quantum wells)
multi-interface novel devices,
165–194
multi-junction

solar cells, 22, 25–34, 42, 64–90, 91, 93, 98, 100, 101, 102, 108, 114, 116, 117, 124, 127, 135–136, 193, 225, 280–281, 283, 285, 305, 320, 326, 327
 characterization, 77–82
 costs, 72–73, 122, 128–135
 efficiency, 16, 33, 35, 60–62, 67, 97, 120, 122, 135, 149, 150, 247–249, 287, 319
 in thermophotovoltaics, 249, 254–255, 261, 264–265,
 mechanical stack, 28, 67–72, 75, 77, 86, 91, 120–122, 135, 251, 255, 265, 274
 monolithic, 28, 67, 72–77, 83–87, 91, 96, 120–122, 125, 135, 251, 263–266
 with quantum wells, 96–97

nitrides, 31, 35, 45
non-imaging optics, 309–315, 319, 321
nuclear energy, 2–4, 8–9, 47

ohmic losses (*see also* series resistance), 35, 37, 42, 75, 82, 86, 254
oil, 2–3
organic-fuels (*see* bio-fuels)

phonon, 57, 60, 196–198, 200–213, 218
phonon-bottleneck, 155, 205–213, 218
photon recycling (*see* self-absorption)

QD-IBSC (*see* quantum dot intermediate band solar cell)
QDs (*see* quantum dots)
quantum confinement, 92, 151–152, 154–156, 199, 201, 207, 240–241

quantum dot intermediate band solar cell, 150–162
quantum dot solar cell, 196–221
quantum dots, 32, 34, 60, 241–242, 326–327
quantum wells, 25, 91–103, 154, 157, 159, 199, 201–209, 241–242
Queisser 1, 56, 61, 63, 66, 148, 197
QW (*see* quantum wells)

recombination, 23, 25, 26, 53, 54, 66, 92, 110–112, 117, 140, 141–149, 159, 172, 175, 180–183, 192–193, 207, 213, 253–257
 radiative, 21, 56–59, 62, 66, 120–121, 153, 206, 209, 247
 surface, 156, 174, 181, 186, 217, 255, 261
relaxation, electron, 53, 62, 141, 157, 196–213, 218
RIGES, 7–9, 12–13
Roosbroeck–Schockley, 153

selection rules, 149, 152–155, 197
self-assembling (*see also* Sranski–Krastanov and Volmer–Weber), 156, 158, 160, 162, 209–210, 213
self-absorption, 111,
series resistance, 69, 71, 75–76, 100, 108–117, 124–126, 134–135, 230, 296–305, 321
Shockley, 1, 56, 61, 63, 66, 133, 136, 148, 153, 196–197, 218
Si (*see* silicon)
SiGe, 101, 240–243, 251
silicon (solar cells and technology), 1, 11, 15–17, 20–31, 35, 43, 64–65, 68, 71, 123, 133, 165, 167–183, 187–189, 193, 225–235, 239, 241–247, 250, 253–255, 274–278, 285, 315, 316

silicone, 27, 40–44

SILO (single optical surface), 307

SK (*see* Stranski–Krastanov)

strain, 92, 93, 94–97, 101–103, 156–158, 170–171, 173, 176–179, 189, 193, 209, 241–242

Stranski–Krastanov 156, 157, 209, 241

superlattices, 25, 27, 34, 151, 167, 199, 201–207, 326

tandem (*see* multi-junction)

thallium sulphide, 20

thermodynamics, 24, 34–35, 50–63, 144, 196, 218, 288, 308–309, 312

thermophotovoltaics (*see* TPV)

thin-film (*see also* CIGS), 17, 22, 31, 43, 45, 64, 80, 235–243, 263, 274, 277, 285, 326

TiO_2, 58, 216

TPV, 42, 44–46, 60, 62, 93, 99–103, 197, 223–243, 246–268, 327
  costs 225, 226, 231, 235, 238, 243, 246, 253, 257, 268

tracking, 36, 42–43, 65, 72, 83–85, 87, 120, 129–131, 286, 292–293, 306

tunnel diodes, 29, 30, 34–35, 67, 73, 75, 76–77, 86, 91, 96, 114, 134, 264–265, 282

up-converters, 61–62, 142

Volmer–Weber, 156–157

VW (*see* Volmer–Weber)

wetting layer, 156–157

wind energy, 3, 9, 11, 12, 15, 293

window layer, 22, 25–27, 37, 45, 70, 113, 116, 256, 257, 261–262, 265